Compendium of Plant Genomes

Series Editor

Chittaranjan Kole, Raja Ramanna Fellow, Government of India, ICAR-National Research Center on Plant Biotechnology, Pusa, New Delhi, India

Whole-genome sequencing is at the cutting edge of life sciences in the new millennium. Since the first genome sequencing of the model plant Arabidopsis thaliana in 2000, whole genomes of about 100 plant species have been sequenced and genome sequences of several other plants are in the pipeline. Research publications on these genome initiatives are scattered on dedicated web sites and in journals with all too brief descriptions. The individual volumes elucidate the background history of the national and international genome initiatives; public and private partners involved; strategies and genomic resources and tools utilized; enumeration on the sequences and their assembly; repetitive sequences; gene annotation and genome duplication. In addition, synteny with other sequences, comparison of gene families and most importantly potential of the genome sequence information for gene pool characterization and genetic improvement of crop plants are described.

Interested in editing a volume on a crop or model plant?

Please contact Prof. C. Kole, Series Editor, at ckoleorg@gmail.com

More information about this series at http://www.springer.com/series/11805

Chang Liu
Editor

The Lingzhi Mushroom Genome

Springer

Editor
Chang Liu
Institute of Medicinal Plant Development
Chinese Academy of Medical Sciences
& Peking Union Medical College
Beijing, China

ISSN 2199-4781 ISSN 2199-479X (electronic)
Compendium of Plant Genomes
ISBN 978-3-030-75709-0 ISBN 978-3-030-75710-6 (eBook)
https://doi.org/10.1007/978-3-030-75710-6

This Springer imprint is published by the registered company Springer Nature Switzerland AG
The registered company address is: Gewerbestrasse 11, 6330 Cham, Switzerland

Preface

Lingzhi, also known as "Magic Mushroom" in Chinese, refers to species from the genus *Ganoderma* P. Karst. The genus consists of mushroom taxa characterized by unique double-wall basidiospores with ornamented endospores. The Dictionary of Fungi recognizes about 80 species. In contrast, 427 name records are in Index Fungorum. *Ganoderma* species are of great importance from three aspects: therapeutic fungal bio-factories, plant pathogens, and bio-bags of ligninolytic enzymes. The therapeutic effects have been described in various myths, folklores, poems, and operas in ancient China.

Researchers have identified the chemical components, determined their bioactivities, and explored their pharmacological effects. More than 600 chemical compounds have been isolated from Lingzhi. A highly diverse range of bioactivities against prolonged list diseases or ill-conditions has been reported. These studies, along with success in large-scale cultivation technology, prompt the wide use of Lingzhi in health-promoting products. There are approximately 180 drugs and 1200 health-promoting products approved by the Chinese Food and Drug Administration, and Lingzhi is the species having the largest number of health-promoting products on the market. Types of products range from drugs to cosmetic products such as facial masks and lotion. The rapidly growing market demand, in turn, prompts further research on all aspects.

Several excellent books have been published, providing a detailed account of Lingzhi's biology, chemistry, industry, pharmacology, and clinical application. However, no book systematically describes the genes and genomes of Lingzhi. The current book was published to fill this blank. It systematically described recent research progresses on whole nuclear-genome sequencing, mitochondrial genome, transcriptomics, proteomics, and genetic transformation system. The book also included one chapter using the bibliometric method to summarize the past 30 years' research trends. In another chapter, we use the meta-analysis method to systematically analyze the data extracted from literature and obtained from unpublished experiments. Lastly, we summarize the current products involving Lingzhi. We hope these chapters, in combination, will provide complete pictures of Lingzhi from research trends to the general applications in the markets. As a result, it will provide valuable information for students, teachers, researchers in academia and industry, and managers from regulatory agencies.

We thank the support from the National Natural Science Foundation of China [81872966], the National Science and Technology Fundamental Resources Investigation Program of China (2018FY100705), and the National Mega-Project for Innovative Drugs of China [2019ZX09735-002], Innovation Funds for Medical Sciences [2016-I2M-3-016, 2017-I2M-1-013]. We appreciate the Series Editor, Prof. Chittaranjan Kole, for providing us this opportunity to publish the book. We are grateful for the professional assistance with the publishing of this book.

Beijing, China Chang Liu

Contents

Contributors

Po Cao School of Information Management, Central China Normal University, Wuhan, China

Yu Cao Institute of Medicinal Plant Development, Chinese Academy of Medical Sciences and Peking Union Medical College, Beijing, China;
Chengdu Institute of Biology, Chinese Academy of Sciences, Chengdu, China

Yuxin Cao Institute of Medicinal Plant Development, Chinese Academy of Medical Sciences, Peking Union Medical College, Beijing, China

Haimei Chen Key Laboratory of Bioactive Substances and Resource Utilization of Chinese Herbal Medicine from the Ministry of Education, Institute of Medicinal Plant Development, Chinese Academy of Medical Sciences, Peking Union Medical College, Beijing, China

Qing Du Institute of Medicinal Plant Development, Chinese Academy of Medical Sciences, Peking Union Medical College, Beijing, China;
Qinghai Nationalities University, Xining, Qinghai, China

Shan He School of Information Management, Central China Normal University, Wuhan, China

Linfang Huang Institute of Medicinal Plant Development, Chinese Academy of Medical Sciences and Peking Union Medical College, Beijing, China

Ailiang Jiang College of Life Sciences, Nanjing Agricultural University, Nanjing, Jiangsu, People's Republic of China

Mei Jiang Key Laboratory of Bioactive Substances and Resource Utilization of Chinese Herbal Medicine From the Ministry of Education, Institute of Medicinal Plant Development, Chinese Academy of Medical Sciences & Peking Union Medical College, Beijing, China

Jingling Li Institute of Medicinal Plant Development, Chinese Academy of Medical Sciences & Peking Union Medical College, Haidian, Beijing, China

Yueyue Liang School of Information Management, Central China Normal University, Wuhan, China

Chang Liu Institute of Medicinal Plant Development, Chinese Academy of Medical Sciences, Peking Union Medical College, Beijing, China

Jingting Liu Institute of Medicinal Plant Development, Chinese Academy of Medical Sciences & Peking Union Medical College, Haidian, Beijing, China

Rui Liu College of Life Sciences, Nanjing Agricultural University, Nanjing, Jiangsu, People's Republic of China

Rui Liu School of Information Management, Central China Normal University, Wuhan, China

Shanfa Lu Institute of Medicinal Plant Development, Chinese Academy of Medical Sciences & Peking Union Medical College, Beijing, China

Yang Ni College of Agriculture, Fujian Agriculture and Forestry University, Fuzhou, China

Ang Ren College of Life Sciences, Nanjing Agricultural University, Nanjing, Jiangsu, People's Republic of China

Liang Shi College of Life Sciences, Nanjing Agricultural University, Nanjing, Jiangsu, People's Republic of China

Liqiang Wang College of Pharmacy, Heze University, Heze, Shandong Province, China

Xin-Cun Wang State Key Laboratory of Mycology, Institute of Microbiology, Chinese Academy of Sciences, Beijing, People's Republic of China

Bin Wu Key Laboratory of Bioactive Substances and Resource Utilization of Chinese Herbal Medicine From the Ministry of Education, Institute of Medicinal Plant Development, Chinese Academy of Medical Sciences & Peking Union Medical College, Beijing, China

Jun-Wei Xu Faculty of Life Science and Technology, Kunming University of Science and Technology, Kunming, China

Yicen Xu College of Horticulture and Landscape Architecture, Southwest University, Chongqing, China

Heyu Yang School of Environmental Science and Engineering, Tianjin University, Tianjin, China

Jiaqi Yu School of Information Management, Central China Normal University, Wuhan, China

Jie Yu College of Horticulture and Landscape Architecture, Southwest University, Chongqing, China

Mingwen Zhao College of Life Sciences, Nanjing Agricultural University, Nanjing, Jiangsu, People's Republic of China

Jing Zhu College of Life Sciences, Nanjing Agricultural University, Nanjing, Jiangsu, People's Republic of China

Abbreviations

A-AMA	Alpha-mannosidase
A3SS	Alternative 3' Splice Site
A5SS	Alternative 5' Splice Site
AACT	Acetyl-CoA C-acetyltransferase
AchE	Acetyl cholinesterase
ADPKD	Autosomal dominant polycystic kidney disease
AIDS	Acquired immune deficiency syndrome
AKI	Acute kidney injury
Akt/PKB/Rac	Serine threonine protein kinase/protein kinase B
ALT	Alanine aminotransferase
Ams	Alveolar macrophages
AS	Alternative Splicing
AST	Aspartate aminotransferase
ATMT	*Agrobacterium tumefaciens*-mediated transformation
bar	Bialaphos resistance gene
BSGLEE	A preparation of triterpenoids from *Ganoderma*
cAMP	Cyclic adenosine 3',5'-monophosphate
CASEG	Clusters of Adjacent and Similarly Expressed Genes.
CAT	Catalase from micrococcus lysodeikticus
CAZy	Carbohydrate-Active Enzymes
CAZymes	Carbohydrate-active enzymes
Cbx	Carboxin-resistance gene
Cd	Cadmium
cDNA-AFLP	cDNA-amplified fragment length polymorphism
CDS	Coding DNA Sequence
CGL	The consortium of *Ganoderma lucidum*
circRNA	Circular RNA
CKD	Chronic kidney disease
COGs	Clusters of Orthologous Groups
CRISPR/Cas9	The clustered regularly interspaced short palindromic repeats (CRISPR)/CRISPR-associated protein 9 (Cas9)
CTL	Cytotoxic T-lymphocytes
Cu	Copper
CYM	Complete yeast extract medium
CYP450	Cytochrome P450
DC	Dendritic cell

DCs	Dendritic cells
DCW	Dry cell weight
DEAE	Diethyl aminoethyl
DEGs	Differentially expressed genes
DNA	Deoxyribonucleic acid
DSBs	Double-stranded breaks
EOG	Eukaryotic Homologous Group Terms
EPS	Exopolysaccharide
ER	Endoplasmic reticulum
ERG	Genes involved in ergosterol biosynthetic pathway
EST	Expressed Sequence Tag
EV71	Enterovirus 71
FABP	Fatty acid binding protein
FIPs	Fungal immunomodulation proteins
FOLy	Fungal Oxidative Lignin enzymes
FOLymes	Fungal oxidative lignin enzymes
FPKM	Fragments per Kb of Transcript per Million Mapped Reads
FPPs	Farnesyl pyrophosphate synthase
FPS	Farnesyl diphosphate synthase
G. lucidum	Ganoderma lucidum
GA	Ganoderic acid
GA-A	Ganoderic acid A
GaLuDB	*Ganoderma lucidum* Database
Lingzhi	*Ganoderma*
GBIF	Global biodiversity information facility
GFF	Generic Feature Format
GL22	A kind of triterpenoid from *Ganoderma leucocontextum*
GLPS	*Ganoderma lucidum* polysaccharides
GLS	β-1,3-glucan synthase
GLSO	*Ganoderma lucidum* spore oil
GLT	*Ganoderma Lucidum* triterpenoids
GM	*Ganoderma* Meroterpenoids
GM-CSF	Granulocyte-macrophage colony-stimulating factor
GMOD	Generic Model Organism Database
GNDT	*Ganoderma*nontriol
GO	Gene Ontology
gpd	glyceraldehyde-3-phosphate dehydrogenase gene.
GSH-Px	Glutathione peroxidase
GSs	*Ganoderma* steroids
GT	*Ganoderma* triterpenoids
GUS	β-glucuronidase
H_2S	Hydrogen sulfide
HCT-116/HT-29	A kind of colon cancer cell
HDL-C	High density lipoprotein cholesterol
HSPs	Heat shock proteins
Hg	Mercury

HL-7702	A kind of human liver cell
HMGR	3-hydroxy-3-methylglutaryl-CoA reductase
hmgr	HMGR gene
HMGS	3-hydroxy-3-methylglutaryl-CoA synthase
Hph	Hygromycin B resistance gene
HR	Homologous recombination
HS	Heat stress
HSV	Herpes simplex virus
IDI	Isopentenyl diphosphate isomerase
IFN-γ	Interferon-γ
IL-1β	Interleukin 1β
InDel	Insertion-deletion
IP3/Ca2+	Inositol triphosphate/Ca2+
iTRAQ	Isobaric tags for relative and absolute quantification
IκB	κB inhibitors
Kan	Geneticin resistance gene
KEGG	Kyoto Encyclopedia of Genes and Genomes
KOGs	Eukaryotic Orthologous Groups
LDL-C	Low density lipoprotein-cholesterol
Lingzhi	*Ganoderma*
LiP	Lignin peroxidase
LLC1	A kind of mouse Lewis lung cancer cells
lncRNA	Long non-coding RNA
LPO	Lipid peroxide
LS	Lanosterol synthase
ls	LS gene
LSSC	Liquid superficial-static culture
Lz-8	Lingzhi-8
MAK	Meiosis-activating kinase
MAPK	Mitogen-activated protein kinase
MCF-7	A kind of human breast cancer
MDA	Malondialdehyde
MCP-1	Monocyte chemotactic protein 1
MDA	Malondialdehyde
MeJA	Methyl Jasmonate
MHC	Major histocompatibility complex
MIP-1α/RHCCL3	Recombinant Human C-C Ligand Motif Chemokine 3
miRNA	micro RNA
MK	Mevalonate kinase
MnP	Manganese peroxidase
MPK	Phosphomevalonate kinase
MS	Mass spectrometry
MVA	Mevalonate/isoprenoid
MVD	Mevalonate pyrophosphate decarboxylase
MVK	Mevalonate kinase
MXE	Mutually Exclusive Exons

ncRNA	Non-coding RNA
NDRG2	n-myc downsream regulator gene 2
NF	Nuclear factor
NGS	Next-generation sequencing
NHEJ	Non-homologous end joining
NK	Natural killer
NMR	Nuclear magnetic resonance
NO	Nitric oxide
Nr	Non-Redundant set
NRPS	Non-Ribosomal Peptide-Synthetase
Nt	Nucleotide
OSC	2, 3-oxidosqualene-lanosterol cyclase
OVCAR-3	A kind of human ovarian cancer cell
PE	Paired-end reads
PGM	Phosphoglucomutase
PI3K/	Phosphatidylinositol 3-kinases
PKC	Protein kinase C
PMT	Polyethylene glycol-mediated transformation
PMVK	Phosphomevalonate kinase
PSG	Polysaccharides from *Ganoderma*
RANTES	Regulated upon activation normal Tcell expressed secreted factor
REMI	Restriction enzyme-mediated integration
rFip-gts	Recombinant FIP-gts (rFip-gts)
RI	Retained intron
rLz-8	Recombinant Lz-8
RNA-seq	RNA-sequencing
ROS	Reactive oxygen species
RPKM or FPKM	Reads (Fragments) Per Kilobase per Million (mapped) reads
SE	Squalene epoxidase
sgRNA	Single-guide RNA
SILAC	Stable-isotope labeling by amino acids in cell culture
SNP	Single nucleotide polymorphism
SOD	Superoxide dismutase
SP	Sulfopropyl
sqs	SQS gene
SQS	Squalene synthase
SR	Super-reads
SRBC	Sheep red blood cells
SSRs	Simple sequence repeats
TC	Hyperlipidemia
TDFs	Transcript-derived fragments
TG	Hyperlipemia
TGF-β/S	Transforming growth factor-β/S
TLR	Toll-like receptor

TNF-α	Tumor necrosis factor-α
UGP	UDP-glucose pyrohposphorylase
ura3	Orotidine 5'-monophosphate decarboxylase gene
Vgb	VHB gene
VHB	*Vitreoscill*a hemoglobin

Abbreviation and Definition

Traditional Medicine	It refers to the knowledge, skills and practices based on the theories, beliefs and experiences indigenous to different cultures, use in the maintenance of health and the prevention, diagnosis, improvement, or treatment of physical and mental illness
TCM	Traditional Chinese Medicine (TCM) is an ancient health and wellness system that's been used in China for thousands of years. TCM practitioners use various mind and body practices (such as acupuncture and tai chi) and TCM products to address health problems
Medicinal Material	Substance used as raw materials in traditional medicine products. More than 600 Medicinal materials are listed in Chinese pharmacopeia (2020 edition)
Health food	It is also called functional foods, nutraceuticals, or dietary supplements. It's a specific kind of food that can regulate human body function. It cannot treat diseases. However, it can be used as special foods for people under certain unhealthy conditions
CFDA	Chinese Food and Drug Administration
Lingzhi Health Food	The health food containing materials of Lingzhi
Lingzhi Drug	The drug-containing materials of Lingzhi
Single Composition Health Food	The health foods containing only one medicinal material
Multiple Composition Health Food	The health foods containing numerous medicinal materials
Health Effect	Effects to maintain or restore physical, mental, or emotional well-being
Lingzhi Fruit Body	The fruit body of the *Ganoderma* fungi used as the source material of Lingzhi
Lingzhi Mycelium	The mycelium of the *Ganoderma* fungi used as the source material of Lingzhi

Lingzhi Spore Powder	The spore powder of the *Ganoderma* fungi used as the source material of Lingzhi. The powder is processed from *Ganoderma* spore
Lingzhi Spore Oil	The spore oil of the *Ganoderma* fungi used as the source material of Lingzhi. The oil is extracted from the spore of *Ganoderma* fungi

1 Lingzhi, An Overview

Qing Du, Yuxin Cao, and Chang Liu

Abstract

Lingzhi mushroom refers to a group of fungal species belonging to the genus *Ganoderma*. During the past 10 years, the genetic studies of Lingzhi have made great strides. The nuclear and mitochondrial genomes of several *Ganoderma* species are publicly available. Transcriptome and proteome studies of Lingzhi have been described and reported. Genetic transformation systems have been developed to validate the gene functions. Several books have been published on the subject of Lingzhi. All of these books give detailed accounts regarding the taxonomic classification, chemical components, biological activities, pharmacological effects, and clinical applications of Lingzhi. However, there is no book discussing in-depth the genetic basis for these biochemical and physiological characteristics. This book intends to fill the research blank. It will describe the recent progress in Lingzhi genome research, and, hopefully, will provide a relationship between the Lingzhi genotype and its phenotype. This chapter serves as an overview of the subsequent chapters. Thus readers who are not familiar with Lingzhi mushroom will gain a solid ground to understand the following chapters' contents.

Q. Du · Y. Cao · C. Liu (✉)
Institute of Medicinal Plant Development, Chinese Academy of Medical Sciences, Peking Union Medical College, Beijing 100193, China
e-mail: cliu6688@yahoo.com

Q. Du
Qinghai Nationalities University, Xining 810007, Qinghai, China

1.1 Introduction

1.1.1 Overall Arrangement of the Book

This chapter serves as a general introduction to the following text. This chapter intends to provide an overview of all verified reported aspects related to Lingzhi mushrooms. Chapter 2 then uses the bibliometric method to give a general account of the Lingzhi mushroom literature. With the rapid development of high throughput DNA sequencing and protein analysis techniques, genomic studies of Lingzhi have blossomed. Chapter 3 summarizes the recent development in the sequencing of the nuclear genome. Chapter 4 described the recent development in the sequencing of the mitochondrial genomes from several Lingzhi species, another genome in a cell whose gene functions are closely related to the oxidative phosphorylation of mitochondria and the energy supply in cells. Chapters 5 and 6 described the progress in the transcriptomics and proteomics of Lingzhi. Both have led to the identification of genes, proteins essential for Lingzhi's biology and biochemistry.

C. Liu (ed.), *The Lingzhi Mushroom Genome*, Compendium of Plant Genomes,
https://doi.org/10.1007/978-3-030-75710-6_1

Chapter 7 focuses on a specific set of molecules, non-coding RNAs, which play crucial roles in the development and various aspects of Lingzhi. These non-coding RNAs include lncRNA, natural antisense RNA, circRNAs, miRNAs, etc.

Since Lingzhi is widely used as a medicine, we discussed the biosynthetic pathway of active ingredients in Chap. 8. Functional validations of genes require a mature genetic transformation system. Therefore, in Chap. 9, the genetic transformation systems of Lingzhi were described in detail. We spend the last three chapters to discuss the clinical applications of Lingzhi mushrooms. Chapter 10 summarizes the application of Lingzhi in cancer treatment. Chapter 11 described the application of meta-analysis to determine the health-benefit effects of Lingzhi, mainly using the immune-regulatory activity of Lingzhi as an example. In the last chapter (Chap. 12), we described the drugs and health products using the main active constituents from Lingzhi as their primary components.

There have been several excellent books on the Lingzhi mushroom (Lin 2019, 2007, 2009; Sissi WG 2011). These books and Journals are considered classic works for the in-depth study of biology, chemistry, and pharmacology of Lingzhi (Zhao 1989; Sanodiya et al. 2009; Lin 2019; Lin and Yang 2019). The current book complements those books by focusing on the genomics of Lingzhi. Through these 12 chapters presented in this book, we hope to show a vivid picture of Lingzhi mushroom, from basic biology to the development of Lingzhi products that benefit human health.

1.1.2 A Brief Account of Researches on Lingzhi, from Ancient China to the Modern Era

Lingzhi is a group of fungi belonging to the genus *Ganoderma*. "Ling" in Chinese means "magic" and "Zhi" means "mushroom." The exact taxonomy classification is not without any debate. Throughout the text, we will use Lingzhi to represent the group of fungi. Whenever it is needed, we will use the complete Latin names for specific species. The most commonly seen Lingzhi species includes *Ganoderma Lucidum* (Leyss. ex Fr.) karst and *Ganoderma sinense* Zhao, Xu et Zhang (Fig. 1.1a and b). Both of them were included in "Chinese Pharmacopoeia" since 2000 (Chinese Pharmacopoeia Commission 2020; Dai and Yang 2008). In Chinese, *G. Lucidum* is also called "Chi-Zhi," and *G. sinense* is called "Zi-Zhi``. Here "Chi" means "red," and "Zi" means "purple," indicating the most obvious phenotypical characteristics of the two mushrooms. The word "Chi-Zhi" was first found in the book "Xijing" written by Zhang Heng of the Eastern Han Dynasty. He wrote, "Shijun is soaked in Chongya, and Zhuke is used to wash Chi-Zhi." (Zhang and Fu 2016). In ancient times, it was called "Yao-Cao" and "Xian-Cao," or 'Shenzhen`` in the classic book named "Shennong materia medica Jing" (Anonymous 1997). Besides, *G. Lucidum*'s shape looks like an umbrella with a cap of kidney-shaped and semicircular or nearly circular shape (Chen 1991). Lingzhi was characterized and classified based on its phenotypical characteristics and potential pharmacological effects. In the compendium of Materia Medica compiled by Li Shizhen, the smell of Lingzhi was distinguished according to "five colors" and "five elements."

Subsequently, it was believed that "the taste may not follow the five colors" (Li 1978). The ancient Chinese people have found the medicinal value of Lingzhi (Zhao 1989; Dai et al. 2009). In ancient times, Lingzhi represents longevity and rejuvenating. It can increase the wisdom and strengthen the body and is considered the "gem" in the treasure house of traditional Chinese medicine (Sanodiya et al. 2009; Zhao and Zhang 1989).

There are many myths, folklores, poems, and operas related to Lingzhi. The myths regarding Lingzhi say it is originated from the spirit of Yao Ji, who is the youngest daughter of Yan Emperor in the works of Shanhaijing. Her spirit was transformed into Lingzhi after her death, and Lingzhi can strengthen the body of people. Yao Ji is the daughter of Yan Emperor Shennong. She is beautiful, naive, and kind, deeply loved by Yan Emperor. Yao Ji grew up in Tiangong, drinking Qiongji jade liquid, eating rare fruit,

Fig. 1.1 The species that are considered representing the Lingzhi (https://www.gbif.org/occurrence/charts?taxon_key=2549730). **a** *Ganoderma Lucidum* and **b** *Ganoderma sinense*

living a happy life. But later, Yao Ji suddenly contracted a strange disease and became weak. She took countless fairy medicine, but still could not cure the peculiar disease. Finally, she left the world. After crying, Emperor Yan buried his daughter in Wu mountain. However, Yao Ji's soul did not disperse, but floated to the mountain, turned into a yellow grass, their leaves growing in pairs, and the fruit like dodder. It is said that the fruit of the grass has magical effects, and the grass absorbs the essence of the sun and moon. In the following text, we will describe the five most well-known myths involving Lingzhi to give you a sense of the deep involvement of Lingzhi in the history of China (Figs. 1.2, 1.3, 1.4, 1.5 and 1.6).

Figure 1.2 is part of a bigger picture called "chaoyuan'. It is a mural created by Ma Junxiang and his son Ma Qi etc., in the Yuan Dynasty. It depicts the story of gods worshiping "Yuanshi Tianzun" (A god in Chinese myth), and the picture shows the spectacular scene of ministers and court ladies paying tribute to Lingzhi. In the pot held in the hand of the palace lady in the picture is the nine-stalked Lingzhi, which means good luck, longevity, and joy.

Figure 1.3 described a tale named "Pirate Immortal Grass Tale of the White Snake." The story originated in the Northern Song Dynasty. Its birthplace is in Xujiagou village at the foot of Hebi Mountain and the shore of the Qi River in Hebi county, Henan province. It records that Bai Suzhen is a white snake that lived for thousands of years and gained human form. When she and her sister Green Snake met scholar Xu Xian in a Broken Bridge, a love affair arised. The white snake expressed love and affection, and the white snake and Xu Xian finally got married. At the Mid-Autumn Festival, the couple invited the moon to have a drink and spent a good night together. The drink was too strong, and the white snake showed her original form of a snake. Xu Xian was scared to death. The white snake and the green snake fought against the mountain fairy boys for the immortal grass: Lingzhi to rejuvenate Xu Xian. Xu Xian survived after taking the Millennium Lingzhi. In the picture, the white lady is holding "Lingzhi immortal grass" and wants to leave after a battle with fairy boys.

Figure 1.4 depicted the tale of Magu. Magu's offering of longevity is a myth that has spreaded widely in China. According to legend, Magu was a simple and beautiful folk woman. Ge Hong of the Jin Dynasty said that she was from Jianchang county in "The Legend of the Immortals," who practiced Taoism in Guyu Mountain, southeast of Mouzhou region. There were 13 clear springs in the mountains, and Magu used this spring water to brew Lingzhi wine. In the 13th years, the wine was matured, and Magu became immortal. On the birthday of Wangmu (The Heaven Queen in Chinese Mythology), Magu took the Lingzhi wine to Yaotai (The Heaven Palace in Chinese

Fig. 1.2 Murals of Yongle Palace in Yuan Dynasty. *Source* National Geographic of China—Lingzhi out of myth

Fig. 1.3 Tale of the White Snake Pirate Lingzhi, An immortal grass. *Source* https://www.zhihu.com/

Mythology) to celebrate Wangmu's birthday. Wangmu was overjoyed and named Magu a "Female Birthday Fairy." This painting depicts Magu and a maid, both dignified and beautiful. Magu holds a fairy stick in one hand, a treasure gourd filled with Lingzhi wine is attached to the end of the stick, and a jade plate in the other hand. The maid behind Magu holds a vase in both hands and puts the snow-white plum and red camellia in the bottle, looking forward with a focused expression. They were going to Yaotai to celebrate Wangmu's birthday.

Figure 1.5 was painted by an emperor of the Ming dynasty named Zhu Jianshen. In the picture, Zhong Kui stared sharply at the flying bats, holding Lingzhi Ruyi, a mascot with the shape of

Fig. 1.4 Magu Xianshou (Magu's birthday offering, collection of Beijing Palace Museum)

Fig. 1.5 Suizhao JiaZhao picture drawn by an Emperor of the Ming dynasty (Collection of Beijing Palace Museum)

Fig. 1.6 Pengzu, an immortal having a long and healthy life. *Source* http://info.service.hc360.com/2018/06/211039538622.shtml

Lingzhi, in one hand and resting the other on the shoulder of the kid. The kid is holding a tray of persimmons and cypress branches in both hands, which means "Bai Shi Ruyi." It is a New Year's picture praying for an auspicious blessing.

Figure 1.6 depicted Peng Zu, a long-lived man in the "Tang" and "Yao" era, hidden in a mountain named "Wuyi." Peng Zu was the first monarch of the Great Peng Kingdom and the great-great-grandson of Zhuan Xu, a descendant of the Yellow Emperor. Because he often took the "Lingzhi immortal grass" from Wuyi Mountain, he looked like a child when he was 760 years old. The history of later generations said that his health preservation method is "eating Lingzhi, drinking waterfall, retreating for health." In the picture, Peng Zu is holding Lingzhi jewelry in his hand, sitting at the foot of Wuyi Mountain, cooking Lingzhi on the stove. He looks young and has a relaxed posture.

These fairy tales demonstrate that "Lingzhi" has become an integral part of Chinese culture.

In history, Lingzhi and its derivative mascot, called "Ruyi," were used to symbolize "national peace and people's peace." Ruyi also represents wishes for Indian Buddhism to pursue the future and look forward to acquiring happiness in the next life. Christians in Western Europe revere Lingzhi as Ruyi. Although Lingzhi has been an essential part of Chinese history for 3000 years, the scientific description of its taxonomy, biology, chemical compositions, and pharmacological effects have not been addressed until recently.

1.1.3 Modern Research on Lingzhi

In recent years, Chen (1991), Zhao and Zhang (1989), Lin (2007, 2009), Lin and Yang (2019), and other experts have made significant contributions to Lingzhi research. They have explored the origin, history, culture, and development of fungus *Ganoderma* from the aspects of archaeology, history, literature, ethnology, linguistics, aesthetics, philosophy, folklore, traditional Chinese medicine, sociology, nutrition, and biochemistry. *G. Lucidum* culture has been discussed in some publications and books such as Chinese Edible Fungi, Microbiology Bulletin, Natural Resources, Bauhinia, and Chinese Science Pictorial (Huang et al. 2010). In the following text, we will briefly introduce the biology, genetics, chemical components, pharmacological effects, clinical applications, and general health-promoting products made of Lingzhi.

1.2 Basic Biology

1.2.1 Taxonomy and General Life Cycle of Lingzhi

As described earlier, there are two definitions of Lingzhi. In the broad sense, Lingzhi refers to many fungal species belonging to *Ganoderma* (Genus), *Ganoderma*taceae (Family), Polyporales (Order), Agaricomycetes (Class), and Basidiomycota (Phylum), Fungi (Kingdom). More than 20 species of *Ganoderma* (Lingzhi) fungi have been studied, including *G. Lucidum, G. sinense, G. japonicum, G. capense, G. australe, G. tsugae, G. applanatum, G. tropicus, G. boniense, G. duropora, G. resinaceum, G. theaecolum, G. cochlear, G. atrum, G. formosanum, G. boninense, G. colossum, G. concinna, G. amboinense, G. pfeifferi, and G. orbiforme*. Two of them, *G. lucidum* and *G. sinense*, are listed as the authentic sources of Lingzhi mushroom in the Chinese Pharmacopeia (Chinese Pharmacopeia Commission 2020). Among the two, *G. lucidum* (Curtis) P. Karst is more widely studied and used. In the following text, we will use *G. lucidum* and Lingzhi interchangeablely.

G. lucidum can grow and develop under artificial cultivation conditions. After inoculation for 1 or 2 days, the mycelium begin to sprout. During the growth of hyphae, the length increases, and a large number of branches grow. After a week, the hyphae can cover the medium surface and then spread deep into the medium. Hyphae absorb water and nutrients on the surface and secrete enzymes to decompose nutrients.

Mycelium grows interlaced to form various hyphal tissues. At this time, the mycelium on the medium's surface developed local swelling, protruding into a tumor-like white bud. With the continuous input of nutrition and water, the bud cells rapidly split and proliferate, extending into rod-like stalks in the form of apical growth. The stem is round and upright or extends up along the bottle wall (Fig. 1.7a and b).

The growth points expanded, connected, and intersected to form a growth circle, and a wheel grew outwards. The lower part of the lid is continuously differentiated and thickened to create a white or yellowish layer. Then appear on the surreal layer many small holes, that is, porous structures. The bud color gradually changed from white to yellowish during the fruiting body's growth and development, and then deepened into yellow, red, and purple (Fig. 1.7c and d). When the light color disappeared around the lid, the lid stopped growing, the surface showed lacquer luster, and many brown spores were dispersed from the tube, forming the mature Lingzhi (Fig. 1.7e and f).

1.2.2 Worldwide Distribution of Wild *G. Lucidum*

In the database GBIF (http://gbif.org, last accessed: January 2021), a total of 5,465 occurrences have been described in GBIF. Among them, 3839 occurrences have coordinates (Fig. 1.8).

Geographically, it shows that *G. lucidum* distributes around the world. However, most occurrences are found in Europe. Time-wise, there is a steady increase in the occurrence of Lingzhi (Fig. 1.9). Across multiple years, the occurrences of Lingzhi rapidly increased from 1950 to now, exceeding all the occurrences in the previous 150 years. Within a year, the Lingzhi can be identified throughout the years. However, it is most likely to be found in September (Fig. 1.9b). This also suggests that the best time for the *G. lucidum* growth and development in the wild is in autumn. In terms of the records' sources, most data are form North European countries (Fig. 1.9c), such as Swedish, Danish, Norwegian, Britain, and Ireland. Most *G. Lucidum* occurrences were found through human observation in terms of types of occurrences, and there are many living specimens. However, this observation cannot be explained as Lingzhi are mostly distributed in these areas. These observations can result from the more active collection and deposition of samples into the museums or herbaria in these regions (Fig. 1.9d).

1.2.3 Cultivation and Industrial Productions.

In the 1950s, the Institute of Microbiology, Chinese Academy of Sciences, successfully cultivated *G. lucidum* for the first time (Li et al. 2016a). Since then, the artificial cultivation of *G. lucidum* in China grows rapidly in response to the demand (Wang et al. 2012, 2014; Steyaert 1980; Yu 2017; Li 2015). Many strains have been developed carrying favorable characteristics (Zheng 2007), such as "Hunong 1" (He et al. 2011), "Liaolingzhi 2" (Liu et al. 2018a), "Xianzhi 2" (Li et al. 2017a), "Lingzhi 2" (Li et al. 2016b) etc. Various cultivation techniques have also been developed (Hapuarachchi et al. 2018; Qiu et al. 2016). There are two general cultivation methods: block wood cultivation and substitute cultivation. For block cultivation, wood logs are usually selected for stumpage cultivation, and fruiting bodies are gradually formed by inoculating mycelium (Wang et al. 2017a; Chaiyasut et al. 2010). For substitute cultivation, suitable sawdust and agricultural and sideline products are used as cultivation materials to make fungus package, and fruiting bodies are gradually formed by inoculating and culturing mycelium (Ma et al. 2007).

Fig. 1.7 Morphology of different growth periods of *Ganoderma lucidum*. **a** the developing fruiting body stage of *Ganoderma lucidum*. **b** Immature *Ganoderma lucidum*. **c** the fruiting body stage of *Ganoderma lucidum*. **d** the fruiting body stage of *Ganoderma lucidum*. **e** the sporophyte stage of *Ganoderma lucidum*. **f** the sporophyte stage of *Ganoderma lucidum*. *Source* https://stock.tuchong.com

The growth and development of *G. lucidum* prefer high temperature and humidity. The optimum temperature is 26–28°C, and the relative humidity of air is 85–95%. The cultivation temperature and humidity should be under control during the period of growing up (Li et al. 2012).

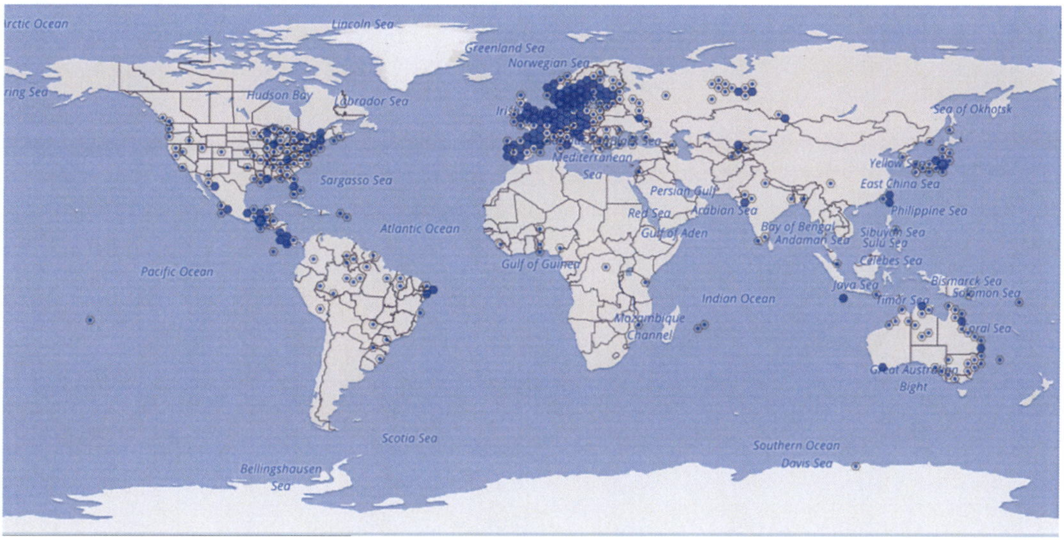

Fig. 1.8 World-wide distribution of *G. lucidum*. The blue dots represent the occurrence of the *G. lucidum* samples. The darkness of the blue dots reflects the number of occurrences

The development of the fruiting body is sensitive to CO_2 (Fu et al. 2009). When the content of CO_2 is 0.1–1.0%, it can promote the growth of the stipe but not the formation of the pileus. The young fruiting bodies tend to grow toward the light. Therefore, various forms of *G. lucidum* can be formed by adjusting the light source and CO_2 concentration in production.

1.3 Genetic Studies

Genome sequencing becomes a standard method for identifying genes and understanding the biosynthesis of active components and their regulations with the rapid development of DNA sequencing technologies. Each lingzhi cell has two different genomes, the nuclear genome and the mitogenome (Yu et al. 2008). The two genomes interact with each other to support the life cycle of Lingzhi (Chen et al. 2012a). The research progress on the nuclear and mitochondrial genomes is described in detail in Chap. 3 and Chap. 4 (Li et al. 2013a). Furthermore, a complementary strategy is to use the RNA-seq technology to obtain the genes encoded in a genome. In Chap. 5, transcriptome studies of Lingzhi are described in detail. Efficient transformation systems are required for the functional validation of particular genes. During the past years, mature transformation systems have been developed to successfully validate lingzhi genes' functions. Furthermore, these technologies set up the stage to conduct transgenic studies in Lingzhi mushrooms. The genetics study is the main subject of this book.

In 2012, Chen et al. used the next-generation sequencing (NGS) technologies, including Roche 454 and Illumina sequencing platforms, to obtain the whole genome sequences of *G. lucidum*. The whole genome of *G. lucidum* is the first published genome of facultative parasitic medicine Edible Fungus in the world. The results show that the size of the whole genome of *G. lucidum* is 39.9 Mb, encoding 12,000 proteins (Li et al. 2013b). Based on this study's results, the biosynthesis pathway of triterpenoids in *G. lucidum* was proposed (Li et al. 2013b; Huang et al. 2013).

Fig. 1.9 The occurrences of lingzhi. **a** by year; **b** per month; **c** by contributing organization; **d** by type

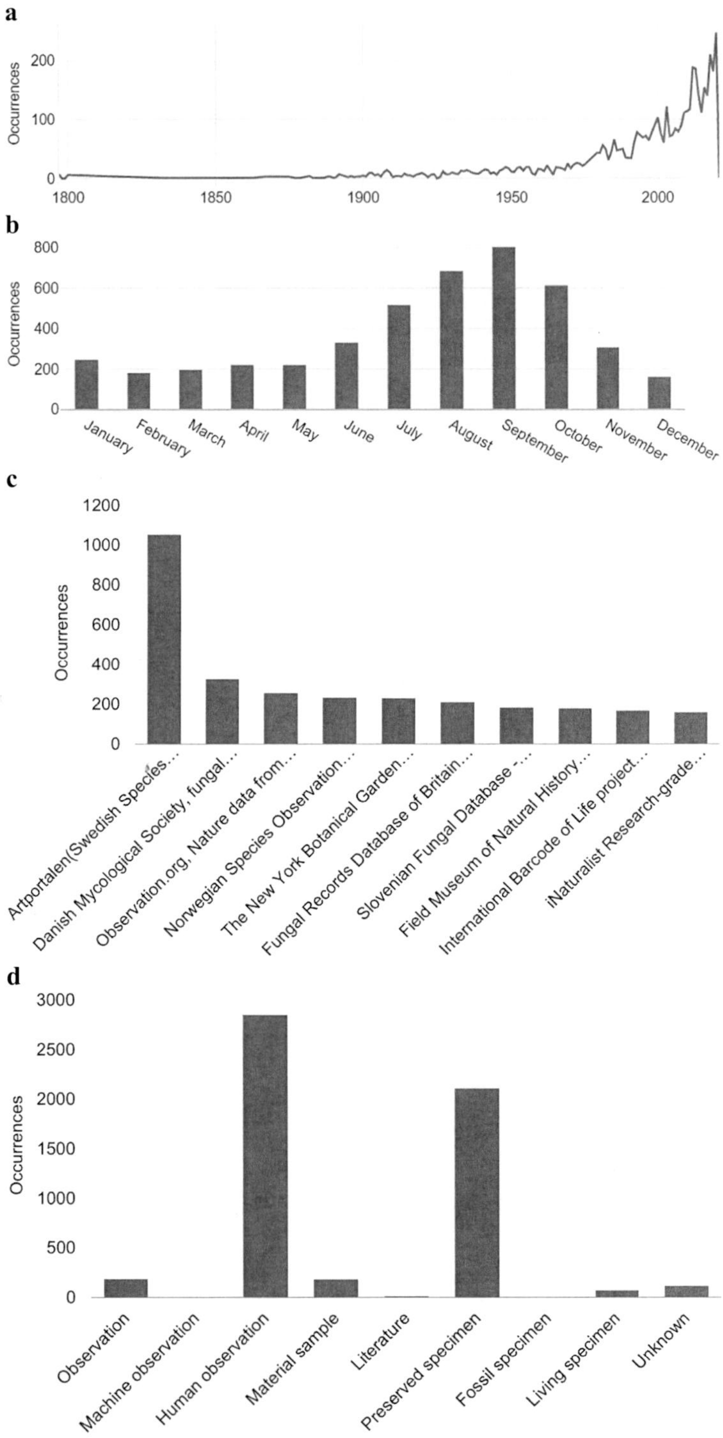

1.4 The Chemical Components of Lingzhi

Over 600 compounds were isolated from the Lingzhi. These include polysaccharides, nucleosides, furans, sterols, alkaloids, triterpenoids, oils, amino acids, protein enzymes, organic germanium, and trace elements (Liu et al. 2012; Gong et al. 2019; He et al. 2010). Their structures are shown in Fig. 1.10. Among them, more than 300 are triterpenes (Zhang et al. 2008; Chen et al. 1991). Triterpenoids are one of the main chemical components of *G. Lucidum* (Ma et al. 2011; Liu et al. 2019a) (Fig. 1.10a). The most abundant group of Triterpenoids is *Ganoderma* acid (GA) (Lin et al. 1988), a kind of tetracyclic triterpenoids (Liu et al. 2020; Ma et al. 2003). It has a variety of structures and is highly oxidative. It can effectively induce the apoptosis of HeLa cells. Furthermore, it can reduce the mitochondrial membrane potential, increase the membrane permeability, and further induce cell apoptosis (Duan et al. 2018). *Ganoderma* triterpenoids can purify the blood and protect liver function (Li et al. 2005). *G. lucidum* polysaccharides (GLPs) are another group of active components (Liu et al. 2014a) (Fig. 1.10e). They have immunomodulatory, hypoglycemic, hypolipidemic, antioxidant, antiaging, and antitumor effects (Boh et al. 2007; Shiao 2003; Ferreira et al. 2015; Xiao et al. 2006). Below, we describe each type of these chemical components.

1.4.1 Triterpenoids

The *Ganoderma* triterpenoids (GTs) are highly oxidized lanostane derivatives with complex chemical structures. They are unique chemical components derived from the genus *Ganoderma* with a variety of activities. At present, more than 300 kinds of GTs have been isolated. The GTs extraction methods include the following: organic solvent extraction, ultrasonic-assisted extraction, microwave-assisted extraction, supercritical fluid extraction (Ahmad 2018). These compounds include tetracyclic (lanostane carbon skeleton) and pentacyclic triterpenoids. According to the numbers of carbons, the triterpenoids can be divided into three groups: C30, C27, and C24. GTs with similar structures but different functions are named by letters, such as *Ganoderma* A, B, C, D, E, G, I, L, Ma, Mb, Mc, Md, Mg, and Ganoderic acid A, B, C, D, E, F, O, etc. (Chen et al. 1991; Duan et al. 2018). The chemical structures of four representative GTs are shown in Fig. 1.10a.

1.4.2 Meroterpenoids

Ganoderma Meroterpenoids (GMs) are hybrid natural products originating from the shikimic acid and mevalonic acid biogenetical pathway (Ko et al. 2008). It is not until 2000 that GMs such as ganomycins A and B were reported from the mature fruiting bodies of *G. pfeifferi* (Zhang and Zhao 2020). With new techniques, about 200 GMs were isolated from the genus *Ganoderma* (Mothana et al. 2000). GMs possessed diverse biological activities, including antioxidant, antifibrotic, anti-allergic, anti-AChE, antimicrobial, cytotoxic activities, etc. (Peng and Qiu 2018). GM is composed of a 1, 2, 4-trisubstituted phenyl group and a mono- or sesquiterpene moiety (C10 or C15 chain or cyclic moiety). According to the characters of structures, GMs could be divided into seven types: (1) GMs with 10- or 15-carbon side chains, (2) GMs with lactone groups, (3) GMs with ether rings, (4) GMs with five- or six-membered carbon rings, (5) GMs with spiro rings, (6) GMs with bridged rings, and (7) dimeric GMs (Mothana et al. 2000). The diverse terpene moiety may be formed through oxidation, cyclization, isomerization, polymerization, etc. An α,β-unsaturated γ-lactone connected to C-1' of phenyl group might be formed through a nucleophilic addition reaction between the carboxyl group at C-10' or C-14' and the ketone carbonyl at C-1'. The ether rings were

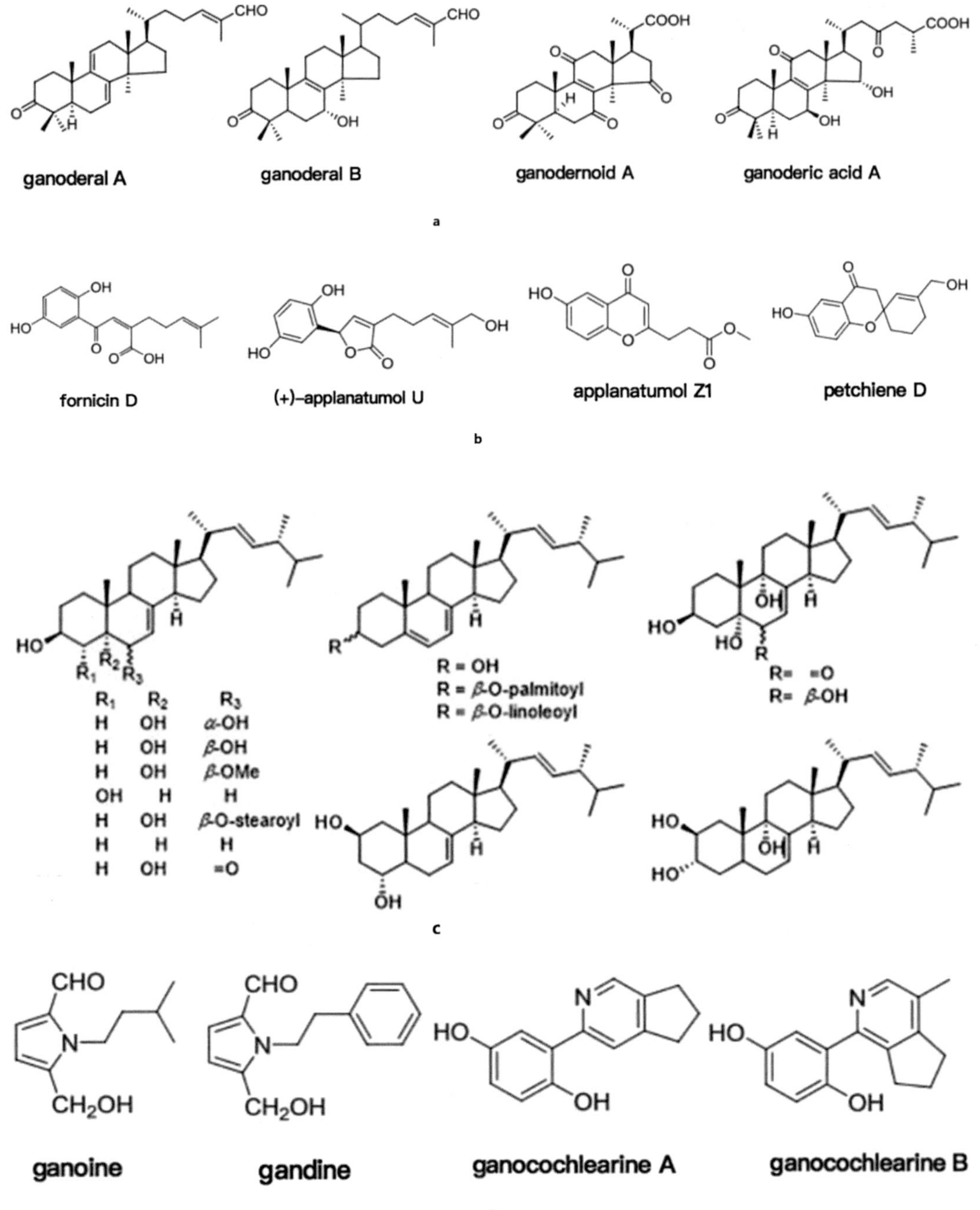

Fig. 1.10 Chemical structures of representative *Ganoderma* triterpenoids (**a**); *Ganoderma* meroterpenoids (**b**); *Ganoderma* steroids (**c**); *Ganoderma* alkaloids (**d**) and *Ganoderma lucidum* polysaccharide (GLP) (**e**)

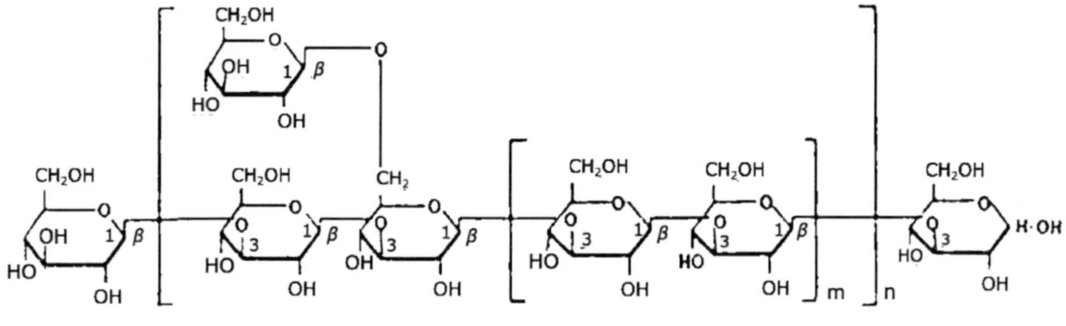

e

Fig. 1.10 (continued)

produced through a nucleophilic substitution reaction in which two hydroxyl groups were involved. The five-, six-, or seven-membered carbon rings were derived from a new C–C bond formation. The chemical structures of the four GMs are shown in Fig. 1.10b.

1.4.3 *Ganoderma* Steroids

Ganoderma Steroids (GSs) include ergosterols, stigmasterols, sitosterol, and daucosterol. Among them, ergosterols are the most commonly found (Qin et al. 2019). More than 30 different ergosterols have been isolated so far, including stigmasta-7, 22-diene-3β, 5α, 6α-triol (Peng and Qiu 2018; Chen et al. 2017), (22E, 24R)-ergosta-7, 22-diene-3β, 5α, 6β-triol (Peng and Qiu 2018; Chen et al. 2017), Ergosta-7, 22-dien-3β-ol (Bao et al. 2018; Gan et al. 1998), 22E-6β-methoxyergosta-7, 22-diene-3β, 5α-diol (Qin et al. 2019; Gan et al. 1998), and so on. The chemical structures of five *Ganoderma* steroids are shown in Fig. 1.10c.

1.4.4 Alkaloids and Nucleosides

In 1990, ganoine and ganodine were firstly isolated from *G. capense* (Seo et al. 2009). Approximately 30 alkaloids have been obtained later on, including: monocyclic and polycyclic alkaloids, purine, pyrimidine (Liu et al. 2012; Chen et al. 2007), and cerebrosides (Yu et al. 1990). Sixteen nucleosides and nucleobases have been studied (Zhao et al. 2015; Phan et al. 2018). The chemical structures of four *Ganoderma* alkaloids and nucleosides are shown in Fig. 1.10d.

1.4.5 *Ganoderma* Polysaccharides

More than 200 different *G. Lucidum* polysaccharides (GLPs) have been isolated and purified from the fruiting bodies, spores, mycelia, and cultivation broth of *G. Lucidum* (Fig. 1.10e). GLP separation methods include fractional precipitation, ion-exchange chromatography, ultrafiltration, etc. (Chen et al. 2012b). Most polysaccharides are composed of homoglucans (α/β-glucans), glycoproteins, and heteropolysaccharides. The GLPs mainly include GLPI, GLPII, GLPIII, GLPIV (Boh et al. 2007), GTM1, GTM2, GTM3, GTM4 (Chen et al. 2007), GLB6, GLB7, GLB10, and GLC1 (Huang et al. 2010; Peng et al. 2005).

1.5 Bioactivities of Lingzhi's Chemical Components

Ganoderma is a medicinal fungus that has been studied systemically for the past 40 years (Qi et al. 1993). Recent studies have identified broad spectra of biological activities in Lingzhi, such as immunomodulatory, anticancer, antidiabetic, antioxidant, anti-atherosclerotic, antitumor, analgesic, anti-inflammatory, antimicrobial, hypolipidemic, hepatoprotective, antiangiogenic, antiosteoporotic, antiaging, antiulcer properties, and estrogenic activities. Below we will briefly describe the most recent advances for each type of activities.

1.5.1 Immunomodulatory Activity

Many works have demonstrated that GLPs have immunomodulatory activity. It could enhance the function of lymphocytes, macrophages, dendritic cells, natural killer cells, etc. A bioactive fraction (named GLIS) purified from the fruiting body of *G. lucidum* stimulated the proliferation of mouse spleen lymphocytes (Bishop et al. 2015). Most of the activated lymphocytes were B cells. GLIS was also found to increase the secretion of immunoglobulin by B cells. Besides, it was found that GLIS activated B cells through the PKC pathway (Bishop et al. 2015). After treatment with GLIS, RAW264.7 macrophages were enlarged and formed pseudopodia. Treating RAW264.7 macrophages by GLIS resulted in a significant increase of NO production, induction of cellular respiratory burst activity, and increased gene expression levels of IL-1β, IL-12p35, and IL-12p40 (Zhang et al. 2002a).

The dendritic cells could be activated by the GPs. Polysaccharides from *G. atrum* (PSG-1) induced the maturation of dendritic cells; the molecule expression of MHC-II, CD80, and CD86; and the production of IL-12 p70, IL-6, IL-10, RANTES, MIP-1α, and MCP-1 in DCs. PSG-1 directly induced DCs maturation via activating MAPK 18 pathways and indirectly stimulated DCs separated by intestinal epithelial cells (Zhang et al. 2010).

The immune activities were also tested in animals. When the GPs were fed to Wistar rats, it increased the degree of toe swelling and enhanced the primary immune response to SRBC (Wang et al. 2017b). GPS significantly upregulated the expression of nuclear factor-κB p65, and secretory immunoglobulin A in the ileum. It markedly improved IFN-γ, IL-2, and IL-4 and decreased the level of diamine oxidase in serum (Zhang et al. 2016). The GLP extracted by ultrasonic/microwave improved the weight of immune organs of immunocompromised mice. A water-soluble extract from *G. lucidum* mycelia (MAK) has an anti-inflammatory effect on murine colitis induced by trinitrobenzene sulfonic acid. The induction of granulocyte–macrophage colony-stimulating factor (GM-CSF) by MAK resulted in its anti-inflammatory effects (Jin et al. 2017).

1.5.2 Antitumor Activity

Modern studies have found that *Ganoderma* can help to treat different types of tumors by regulating immunity. The main bioactive components are polysaccharides, triterpenoids, and organic germanium. Jiao evaluated the anticancer properties of *G. Lucidum* spore oil (GLSO) in vitro and in vivo. He found that GLSO could inhibit the expression of Bax and caspase-3 in MDA-MB-231 cells and tumor growth in vivo by inducing apoptosis (Huang and Ning 2010). Das found that ganoderic acid A/DM could promote the expression of apoptotic factors by inhibiting the growth of NDRG2 and changing the intracellular signaling pathway, reducing the proliferation and spread of tumor cells and playing a particular role in the treatment of meningioma (Jiao et al. 2020). Smina found that *G. lucidum* total triterpenes could promote human breast cancer cells MCF-7 apoptosis, reduce the number of tumors, and extend the tumor incubation period (Das et al. 2019). However, GLPs could not directly inhibit the growth of tumor cells; they exert the antitumor effect by enhancing the host's defense system (Smina et al. 2017). In another study, GLIS markedly increased macrophage phagocytosis and raised the macrophage-mediated tumor cytotoxicity. Treating mice with GLIS caused inhibition of mouse sarcoma S180 tumor growth by 60% in vivo (Ji et al. 2007). Finally, GLP-Au exhibited strong inhibitory effects on tumor growth and pulmonary metastasis combined with doxorubicin (Zhao et al. 2009).

1.5.3 Hypoglycemic Activities

GLP was hypoglycemic in vivo, which could enhance insulin sensitivity or increase insulin levels. In a T2DM rat model, the polysaccharides significantly improved the glucose and lipid metabolism-related parameters, which seemed to enhance insulin sensitivity, increase glycogen synthesis, and facilitate glucose transportation by regulating the PI3K/Akt pathway (Zhang et al. 2019). GLP exerts its hypoglycemic activity by increasing plasma insulin levels and decreasing plasma sugar levels in mice (Liu et al. 2019b). GLP was used in patients with type II diabetes mellitus and appeared effective and safe in lowering blood glucose concentration (Ma et al. 2015). The possible mechanism was to reverse alloxan-induced islet viability loss by inhibiting the free radicals production, increasing serum insulin, and reducing serum glucose levels (Xiao et al. 2017).

1.5.4 Hepatoprotective Effect

G. Lucidum spore has an effect on [Cd(II)]-induced hepatotoxicity in mice. It protected against Cd(II)-induced liver injury in a dose-dependent manner (Gao et al. 2004). It decreased Cd(II) accumulation in hepatic nuclei, mitochondria, and microsomes; and induced the expression of hepatic metallothionein-1 mRNA eight fold (Zhang et al. 2003). *G. atrum* polysaccharides (PSG) can be used for managing drug-induced liver injury (Jin et al. 2013a). These effects may relate in part to GLP's antioxidant properties. Treatment with *G. lucidum* aqueous extracts significantly decreased serum ALT, and AST levels, significantly increased SOD and CAT activities, and decreased the MDA content in the liver compared with the α-AMA control group (Fan et al. 2018; Zhang et al. 2002b).

1.5.5 Anti-hyperlipidemia Effect

Animal feeding experiments showed that *G. lucidum* had hypolipidemic effects. Oral administration of the polysaccharide from *G. atrum* (PSG-1) at 200 or 400 mg/kg body weight significantly reduced fasting blood glucose. PSG-1 also decreased total serum cholesterol, triglyceride, low-density lipoprotein cholesterol, free fatty acid, and insulin resistance (Wu et al. 2013a). After the rats were treated with GLPs at dosages of 200, 400, and 800 mg/kg, the serum TC, TG, HDL-C, LDL-C, GSH-Px, SOD, and LPO significantly decreased (Zhu et al. 2013). Oral administration of GLPs for 40 days resulted in a dose-dependent significant reduction of TC and TG (Chen et al. 2005).

1.5.6 Regulation of Intestinal Flora

Recently, the effect of GPs on intestinal flora has attracted more attention. GPs can improve intestinal barrier function, regulate intestinal immunity, and modulate intestinal microbiota. Different metabolites associated with the improvement of intestinal immunological function and intestinal microbiota regulations were also identified. The results provided a potential mechanism of health-beneficial properties of GLPs (Yang et al. 2010). The GLP strain S3 (GLPS3) markedly alleviated pancreatitis in mice by decreasing lipase, AMs, IFN-γ, and TNF-α level and increasing SOD and total antioxidant activity. Furthermore, GLPS3 altered intestinal microbiota's composition and diversity, primarily via decreasing the relative abundance of the phylum Bacteroidetes and increasing the relative abundance of the phylum Firmicutes. At the genus level, supplementation of GLPS3 increased the relative abundance of beneficial bacteria such as Lactobacillales, Roseburia, and Lachnospiraceae (Jin et al. 2019).

1.6 Pharmacological Effects of Lingzhi Preparations

From the fruiting bodies, mycelium, and spores of *G. lucidum*, various chemical components with pharmacological activities can be extracted as water and fat-soluble fractions (Liu et al. 2012; He et al. 2010). The water-soluble components have antitumor, antiaging, and immune regulatory effects (Boh et al. 2007). In contrast, the fat-soluble components have the functions of lowering blood pressure, reducing blood viscosity, reducing thrombosis, and promoting blood circulation, etc. (Liu et al. 2014a; Ma et al. 2015).

1.6.1 Antitumor Effect

Both GLPs and GLTs have anticancer or antitumor effects (Li et al. 2016c; Wachtel-Galor et al. 2011; Xu et al. 2011). Wu et al. reviewed GLTs acting as an anticancer constituent (Sohretoglu and Huang 2018). Yan et al. studied the synthesis and antitumor pharmacological activity of GLTs (Wu et al. 2013b). Gao et al. reported that Ganopoly, an aqueous polysaccharide fraction extracted from *G. lucidum*, had an antitumor activity with a broad spectrum of immunomodulating activities (Yan et al. 2017). Hsu et al. found WSG, a water-soluble GLP, effectively inhibited cell viability and mobility, and lung cancer cells' growth (Gao et al. 2005). WSG also inhibited phosphorylation of multiple intracellular signaling molecules to reduce the size of metastatic nodules in the lungs and to prolong the survival of LLC1-bearing mice. Xie et al. prepared two fractions: a GLP called GLE-1 and a triterpenoid fraction without polysaccharides called GLE-2. He found that both GLE-1 and GLE-2 significantly inhibited SW 480 human colorectal cancer cells' proliferation (Hsu et al. 2020). Li et al. found that BSGLEE, a preparation of triterpenoids from *Ganoderma*, effectively inhibited colorectal cancer carcinogenesis. The mechanism involved the induction of apoptosis, inhibition of migration, and promotion of cell cycle arrest (Xie et al. 2006).

Jedinak et al. studied the *Ganoderma* nontriol (GNDT), a purified triterpene from *G. lucidum*. He found that GNDT inhibited the proliferation of HCT-116 and HT-29 colon cancer cells without a significant effect on cell viability (Li et al. 2017b). GNDT suppressed tumor growth in a xenograft model of human colon adenocarcinoma cells HT-29 implanted in nude mice without any side-effects and inhibited the expression of cyclin D1 in tumors. The novel natural triterpene GL22 isolated from *G. leucocontextum* significantly inhibits the growth of the liver cancer cell line in vitro and of tumor xenografts in vivo through targeting lipid metabolism via manipulating FABPs as a cancer treatment strategy (Andrej et al. 2011). Moreover, Ivette et al. reported that some *Ganoderma* species had great potential as a natural therapeutic for breast cancer (Liu et al. 2018b). Hsieh et al. found that *G. lucidum* preparations elicited its antitumor effects by suppressing cell growth and inducing antioxidative or detoxification activity in human ovarian OVCAR-3 cells (Ivette et al. 2017).

1.6.2 Immunoregulation Effect

Ganoderma can be used for cancer immunotherapy by regulating the immune system (Hsieh and Wu 2011; Cao et al. 2018). Guo et al. reported that a water-soluble polysaccharide named GSG is an effective immunomodulator. It contained antitumor activity against Lewis lung cancer in mice (Jiang et al. 2010). Bao et al. isolated and identified three immune-activating polysaccharides based on their activities in enhancing the proliferation of T- and B-lymphocytes in vitro in mice (Guo et al. 2009). Pan K et al. studied the extraction of GLPs. They found that GLP could enhance immunity and antioxidant activities in gastric cancer rats by reducing the levels of serum IL-6 and TNF-α levels. This, in turn, upregulated serum IL-2, IL-4, and IL-10 and activated serum and gastric tissue SOD, CAT, and GSH-Px (Bao et al. 2002).

Hem et al. found the immunosuppressive effects of LZ-8 in an allogeneic mouse skin transplantation model (Pan et al. 2013). Lin et al. found that LZ-8 can accelerate wound healing in rat liver tissues after monopolar electrosurgery. The molecular mechanism involved decreasing the expression and minimal dispersion of NF-KappaB, reducing caspase-3 activity, and the apoptosis of the liver tissue (Hem et al. 1996). LZ-8 could be developed as an adjuvant in the DNA vaccines (Lin et al. 2014). Li et al. characterized 109 constituents from *G. lucidum* spore. They found positive correlations between triterpenoid contents and the zebrafish-based bioassay results (Lin et al. 2014). The molecular mechanisms involved the alleviation of macrophage deficiency and significant enhancement of the phagocytic function of macrophages.

1.6.3 Protection and Treatment of Organ Damage

Liver injury cannot be completely cured in the world. It tends to develop into liver cancer. *G. lucidum* demonstrated hepatoprotective effects in different liver diseases, including hepatocellular carcinoma, etc. (Chu et al. 2011). Jin et al. found that *G. lucidum* spore powder has a protective effect on cadmium (Cd)-induced liver injury in mice (Li et al. 2020). Liu et al. found that *G. theaecolum* triterpenoids exhibited hepatoprotective activities against DL-galactosamine-induced cell damage in HL-7702 cells (Qiu et al. 2019). Furthermore, triterpenoids and polysaccharide peptides-enriched *G. Lucidum* preparation have antioxidation, antiaging, and hepatoprotective effects (Jin et al. 2013b).

Many lines of evidence showed that GLPs and GTs possessed potent inhibitory activities on CKD and AKI (Liu et al. 2014b). The molecular mechanism might involve the inhibition of the TGF-β/S mad signaling pathways (Liu et al. 2014b). Meng et al. reported that ganoderic acid A (GA-A) might be the main ingredient of GTs as a potential therapeutic reagent for treating autosomal dominant polycystic kidney disease (ADPKD) (Chiu et al. 2017). KUOK et al. found that GAs are the bioactive components against cardiac insults created with isoproterenol on mice (Geng et al. 2020). GAs dissipated the cellular reactive oxygen species and prevented cell death through testing the cultured cardiomyoblast H9c2 cells against the insult of H_2O_2 (Geng et al. 2020). GTs can improve the therapeutic effect of frostbite in rats (Meng et al. 2020). *G. lucidum* polysaccharides are thought to prevent the T/D-induced I/R injury by reducing the apoptotic effect in testicular torsion-detorsion (Kuok et al. 2013).

1.6.4 Antiaging Effect and Antioxidant Activity

It is reported that polysaccharides, triterpenes, peptides, and polysaccharide-peptides of *G. lucidum* have antiaging functions. The molecular mechanisms involve antioxidation, immunomodulation, lifespan extension, promotion of stem/progenitor cell survival, and anti-neurodegeneration (Shen et al. 2016). The terpene fraction of ganoderic acids A, B, C, and D, lucidenic acid B, and *Ganoderma* nontriol was found to possess the highest antioxidative effect against erythrocyte membrane oxidation and lipid peroxidation (Dogan and Ipek 2020). Triterpenes and aromatic meroterpenoids from *G. lucidum* have antioxidant potencies, radical scavenging activities, and neuroprotective activities against H_2O_2 (Wang et al. 2017c). Activities of the antioxidant enzymes in blood and tissue were increased by the administration of total triterpenes (Zhu et al. 1999). The mycelia protein extracts of Lingzhi partially purified by Diethyl aminoethyl (DEAE)-Sepharose column and Sulfopropyl (SP)-Sepharose column could protect DNA damage by hydroxyl radicals. As a result, it has the potential to be used as antioxidant and antibacterial agents (Wang et al. 2019). GLPs have been used in promoting skin wound healing, mitigating postburn infection, and preventing skin flap ischemia–reperfusion injury. *Ganoderma* extracts have also been used in skincare because of their roles in skin photoaging and skin whitening (Smina et al. 2011). Chitosan preparations extracted from *G. lucidum* have a better antioxidant, cytotoxic, and antimicrobial activity (Sa-Ard et al. 2015).

1.6.5 Antibacterial and Anti-Inflammatory Effects

SA-ARD et al. reported that *G. lucidum* extracts showed antibacterial activity (Wang et al. 2019). Three triterpenoids and five steroids from the fruiting bodies of *G. mastoporum* collected in Vietnam were characterized. They inhibited superoxide anion generation and elastase release by human neutrophils in treating inflammatory diseases (Yin et al. 2019). Another study found that GTs might be useful for Alzheimer's disease treatment (Savin et al. 2020). Lanostane triterpenoids isolated from *G. lucidum* could possess an inhibitory effect against NO production in vitro and has anti-inflammatory effects (Thang et al. 2013; Patocka 1999).

1.6.6 Antiviral and Other Effects

Two GLTs: GLTA and GLTB can bind the viral capsid protein at a hydrophobic pocket to prevent EV71 infection. The mechanism involves interacting with the viral particle to block the adsorption of the virus to the cells (Koo et al. 2019). Two proteins bound polysaccharides were isolated from water-soluble substances of *G. lucidum* by EtOH precipitation and DEAE–cellulose column chromatography. The results showed that they have antiviral activities against herpes simplex virus type 1 (HSV-1) and type 2 (HSV-2) through complex interactions of viruses with cell plasma membranes (Wu et al. 2019).

In addition to the above effects described, *G. lucidum* has many other effects, including: antidiabetic effect (Liu et al. 2019b; Ma et al. 2015; Zhang et al. 2014), antihypoglycemic, antihypertensive and hypolipidemic effect (Zhang et al. 2019; Zhu et al. 2013), nervous system effect (Wang et al. 2017c), antiatherosclerotic effect (Eo et al. 2000; Li et al. 2011; Handayani et al. 2016) antiallergic activities (Wu et al. 2012), antiradiation and antimutation effect (Liu et al. 2006; Nurhuda et al. 2013), endocrine effect, (Chen et al. 2015), reducing the side effects of radiotherapy and chemotherapy, etc. (Lakshmi et al. 2003).

1.7 Clinical Application

The clinical application of Lingzhi preparations has been described in detail. It shows that *G. lucidum* preparations have a wide range of applications in a series systems such as nervous system, gastric and intestinal system, respiratory system, and cardiovascular system (Lin and Yang 2019). The list of potential diseases that Lingzhi can treat includes: chronic bronchitis, bronchial asthma, hypertension, coronary heart disease, diabetes, liver disease, peptic ulcer, acute ischemic cerebrovascular disease, tumors, various allergic diseases, allergic or autoimmune diseases. These further include allergic rhinitis, lupus erythematosus, a variety of intractable skin diseases, glomerulonephritis, nephrotic syndrome, Alzheimer's disease, hyperlipidemia, climacteric syndrome, and benign prostatic hyperplasia, allergic asthma, hyperthyroidism, angina pectoris, leukopenia, AIDS, nephrotic syndrome, toxic diffuse goiter, localized scleroderma, dermatomyositis, polymyositis, alopecia areata, psoriasis, progressive muscular dystrophy and atrophic ankylosis, Behcet's syndrome, brain hypoplasia, retinitis pigmentosa, Keshan disease, herpes zoster, condyloma acuminatum, and other diseases (Ma et al. 2011; Li et al. 2016c). The detailed discussion is beyond the scope of this chapter. Interested readers are advised to review the cited articles.

1.8 Legal Status of Lingzhi Mushroom

Lingzhi preparations have been used as a medicine, health food, and ordinary food (Zhao et al. 2012). The preparations can derive from different tissues or developmental stages, such as fruiting body, mycelium, and spore powder. At present, more than 200 factories produce pharmaceutical, health products, and cosmetics from Lingzhi (Jianhua and Peng 2019; Liu et al. 1991). The complete list can be found on the website (https://www.nmpa.gov.cn/). The typical drug formulations are capsule, tablet, syrup, vimum, oral liquid, particles, and granules, representing 94.7% of all products (Fig. 1.11).

More than 800 kinds of health care products containing Lingzhi are registered on the market in China, including many high-end products manufactured from the USA, Korea, and Japanese (Fig. 1.12). These products are of diverse types, including facial mask (Fig. 1.12a), Cleanser (Fig. 1.12b), Tonic lotion (Fig. 1.12c), Series of

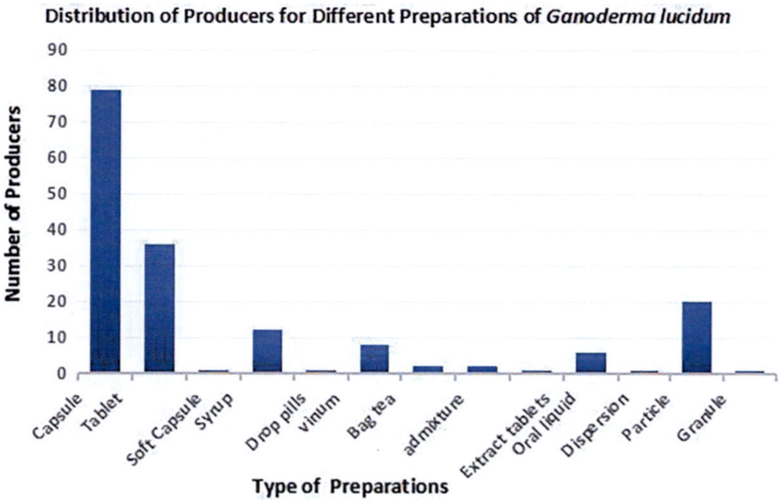

Fig. 1.11 Types of *Ganoderma Lucidum* health products approved by CFDA in China

Fig. 1.12 Examples of cosmetics products claim to contain components of Lingzhi manufactured in America, Korea, Japan and China. **a** Mask (Korean Beauty Co. Ltd). *Source* https://www.rnw.com/info/mask/76.html. **b** Cleanser. *Source* https://item.jd.com/60431509908.html. **c** Toning Lotion (Japanese An Yutang Corporation). *Source* https://item.jd.com/100002671444.html. **d** A series of products with the brand ORIGINS. *Source* https://www.origins.com.cn/dr-weil-mega-mushroom2018. **e** Eye cream (Yichang Tianmei International Cosmetics Co. Ltd). *Source* https://item.jd.com/10021965084553.html#product-detail

products (Fig. 1.12d), and essence Mile (Fig. 1.12d). Thus it is of no question that Lingzhi supports a large number of health products, including cosmetics (Hyde et al. 2010; Li et al. 2019). Details will be described in Chap. 12.

1.9 Concluding Remarks

Lingzhi has a long history of applications in human healthcare. The past 60 years have witnessed extensive studies of the chemical components, biological activities, pharmacological effects of Lingzhi and its preparations. With the rapid development of technologies, metabolic engineering, synthetic biology, and transgenic approaches are ready to breed better lingzhi strains. This book intends to cover the progress in this new frontier of Lingzhi research.

Acknowledgements We would like to thank Miss Silin Zhu and Sihui Sun for their help with the preparation of the figures. This study is supported by National Mega-Project for Innovative Drugs of China [2019ZX09735-002], National Science & Technology Fundamental Resources Investigation Program of China [2018FY100705], National Science Foundation of China Funds [81872966], and Chinese Academy of Medical Sciences, Innovation Funds for Medical Sciences (CIFMS) [2016-I2M-3-016, 2017-I2M-1-013].

References

Ahmad MF (2018) *Ganoderma lucidum*: persuasive biologically active constituents and their health endorsement. Biomed Pharmacother 107:507–519

Andrej J, Thyagarajan-Sahunita A, Jiang JJ, Sliva D (2011) Ganodermanontriol, a lanostanoid triterpene from *Ganoderma lucidum*, suppresses growth of colon cancer cells through ß-catenin signaling. Int J Oncol 38(3):761–767

Anonymous. (1997) Shen Nong Ben Cao Jing (Shennong Materia Medica). Liaoning Science and Technology Press, Shenyang

Bao XF, Wang XS, Dong Q, Fang JN, Li, (2002) Structural features of immunologically active polysaccharides from *Ganoderma lucidum*. Phytochemistry 59(2):175–181

Bao FY, Yang KY, Wu CR, Gao SY, Wang PH, Chen LX, Li H (2018) New natural inhibitors of hexokinase 2 (HK2): steroids from *Ganoderma sinense*. Fitoterapia 125:123–129

Bishop KS, Kao Chi HJ, Xu YX, Glucina MP, Paterson RRM, Ferguson LR (2015) From 2000 years of *Ganoderma lucidum* to recent developments in nutraceuticals. Phytochemistry 114:56–65

Boh B, Berovic M, Zhang J, Lin ZB (2007) *Ganoderma lucidum* and its pharmaceutically active compounds. Biotechnolo Annu Rev 13:265–301

Cao Y, Xu XW, Liu SJ, Huang LF, Gu J (2018) *Ganoderma*: a cancer immunotherapy review. Front Pharmacol 9:1217

Chaiyasut C, Kruatama C, Sirilun S (2010) Breaking the spores of *Ganoderma lucidum* by fermentation with lactobacillus Plantarum. Afr J Biotechnol 9(43):7379–7382

Chen SY (1991) Textural research on a list of lost books on glossy *Ganoderma* in ancient China. China Hist Mat Sci Technol 12(3):70–80

Chen RY, Wang YH, Yu DQ (1991) Studies on the chemical constituents of the spores from *Ganoderma lucidum*. Acta Bot Sin 33(1):65–68

Chen WQ, Luo SH, Li HZ, Yang H (2005) Effects of *Ganoderma lucidum* polysaccharides on serum lipids and lipoperoxidation in experimental hyperlipidemic rats. Zhongguo Zhong Yao Za Zhi 30(17):1358–1360

Chen Y, Xie MY, Gong XF (2007) Microwave-assisted extraction used for the isolation of total triterpenoid saponins from *Ganoderma atrum*. J Food Eng 81 (1):162–170

Chen SL, Liu C, Zhu YJ, Nelson DR, Zhou SG, Li CF, Wang LZ, Guo X, Sun YZ, Luo HM, Li Y, Song JY, Henrissat B, Levasseur A, Qian J, Li JQ, Luo X, Shi LC, He L, Xiang L, Xu XL, Niu YY, Li QS, Han MV, Yan HX, Zhang J, Chen HM, Lv AP, Wang Z, Liu MZ, Schwartz DC, Sun C (2012a) Genome sequence of the model medicinal mushroom *Ganoderma Lucidum*. Nat Commun 3:913

Chen Y, Bicker W, Wu JY, Xie MY, Lindner W (2012b) Simultaneous determination of 16 nucleosides and nucleobases by hydrophilic interaction chromatography and its application to the quality evaluation of *Ganoderma*. J Agric Food Chem 60(17):4243–4252

Chen ML, Hsieh CC, Chiang BL, Lin BF (2015) Triterpenoids and polysaccharide fractions of *Ganoderma tsugae* exert different effects on antiallergic activities. Evid Based Complement Alternat Med 2015:754836

Chen XQ, Chen LX, Li SP, Zhao J (2017) A new nortriterpenoid and an ergostane-type steroid from the fruiting bodies of the fungus *Ganoderma resinaceum*. J Asian Nat Prod Res 19(12):1239–1244

Chiu HF, Fu HY, Lu YY, Han YC, Shen YC, Venkatakrishnan K, Golovinskaia O, Wang CK (2017) Triterpenoids and polysaccharide peptides-enriched *Ganoderma lucidum*: a randomized, double-blind placebo-controlled crossover study of its antioxidation and hepatoprotective efficacy in healthy volunteers. Pharm Biol 55(1):1041–1046

Chu CL, Dz C, Lin CC (2011) A novel adjuvant Ling Zhi-8 for cancer DNA vaccines. Hum Vaccin 7 (11):1161–1164

Committee CP (2020) The pharmacopoeia of the Peoples Republic of China (Part 1). China Medical Science and Technology Press, Beijing

Dai YC, Yang ZL (2008) A revised checklist of medicinal fungi in China (in Chinese). Mycosystema 27:801–824

Dai YC, Yang ZL, Cui BK, Yu CJ, Zhou LW (2009) Species diversity and utilization of medicinal mushrooms and fungi in China (Review). Int J Med Mushrooms 11:287–302

Das A, Alshareef M, Henderson FJ, Martinez Santos JL, Vandergrift WA, Lindhorst SM, Varma AK, Infinger L, Pate SJ, Cachia D (2019) Ganoderic acid A/DM-induced NDRG2 over-expression suppresses high-grade meningioma growth. Clin Transl Oncol 22 (7):1138–1145

Dogan G, Ipek H (2020) The protective effect of *Ganoderma lucidum* on testicular torsion/detorsion-induced ischemia-reperfusion (I/R) injury. Acta Cir Bras 35(1):e202000103

Duan XY, Fan LY, Ma QY, Niu XJ, Guo J, Xu LR (2018) Optimization on Extraction and Purification Technology of Total Triterpenoids from *Ganoderma lucidum*. Traditional Chin Med Res 31(11):59–63

Eo SK, Kim YS, Lee CK, Han SS (2000) Possible mode of antiviral activity of acidic protein-bound polysaccharide isolated from *Ganoderma lucidum* on herpes simplex viruses. J Ethnopharmacol 72(3):475–481

Fan S, Huang XJ, Wang SN, Li C, Zhang ZH, Xie MY, Nie SP (2018) Combinatorial usage of fungal polysaccharides from Cordyceps sinensis and *Ganoderma atrum* ameliorate drug-induced liver injury in mice. Food Chem Toxicol 119:66–72

Ferreira IC, Heleno SA, Reis FS, Stojkovic D, Queiroz MJ, Vasconcelos MH, Sokovic M (2015) Chemical features of *Ganoderma* polysaccharides with antioxidant, antitumor and antimicrobial activities. Phytochemistry 114:38–55

Fu YJ, Liu W, Zu YG, Shi XG, Liu ZG, Schwarz G, Efferth T (2009) Breaking the spores of the fungus *Ganoderma lucidum* by supercritical CO_2. Food Chem 112(1):71–76

Gan KH, Kuo SH, Lin CN (1998) Steroidal constituents of *Ganoderma applanatum* and *Ganoderma neojaponicum*. J Nat Prod 61(11):1421–1422

Gao YH, Lan J, Dai XH, Ye JX, Zhou SF (2004) A phase I/II study of Ling Zhi mushroom *Ganoderma lucidum* (W.Curt.:Fr.) Lloyd (Aphyllophoromycetideae) extract in patients with coronary heart disease. Int J Med Mushrooms 6(1):327–334

Gao YH, Gao H, Chan E, Tang WB, Xu AL, Yang HY, Huang M, Lan J, Li XT, Duan W, Xu CJ, Zhou SF (2005) Antitumor activity and underlying mechanisms of Ganopoly the Refined Polysaccharides Extracted from *Ganoderma lucidum* in mice. Immunol Invest 34 (2):171–198

Geng XQ, Zhong DD, Su LM, Lin ZB, Yang BX (2020) Preventive and therapeutic effect of Ganoderma lucidum on kidney injuries and diseases. Adv Pharmacol 87:257–276

Gong T, Yan R, Kang J, Chen R (2019) Chemical Components of *Ganoderma*. Adv Exp Med Biol 1181:59–106

Guo L, Xie JH, Ruan YY, Zhou L, Zhu HY, Yun XJ, Jiang Y, Lü L, Chen KL, Min ZH, Wen YM, Gu JX (2009) Characterization and immunostimulatory activity of a polysaccharide from the spores of *Ganoderma lucidum*. Int Immunopharmacol 9(10):1175–1182

Handayani O, Siwi K, Widya A, Ubaidillah N, Vitriyaturidda V, Failasufi M, Ramadhan F, Wulandari H, Waranugraha Y, Hayuningputri D (2016) PS 16–11 *Ganoderma lucidum* polysaccharide peptides: a potent protective endothelial vascular and anti-lipid in atherosclerosis. J Hypertens 34(1):e468

Hapuarachchi KK, Elkhateeb WA, Karunarathna SC, Cheng CR, Bandara AR, Kakumyan P, Hyde KD, Daba GM, Wen TC (2018) Current status of global *Ganoderma* cultivation, products, industry and market. Mycosphere 9(5):1025–1052

He JZ, Shao P, Ni HD, Chai NJ, Sun PL (2010) Study on the structure and constituents of polysaccharide from *Ganoderma lucidum*. Guang Pu Xue Yu Guang Pu Fen Xi 30(1):123–127

He JF, Li ZQ, Zhang LM (2011) Breeding of *Ganoderma* cultivar Hunong 1 and its characteristics. Edib Fungi 33(5):21–22

Hem LGVD, Vliet JAVD, Kino K, Hoitsma AJ, Tax WJ (1996) Ling-Zhi-8: a fungal protein with immunomodulatory effects. Transplant Proc 28(2):958–959

Hsieh T-C, Wu JM (2011) Suppression of proliferation and oxidative stress by extracts of *Ganoderma lucidum* in the ovarian cancer cell line OVCAR-3. Int J Mol Med 28(6):1065–1069

Hsu WH, Qiu WL, Tsao SM, Tseng AJ, Lu MK, Hua WJ, Cheng HC, Hsu HY, Lin TY (2020) Effects of WSG, a polysaccharide from *Ganoderma lucidum*, on suppressing cell growth and mobility of lung cancer. Int J Biol Macromol 165(Pt A):1604–1613

Huang SQ, LI JW, Wang Z, Pan HX, Chen JX, Ning ZX, (2010) Optimization of alkaline extraction of polysaccharides from *Ganoderma lucidum* and their effect on immune function in mice. Molecules 15(5):3694–3708

Huang SQ, Ning ZX (2010) Extraction of polysaccharide from *Ganoderma lucidum* and its immune enhancement activity. Int J Biol Macromol 47(3):336–341

Huang NL, Lin ZB, Chen GL (2010) Medicinal and edible fungi. Shanghai Scientific and Technological Literature Press, Shanghai, Shanghai

Huang YH, Wu HY, Wu KM, Liu TT, Liou RF, Tsai SF, Shiao MS, Ho LT, Tzean SS, Yang UC (2013) Generation and analysis of the expressed sequence tags from the mycelium of *Ganoderma lucidum*. PloS ONE 8(5):1–13

Hyde KD, Bahkali AH, Moslem MA (2010) Fungi-an unusual source for cosmetics. Fung Divers 43:1–9

Ivette JSA, Yaliz LA, Raysa RA, Michelle MMM (2017) *Ganoderma* spp.: a promising adjuvant treatment for breast cancer. Medicines 4(1):15

Ji Z, Tang QJ, Zhang JS, Yang Y, Jia W, Pan YJ (2007) Immunomodulation of RAW264.7 macrophages by GLIS, a proteopolysaccharide from *Ganoderma Lucidum*. J Ethnopharmacol 112(3):445–450

Jiang MH, Zhu L, Jiang JG (2010) Immunoregulatory actions of polysaccharides from Chinese herbal medicine. Expert Opin Ther Targets 14(12):1367–1402

Jiao CW, Chen W, Tan XP, Liang HJ, Li JY, Yun H, He CY, Chen JM, Ma XW, Xie YZ, Yang BB (2020) *Ganoderma lucidum* spore oil induces apoptosis of breast cancer cells in vitro and in vivo by activating caspase-3 and caspase-9. J Ethnopharmacol 247:112256

Jin H, Jin F, Jin JX, Xu J, Tao TT, Liu J, Huang HJ (2013a) Protective effects of *Ganoderma lucidum* spore on cadmium hepatotoxicity in mice. Food Chem Toxicol 52(2):171–175

Jin H, Jin F, Jin JX, Xu J, Tao TT, Liu J, Huang HJ (2013b) Protective effects of *Ganoderma lucidum* spore on cadmium hepatotoxicity in mice. Food Chem Toxicol 52:171–175

Jin M, Zhu Y, Shao D, Zhao K, Xu C, Li Q, Yang H, Huang Q, Shi J (2017) Effects of polysaccharide from mycelia of *Ganoderma lucidum* on intestinal barrier functions of rats. Int J Biol Macromol 94(Pt A):1–9

Jin ML, Zhang H, Wang JJ, Shao DY, Yang H, Huang QS, Shi JL, Xu CL, Zhao K (2019) Response of intestinal metabolome to polysaccharides from mycelia of *Ganoderma lucidum*. Int J Biol Macromol 122:723–731

Jianhua X, Peng L (2019) Researches and application of *Ganoderma* spores powder. Adv Exp Med Biol 1181:157–186. https://doi.org/10.1007/978-981-13-9867-4_6

Ko HH, Hung CF, Wang JP, Lin CN (2008) Anti-inflammatory triterpenoids and steroids from *Ganoderma lucidum* and G. tsugae. Phytochemistry 69 (1):234–249

Koo MH, Chae HJ, Lee JH, Suh SS,Youn UJ (2019) Anti-inflammatory lanostane triterpenoids from *Ganoderma lucidum*. Nat Prod Res 1–8

Kuok QY, Yeh CY, Su BC, Hsu PL, Ni H, Liu MY, Mo FE (2013) The triterpenoids of *Ganoderma tsugae* prevent stress-induced myocardial injury in mice. Mol Nutr Food Res 57(10):1892–1896

Lakshmi B, Ajith TA, Sheena N, Nidhi Gunapalan KK, Janardhanan, (2003) Antiperoxidative, anti-inflammatory, and antimutagenic activities of ethanol extract of the mycelium of *Ganoderma lucidum* occurring in South India. Teratog Carcinog Mutagen 1:85–97

Li SZ (1978) Ben Cao Gang Mu (Compendium of Materia Medica). People's Medical Publishing House, Beijing

Li MY (2015) Research and development of quality standard and specifications of *Ganoderma lucidum* from an industry chain perspective. Edible Med Mushroom 23(5):276–279

Li CH, Chen PY, Chang UM, Kan LS, Fang WH, Tsai KS, Lin SB (2005) Ganoderic acid X, a lanostanoid triterpene, inhibits topoisomerases and induces apoptosis of cancer cells. Life Sci 77:252–265

Li FL, Zhang YM, Zhong ZJ (2011) Antihyperglycemic Effect of *Ganoderma lucidum* Polysaccharides on Streptozotocin-Induced Diabetic Mice. Int J Mol Sci 12(9):6135–6145

Li PF, Yao SJ, Qiu YM (2012) The optimum study on *Ganoderma lucidum* fermentation process. Food Sci Technol 37(2):33–35

Li JQ, Zhang JH, Chen HM, Chen XD, Lan J, Liu C (2013a) Complete Mitochondrial Genome of the Medicinal Mushroom *Ganoderma lucidum*. PLoS ONE 8(8):1–12

Li QS, Xu J, Zhe YJ, Sun C, Song JY, Chen SL (2013b) Estimation of the genome size of *Ganoderma Lucidum* based on a flow cytometric analysis. Mycosystema 32 (5):899–906

Li QY, Zhong YY, Chen YX, Zhou WX, Zeng ZJ, Song B (2016) Research advances on the production technology of *Ganoderma* in China (in Chinese). Edible Fungi China 35(1):8–12+20

Li ZQ, Cai WM, Han HP, Ye XJ, Jin QL, Li C, Li XJ (2016b) Breeding of *Ganoderma* cultivar "Lingzhi 2" and its characteristics. Edib Med Mushrooms 24 (2):118–119

Li KK, Zhuo C, Teng CY, Yu SM, Wang X, Hu Y, Ren GM, Yu M, Qu JJ (2016) Effects of *Ganoderma lucidum* polysaccharides on chronic pancreatitis and intestinal microbiota in mice. Int J of Biol Macromol 93(Part A):904–912

Li JM, Xu J, Wang XJ, Zhang GL, Li MY, Shao QS, Pan QX (2017) Breeding of *Ganoderma* var. "Xianzhi 2 and its characteristics". Mod Chinese Med 19 (3):324–345

Li K, Na K, Sang TT, Wu KK, Wang Y, Wang XY (2017) The ethanol extracts of sporoderm-broken spores of *Ganoderma lucidum* inhibit colorectal cancer in vitro and in vivo. Oncol Rep 38(5):2803–2813

Li ZH, Shi YQ, Zhang XH, Xu J, Wang HB, Zhao L, Wang Y (2020) Screening Immunoactive compounds of *Ganoderma lucidum* spores by mass spectrometry molecular networking combined within vivo zebrafish assays. Front Pharmacol 11:287

Li L-D, Mao P-W, Shao K-D, Bai X-H, Zhou X-W (2019) *Ganoderma* proteins and their potential applications in cosmetics. Appl Microbiol Biotechnol 103 (23–24):9239–9250. https://doi.org/10.1007/978-981-13-9867-4_1

Lin ZB (2007) Modern research on *Ganoderma lucidum* (in Chinese), 3rd edn. Peking University Medical Press, Beijing

Lin ZB (2009) Lingzhi from mystery to science, 1st edn. Peking University Medical Press, Beijing

Lin ZB, Yang XB.(2019) *Ganoderma* and health (Biology, chemistry and industry; pharmacology and

clinical application). Springer Nature, Singapore Pte Ltd.
Lin LJ, Shiao MS, Yeh SF (1988) Seven new triterpenes from *Ganoderma lucidum*. J Nat Prod 51(5):918–924
Lin HJ, Chang YSS, Lin LH, Huang CF, Wu CY and Ou KL (2014) An Immunomodulatory Protein (Ling Zhi-8) from a *Ganoderma lucidum* Induced Acceleration of Wound Healing in Rat Liver Tissues after Monopolar Electrosurgery. Evid Based Complement Alternat Med 916531:1–12
Lin ZB (2019) *Ganoderma* (Lingzhi) in traditional Chinese medicine and Chinese culture. Adv Exp Med Biol 1181:1–13. https://doi.org/10.1007/978-981-13-9867-4_1
Liu XM, Liu FP, Lin YH, Xu WJ (1991) Application of *Ganoderma* in drugs and supplements. Sub Trop Plant Sci 20(2):48–50
Liu JY, Jin L, Jiang ZJ (2006) Effect of *Ganoderma lucidum* spores oil on reducing the toxicity of radiochemotherapy and enhancing the function of immune system in mice. LiShiZhen Med Mater Med Res 17(11):2179–2181
Liu DB, Gong J, Dai WK, Kang XC, Huang Z, Zhang HM, Liu W, Liu L, Ma JP, Xia ZL, Chen YX, Chen YW, Wang DP, Ni PX, Guo AY, Xiong XY (2012) The Genome of *Ganoderma lucidum* Provide Insights into Triterpense Biosynthesis and Wood Degradation. PLoS ONE 7(5):1–14
Liu LY, Chen H, Liu C, Wang HQ, Kang J, Yan L, Chen RY (2014) Triterpenoids of *Ganoderma theaecolum* and their hepatoprotective activities. Fitoterapia 98:254–259
Liu LY, Chen H, Liu C, Wang HQ, Kang J, Li Y, Chen RY (2014) Triterpenoids of *Ganoderma theaecolum* and their hepatoprotective activities. Fitoterapia 98:254–259
Liu YY, Song Y, Liu N, Liu JJ (2018) Traits comparison between new *Ganoderma* cultivar 'Liaolingzhi 2' and main cultivars in Liaoning. North Hortic 42(4):165–168
Liu G, Kai W, Kuang S, Cao RB, Bao L, Liu R, Liu HW, Sun CM (2018) The natural compound GL22, isolated from *Ganoderma* mushrooms, suppresses tumor growth by altering lipid metabolism and triggering cell death. Cell Death Dis 9(6):689
Liu W, Hu XZ, Zhu L, Gan JH, Lu Y, Tao NP, Wang XC, Xu CH (2019) Recent progress in research and application of *Ganoderma lucidum* triterpenoids. Food Sci Technol 40(5):309–315
Liu YP, Li YM, Zhang WL, Sun MZ, Zhang ZS (2019) Hypoglycemic effect of inulin combined with *Ganoderma lucidum* polysaccharides in T2DM rats. J Funct Foods 55:381–390
Liu MT, Chen LX, Zhao J, Li SP (2020) *Ganoderma* spore powder contains little triterpenoids. Chin Med 15:111
Ma L, Wu F, Chen RY (2003) Analysis of triterpene constituents from *Ganoderma lucidum*. Acta Pharm Sin 38(1):50–52
Ma JJ, Fu ZY, Ma PY, Su YL, Zhang QJ (2007) Breaking and characteristics of *Ganoderma lucidum* spores by high speed centrifugal shearing pulverizer. J Wuhan Univ Technol (Mater Sci Ed) 22(4):617–621
Ma BJ, Ren W, Zhou Y, Ma JC, Ruan Y, Wei CN (2011) Triterpenoids from the spores of *Ganoderma lucidum*. N Am J Med Sci 3(11):495–499
Ma HT, Hsieh J-F, Chen ST (2015) Antidiabetic effects of *Ganoderma lucidum*. Phytochemistry 114:109–113
Meng J, Wang SZ, He JZ, Zhu S, Huang BY, Wang SY, Li M, Zhou H, Lin SQ, Yang BX (2020) Ganoderic acid A is the effective ingredient of *Ganoderma* triterpenes in retarding renal cyst development in polycystic kidney disease. Acta Pharmacol Sin 41 (6):782–790
Mothana RA, Jansen R, Julich WD, Lindequist U (2000) Ganomycins A and B, new antimicrobial farnesyl hydroquinones from the basidiomycete *Ganoderma pfeifferi*. J Nat Prod 63(3):416–418
Nurhuda MA, Noorlidah A, Norhaniza A (2013) Anti-angiotensin converting enzyme(ACE)proteins from mycelia of *Ganoderma lucidum* (Curtis) P Karst. BMC Complementary Alternat Med 13:256
Pan K, Jiang QG, Liu GQ, Miao XY, Zhong DW (2013) Optimization extraction of *Ganoderma lucidum* polysaccharides and its immunity and antioxidant activities. Int J Biol Macromol 55:301–306
Patocka J (1999) Anti-inflammatory triterpenoids from mysterious mushroom *Ganoderma lucidum* and their potential possibility in modern medicine. Acta Medica 42(4):123–125
Peng XR, Qiu MH (2018) Meroterpenoids from *Ganoderma* species: a review of last five years. Nat Prod Bioprospect 8(3):137–149
Peng YF, Zhang L, Zeng F, Kennedy JF (2005) Structure and antitumor activities of the water-soluble polysaccharides from *Ganoderma tsugae* mycelium. Carbohyd Polym 59(3):385–392
Phan CW, Wang JK, Cheah SC, Naidu M, David P, Sabaratnam V (2018) A review on the nucleic acid constituents in mushrooms: nucleobases, nucleosides and nucleotides. Crit Rev in Biotechnol 38(5):762–777
Qi C, Li RZ, He YQ, Lei LS, Lin ZB, Shen CY, Lu CS (1993) Studies on antiaging polysaccharides GLB GLC of *Ganoderma lucidum*. Beijing Yike Daxue Xuebao 25(4):303–305
Qin FY, Yan YM, Tu ZC, Cheng YX (2019) (±) Cochlearoids N-P: three pairs of phenolic meroterpenoids from the fungus *Ganoderma* cochlear and their bioactivities. J Asian Nat Prod Res 21(6):542–550
Qiu SB, Zhou M, Li QF, Chu F (2016) Experimental study on the toxicity of broken *Ganoderma* spore powder capsules. Lab Anim Sci 33(3):27–31
Qiu ZW, Zhong DD, Yang BX (2019) Preventive and therapeutic effect of *Ganoderma* (Lingzhi) on Liver Injury. Adv Exp Med Biol 1182:217–242
Sa-Ard P, Sarnthima R, Khammuang S, Kanchanarach W (2015) Antioxidant, antibacterial and DNA protective

activities of protein extracts from *Ganoderma lucidum*. J Food Sci Technol 52(5):2966–2973
Sanodiya BSTG, Baghel RK, Prasad GB, Bisen PS (2009) *Ganoderma lucidum*: a potent pharmacological macrofungus. Curr Pharma Biotech 10(8):717–742
Savin S, Craciunescu O, Oancea A, Ilie D, Ciucan T, Loredana SA, Agnes T, Alina N, Calin D, Florin O (2020) Antioxidant, cytotoxic and antimicrobial activity of chitosan preparations extracted from *Ganoderma lucidum* Mushroom. Chem Biodivers 17(7): e2000175
Seo HW, Hung TM, Na MK, Jung HJ, Kim JC, Choi JS, Kim JH, Lee HK, Lee IS, Bae KH, Hatorri M, Min BS (2009) Steroids and triterpenes from the fruit bodies of *Ganoderma lucidum* and their anti-complement activity. Arch Pharm Res 32(11):1573–1579
Shen CY, Shen BD, Shen G, Li JJ, Zhang FC, Xu PH, Li XL, Cheng L, Qiu L, Han J, Yuan HL (2016) Therapeutic effects of nanogel containing triterpenoids isolated from *Ganoderma lucidum* (GLT) using therapeutic ultrasound (TUS) for frostbite in rats. Drug Deliv 23(8):2643–2650
Shiao MS (2003) Natural products of the medicinal fungus *Ganoderma lucidum*: occurrence, biological activities, and pharmacological functions. Chemi Rec 3(3):172–180
Smina TP, Mathew J , Janardhanan KK, Devasagayam TPA (2011) Antioxidant activity and toxicity profile of total triterpenes isolated from *Ganoderma lucidum* (Fr.) P. Karst occurring in South India. Environ Toxicol Pharmacol 32(3):438–446
Smina TP, Nitha B, Devasagayam TPA, Janardhanan KK (2017) *Ganoderma lucidum* total triterpenes induce apoptosis in MCF-7 cells and attenuate DMBA induced mammary and skin carcinomas in experimental animals. Mutat Res 813:45–51
Sohretoglu D, Huang S (2018) *Ganoderma lucidum* Polysaccharides as an anticancer agent. Anticancer Agents Med Chem 18(5):667–674
Steyaert RL (1980) Study of some *Ganoderma* species. Bull Jard Bot Natl Belg 50(1, 2):135–186
Thang TD, Kuo PC, Hwang TL, Yang ML, Ngoc NTB, Han TTN, Lin CW, Wu TS (2013) Triterpenoids and steroids from *Ganoderma mastoporum* and their inhibitory effects on superoxide anion generation and elastase release. Molecules 18(11):14285–14292
Wachtel-Galor S, Yuen J, Buswell JA (2011) *Ganoderma lucidum* (Lingzhi or Reishi): a medicinal mushroom. In: Wachtel-Galor S (ed) Herbal medicine: biomolecular and clinical aspects benzie IFF. CRC Press/Taylor & Francis: Boca Raton (FL)
Wang XC, Xi RJ, Li Y, Wang DM, Yao YJ (2012) The species identity of the widely cultivated *Ganoderma*, '*G. Lucidum*' (Ling-Zhi), in China. PLoS ONE 7: e40857
Wang XD, Zhang Y, Jiang HY, Luo Y, Zhang X, Li Y (2014) Cultivation of Zi-Zhi and bonsai production (in Chinese). Edible Fungi China 36(4):51–52
Wang JJ, Ke R, Zhang SY (2017) Breaking the sporoderm of *Ganoderma lucidum* spores by combining chemical reaction with physical actuation. Nat Prod Res 31(20):2428–2434
Wang H, Yu Q, Nie SP, Xiang QD, Zhao MM, Liu SY, Xie MY, Wang SQ (2017) Polysaccharide purified from *Ganoderma atrum* induced activation and maturation of murine myeloid-derived dendritic cells. J Food Chem Toxicol 108(Pt B):478–485
Wang J, Cao B, Zhao HP, Feng J (2017) Emerging Roles of *Ganoderma lucidum* in Anti-Aging. Aging Dis 8 (6):691–707
Wang CF, Liu XM, Lian CL, Ke JY, Liu JQ (2019) Triterpenes and aromatic meroterpenoids with antioxidant activity and neuroprotective effects from *Ganoderma lucidum*. Molecules 24(23):1–11
Wu F, Meng GL, Chang SS, Xu JL (2012) The anti-atherosclerotic effect of *Ganoderma lucidum* Polysaccharides via down-regulation of vascular NADPH oxidases expression in atherosclerosis rats. Chin Pharmacol Bull 28(7):944–947
Wu X, Zeng J, Hu J, Liao Q, Zhou R, Zhang P, Chen Z (2013) Hepatoprotective effects of aqueous extract from Lingzhi or Reishi medicinal mushroom *Ganoderma lucidum* (higher basidiomycetes) on α-amanitin-induced liver injury in mice. Int J Med Mushrooms 15(4):383–391
Wu GS, Guo JJ, Bao JL, Li XW, Chen XP, Lu JJ, Wang YT (2013) Anticancer properties of triterpenoids isolated from *Ganoderma lucidum*-a review. 22(8):981–992
Wu YL, Han F, Luan SS, Ai R, Zhang P, Li H, Chen LX (2019) Triterpenoids from *Ganoderma lucidum* and their potential anti-inflammatory effects. J Agric Food Chem 67(18):5147–5158
Xiao ZJ, Wang JJ, Lian B (2006) Actuality of the exploitation and research on production of *Ganoderma lucidum*. Food Sci 27(12):837–842
Xiao C, Wu QP, Zhang JM, Xie YZ, Cai W, Tan JB (2017) Antidiabetic activity of *Ganoderma lucidum* polysaccharides F31 down-regulated hepatic glucose regulatory enzymes in diabetic mice. J Ethnopharmacol 196:47–57
Xie JT, Wang CZ, Wicks S, Yin JJ, Kong J, Li J, Li YC, Yuan CS (2006) *Ganoderma lucidum* extract inhibits proliferation of SW 480 human colorectal cancer cells. Exp Oncol 28(1):25–29
Xu ZT, Chen XP, Zhong ZF, Chen LD, Wang YT (2011) *Ganoderma lucidum* Polysaccharides: immunomodulation and potential antitumor activities. Am J Chin Med 39(1):15–27
Yan SY, Wu ZC, Xin GY, Xiao X, Huang MM, Meng XX (2017) Progresses on synthesis and antitumor pharmacological activity of *Ganoderma lucidum* Triterpenoids. Life Sci Res 21(5):454–457
Yang Q, Wang SW, Xie YH, Sun JY, Wang JB (2010) HPLC analysis of *Ganoderma lucidum* polysaccharides and its effect on antioxidant enzymes activity and

Bax, Bcl-2 expression. Int J Biol Macromol 46 (2):167–172

Yin ZM, Yang BX, Ren HW (2019) Preventive and therapeutic effect of *Ganoderma* (Lingzhi) on skin diseases and care. Adv Exp Med Biol 1182:311–321

Yu WY (2017) Study on introduction and cultivation techniques of *Ganoderma lucidum* (in Chinese). Zhejiang University, Hangzhou

Yu JG, Chen RY, Yao ZX, Zhai YF, Yang SL, Ma JL (1990) Studies on constituents of *Ganoderma capense* IV. The chemical structures of ganoine, ganodine and ganoderpurine. Yao Xue Xue Bao 25(8):612–616

Yu ZH, Gao YG, Zeng ZQ, Su K, Xiao L, Luo MF (2008) Advances in studies of fungal genomics. Mycosystema 31(5):464–470

Zhang H, Fu X (Christian era 119) The Eastern Han Dynasty

Zhang F, Zhao YM (2020) Divergent Total Syntheses of Six *Ganoderma* Meroterpenoids: A Bioinspired Two-Phase Strategy. Synlett 31:A–D

Zhang J, Tang Q, Zimmerman Kordmann M, Reutter W, Fan H (2002) Activation of B lymphocytes by GLIS, a bioactive proteoglycan from *Ganoderma lucidum*. Life Sci 71(6):623–638

Zhang GL, Wang YH, Ni W, Teng HL, Lin ZB (2002) Hepatoprotective role of *Ganoderma lucidum* polysaccharide against BCG-induced immune liver injury in mice. World J Gastroenterol 8(4):728–733

Zhang HN, He JH, Yuan L, Lin ZB (2003) In vitro and in vivo protective effect of *Ganoderma lucidum* polysaccharides on alloxan-induced pancreatic islets damage. Life Sci 73(18):2307–2319

Zhang XQ, Pang GL, Cheng Y, Wang Y, Ye WC (2008) Chemical constituents of the spores of *Ganoderma lucidum*. Zhong Yao Cai 31(1):41–44

Zhang J, Tang Q, Zhou C, Jia W, Silva LD, Nguyen LD, Reutter W, Fan H (2010) GLIS, a bioactive proteoglycan fraction from *Ganoderma lucidum*, displays anti-tumour activity by increasing both humoral and cellular immune response. Life Sci 87(19–22):628–637

Zhang WJ, Tao JY, Yang XP, Yang ZL, Zhang L, Liu HS, Wu KL, Wu JG (2014) Antiviral effects of two *Ganoderma lucidum* triterpenoids against enterovirus 71 infection. Biochem Biophys Res Commun 449(3):307–312

Zhang J, Gao X, Pan Y, Xu N, Jia L (2016) Toxicology and immunology of *Ganoderma lucidum* polysaccharides in Kunming mice and Wistar rats. Int J Biol Macromol 85:302–310

Zhang SL, Pang GB, Chen C, Qin JZ, Yu H, Liu YM, Zhang XH, Song ZT, Zhao J, Wang FJ, Wang YY, Zhang LHW (2019) Effective cancer immunotherapy by *Ganoderma lucidum* polysaccharide-gold nanocomposites through dendritic cell activation and memory T cell response. Carbohydr Polym 205:192–202

Zhao JD (1989) The *Ganoderma*taceae in China. Bibliotheca Mycologica, Band132. J. Cramer 1–176

Zhao JD, Zhang XQ (1989) Textual research on the classification of "six zhi" described by Chinese ancient books. Microbiol Bull 3:180–181

Zhao LY, Dong YH, Chen GT, Hu QH (2009) Extraction, purification, characterization and antitumor activity of polysaccharides from *Ganoderma lucidum*. Carbohydr Polym 80(3):783–789

Zhao H, Zhang QY, Zhao L, Huang X, Wang JC, Kang XM (2012) Spore powder of *Ganoderma lucidum* improves cancer-related fatigue in breast cancer patients undergoing endocrine therapy: a pilot clinical trial. Evid Based Complement Alternat Med 809614:1–8

Zhao ZZ, Chen HP, Teng T, Li ZH, Dong ZJ, Liu JK (2015) Lucidimine A-D, four new alkaloids from the fruiting bodies of *Ganoderma lucidum*. J Asian Nat Prod 17(12):1160–1165

Zheng LY (2007) Study on the genetic diversity of *Ganoderma* strains and their difference in medication efficacy (in Chinese). Sichuan University

Zhu M, Chang Q, Wong LK, Chong FS, Li RC (1999) Triterpene antioxidants from *Ganoderma lucidum*. Phytother Res 13(6):529–531

Zhu KX, Nie SP, Li C, Lin SL, Xing MM, Li WJ, Gong DM, Xie MY (2013) A newly identified polysaccharide from *Ganoderma atrum* attenuates hyperglycemia and hyperlipidemia. Int J Biol Macromol 57:142–150

The Trend of *Ganoderma Lucidum* Research (1936–2019)

2

Yicen Xu and Jie Yu

Abstract

Ganoderma lucidum, as the symbol of traditional Chinese medicine, is one of the most economically important medicinal fungi and has received attention worldwide. The number of scientific research publications about *G. lucidum* increased rapidly. However, there have been few research reviews concerning the research trend of *G. lucidum.* Here we used Biblioshiny to analyze 3,286 documents from 1936 to 2019. These documents were screened and analyzed using four software tools: R-package bibliometrix, HisCite, Citespace, and Bibliometric Online Analysis Platform. We presented the performance of relevant sources, authors, institutions, and countries. We analyzed the most highly cited documents of *G. lucidum* to define the research hotspots and research trends in this field.

2.1 Introduction

As the public becomes more aware of drug safety, people prefer natural medicinal products derived from plants and fungi because they have the distinct advantage of being less toxic and having fewer side effects (Tan et al. 2011). Biological natural products are secondary metabolites from living organisms. These sources can be complex mixtures derived from raw materials or single compounds (Lyu et al. 2020; Liu et al. 2020). Several *Ganoderma* species, represented by *Ganoderma lucidum* (Curtis) P. Karst. are well-known medicinal fungi. They have greater advantages over other chemical drugs. *Ganoderma* comes from the Greek word, "gano" meaning shiny, and "derma" meaning skin. For the first time, the mushroom was called *Ganoderma* by mycologist Petter Adolf Karsten in Finland in 1881 (Karsten 1881). *Ganoderma* species is widely distributed in tropical and subtropical regions of Africa, America, Asia, Oceania, and Europe (Zhou et al. 2016; Sun et al. 2020). There are more *Ganoderma* species in southern China, especially in the south of tropical and subtropical regions (Zhao et al. 2018). It mainly regulates human immune function, prolongs life span, has antitumor and neuroprotective functions (Batra 2019). This is one of the reasons for the growing demand for *Ganoderma* products. The main medicinal part of *Ganoderma* is the fruiting body, but its spore powder and mycelium can also be used medicinally. In the United States, *Ganoderma* is available as a dietary supplement and is listed in the United States Pharmacopoeia of Dietary Supplements and Herbs (Nishita et al. 2017; Paterson et al. 2006). With the industry's development, the

Y. Xu · J. Yu (✉)
College of Horticulture and Landscape Architecture, Southwest University, Chongqing 400716, China
e-mail: yujie1982@swu.edu.cn

C. Liu (ed.), *The Lingzhi Mushroom Genome*, Compendium of Plant Genomes,
https://doi.org/10.1007/978-3-030-75710-6_2

spores and mycelium of *Ganoderma* have been used as new medicinal ingredients. *Ganoderma* spores have been shown to have good antitumor and antioxidant activity (Min et al. 1998; Heleno et al. 2012). The main compounds from *Ganoderma* species include triterpenes, polysaccharides, terpenoids, sesquiterpenoids, steroids, and alkaloids. Across different countries, triterpenes and polysaccharides are considered the main active ingredients (Soccol et al. 2016; Gao et al. 2003a). The fruit bodies, cultured mycelia, and spores of *G. lucidum* have been reported to have therapeutic effects on chronic liver disease, hypertension, hyperglycemia, and neonatal disease (Franz 1989; Furusawa et al. 1992; Shiao et al. 1994). The main active ingredients of *G. lucidum* are triterpenoid and polysaccharides (Paterson et al. 2006; Shaoping et al. 2013).

Literature reviews document and quantify trends in scientific publications. Bibliometrics is a scientific, objective, and convenient literature review method. Pritchard first proposed bibliometrics in 1969 as a pattern-based quantitative method for research and development (Pritchard 1969). It is defined as a mathematical and statistical method that quantitatively analyzes trends across multiple scientific disciplines. This method is often used as a powerful tool to evaluate many scientific publications in different research areas (Ge et al. 2017; Torres et al. 2015; Barbero-Sierra et al. 2015). Its scope extends from the statistical analysis of the single relationship among countries, institutions, authors, and other variables to revealing potential multivariable relationships by establishing and analyzing contact networks (Zhang et al. 2017; Yu et al. 2017; Li et al. 2017).

Several sophisticated software tools have been developed for bibliometrics analysis, including HistCite (Li et al. 2016; Wanqi et al. 2018), Citespace (Gao et al. 2016; Wanqi et al. 2017) and Bibliometrics Online Analysis Platform (BOAP, https://bibliometric.com/). These tools can be used for qualitative and quantitative analysis of data collected from many publications. Histsite allows users to analyze citations in an unfamiliar field to quickly map the trends in this field, identify the most frequently researched research issues, and reveal relevant geographic regions, institutions, and specific research scientists who are leaders in the research field. CiteSpace allows users to understand the history and trends of the field through visual analysis of published research. BOAP is a web application for identifying metrics related to bibliometric analysis. CiteSpace and BOAP have similar functionality, but the former has more modules, and the latter is faster and easier to use. R-package bibliometrix is a bibliometrics analysis tool based on the R language (Aria and Cuccurullo 2018). Bibliometrix provides a workflow that can be used automatically. The bibliometrix is flexible and can be integrated with other R-packages. After successful installation, it provides a web interface based on Biblioshiny, through which users can obtain basic statistics of literature information and the cooperation of various indicators. It is easy to use and fast to produce results. Because these tools have complementary theoretical and functional modules, multiple tools should be used together for the most comprehensive literature analysis.

The health-promoting effects of *G. lucidum* have attracted international research interests on diverse subjects. Related publications have rapidly increased in recent years. However, there are no studies that carried out a comprehensive review on this field. In this study, we use sophisticated tools to carry out a bibliometric analysis. Here, we analyze the past development history and identify future research directions from the following aspects: global publication output, country, author, journal, most local citation references, the conceptual structure, the intellectual structure, and the social structure.

2.2 Data Collection and Analytical Method

Data were retrieved from the Web of Science Core Collection database, using "*Ganoderma lucidum*" as the keywords. We set the retrieval time from 1900 to 2019. A total of 3286 records were obtained between 1936 and 2019. Every record contains the authors' information, title, source, abstract, keywords, and cited references. The data were analyzed using the four software tools described below.

The manually processed data were visually analyzed using the R-package bibliometrix. Besides, Histsite and Citespace are used for analyzing comprehensive indicators. BOAP is used for cooperative relationship analysis.

2.3 Results

2.3.1 Descriptive Analysis (Country, Author, Source)

The 3,286 documents from 1936 to 2019 include 381 sources, 25.34 average citations per document, 2.727 average citations per year per document, 79,529 references, 9,339 authors, 17,734 author appearances, 9,263 authors of multi-authored documents, 76 authors of single-authored documents. Documents also contain 6,898 author's keywords (DE) and 6,074 keywords plus (ID). The number of documents per author is 0.352. The number of authors per document is 2.84. The majority of these papers are articles (3,021 records, 91.94%), and the remaining ones are reviews (265 records, 8.06%).

As shown in Fig. 2.1, the number of publications increases gradually, and the annual growth rate is 4.73%. It can be divided into three stages, which are stage I: 1936–1992; stage II: 1993–2007; and stage III: 2008–2019. Stage I (1936–1992) was the initial period when less than ten papers were published annually except 1986, 1988, and 1991 which have 12, 15, and 12 publications, respectively. Stage II (1993–2007) is the rising period; the number of papers is higher than before, but not more than 100. Stage III (2008–2019) is the rapid development period that more scholars began to focus on *G. lucidum* research.

During 1936–2019, a total of 87 countries contributed to publications. Figure 2.2 shows the geographical distribution of the country's scientific production and the total publications (TP), total citations (TC), and average citations (AC) of the top 10 most productive countries by the corresponding authors. China produced the most studies with 1,653 publications, account for more than 50%. In contrast, India ranks in second place, has a much smaller number than China. And China shows the highest TC of 41,214. Although China has the highest number of publications and total citations, the average citation is lower than other countries, only 24.93. It indicated that the number of published articles in China is high, but the number of high-level articles is relatively small, and the research field's influence is relatively weak. In terms of the number of publications, Korea ranks the last, but its average citation is in the first place of the top 10 productive countries with 34.64. The USA with 31.93 is second only to Korea.

Figure 2.3a presents the top 10 productive authors in the *G. lucidum* research area. The most productive author is Zhong JJ, with 58 publications from 2002 and has the highest total citations, 2591 and H-index 28. We analyzed the numbers of articles published yearly by these ten prolific authors (Fig. 2.3b). The most productive author Zhong JJ published articles cover the longest time, from 2002 to 2018. And Zhong JJ has published most articles and obtained the highest citations in 2010, 10 and 36.08, respectively. Besides, Shi L and Chen Y started *G. lucidum* research earlier in 1997. Their articles have been published continuously since 2008. It's worth noting that Chen Y at the 10th productive author but has the most articles per year and the highest citations among all authors listed. In 2012, Chen (2012) published 12 articles per year. Besides, Chen Y only published five articles in 2015, but got the highest citations per year: 59.57. Among them, the title "*Ganoderma lucidum* reduces obesity in mice by modulating the composition of the gut microbiota" is a highly cited paper. This research suggested that *G. lucidum* and its high molecular weight polysaccharides can be used as prebiotic agents to prevent intestinal disorders and obesity-related metabolic disorders in obese individuals (Chang et al. 2015).

A total of 3,286 papers were published in 831 different sources by 2019. The top 10 journals account for 20.48% of the total articles analyzed

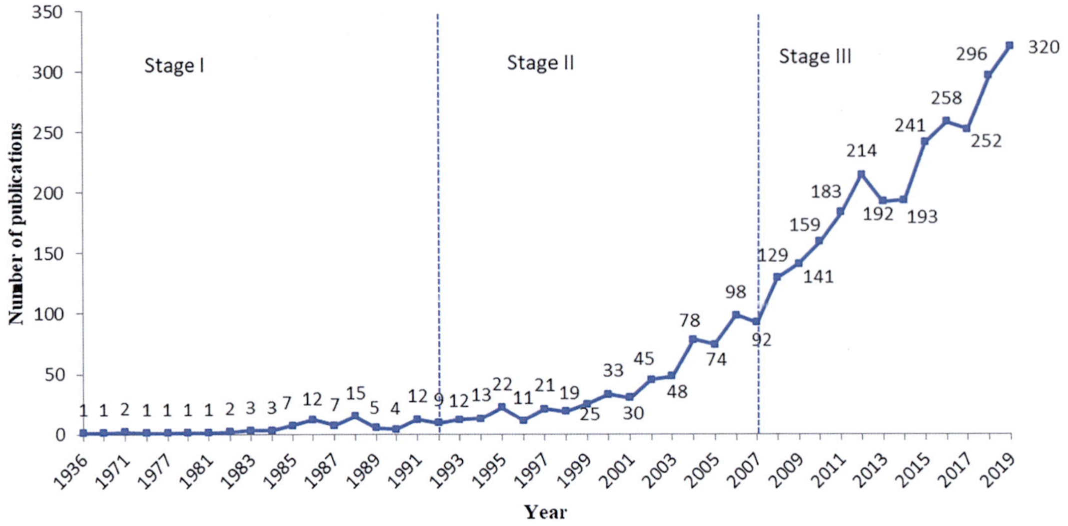

Fig. 2.1 Annual publications in the *Ganoderma lucidum* research area from 1936 to 2019

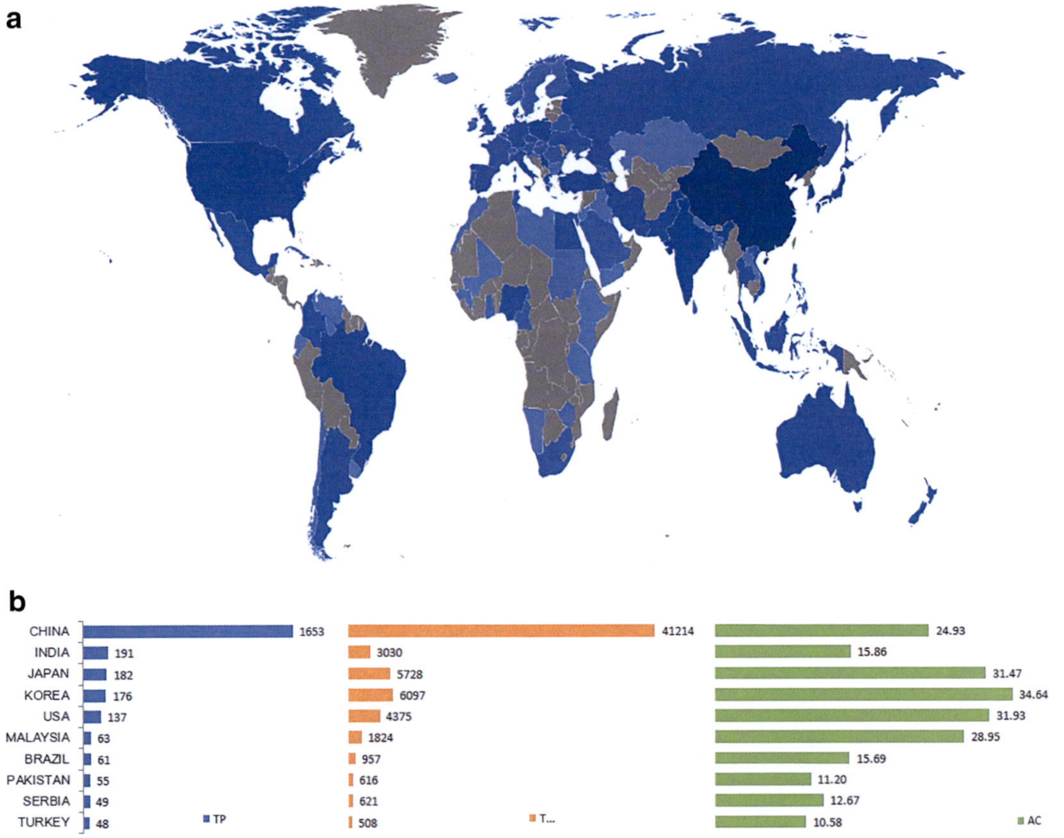

Fig. 2.2 Characteristics of countries/regions in *Ganoderma lucidum* research area from 1936 to 2019. **a** Geographical distribution of research countries/regions. **b** The top 10 relevant countries by the corresponding author

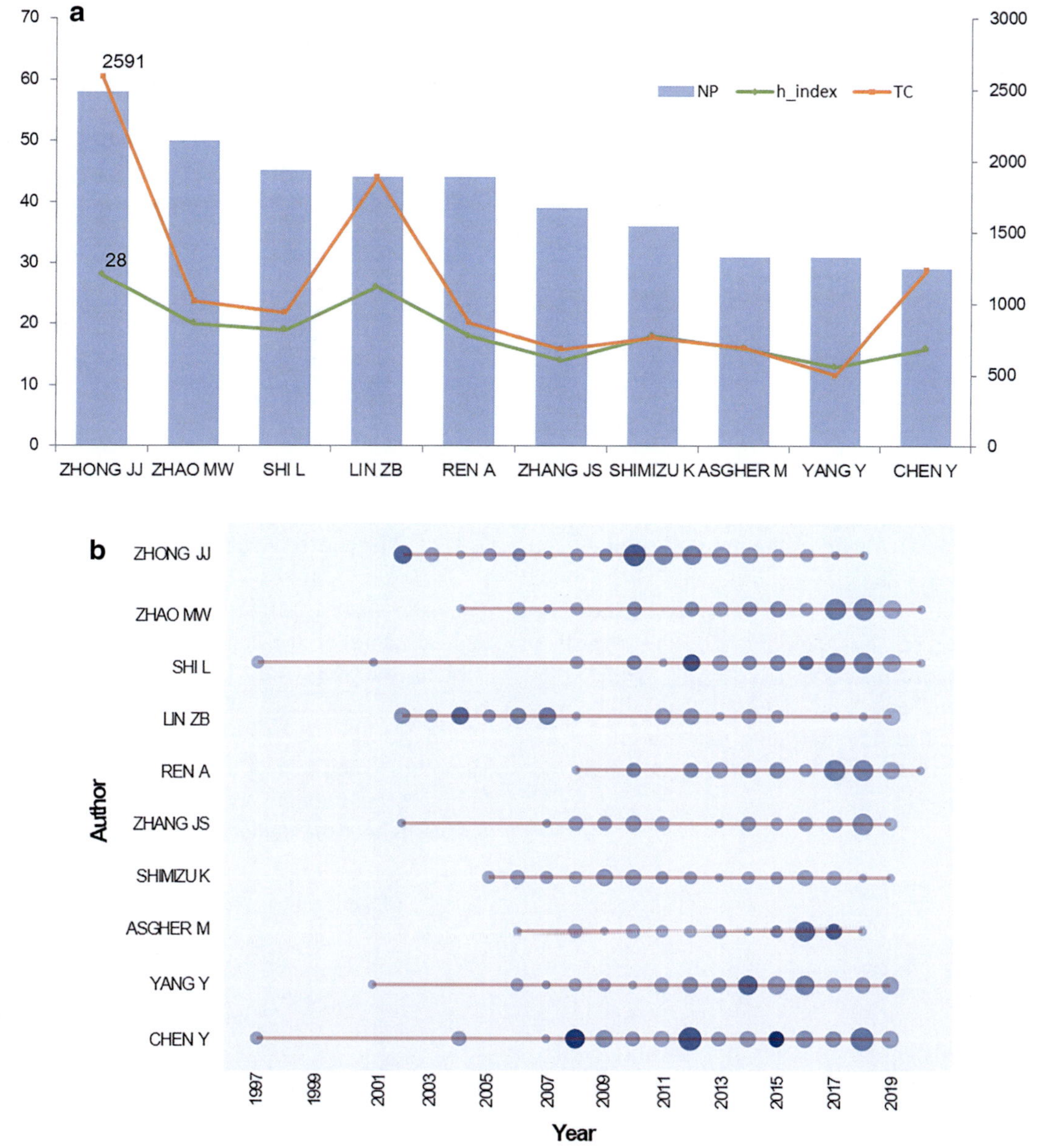

Fig. 2.3 The performance of the authors in *G. lucidum* research from 1936 to 2019. **a** The top 10 productive authors in the *G. lucidum* research area. **b** Top-authors' production over time. The size of the circle represents the number of articles. The depth of the color of the circle represents the number of citations per year

(Table 2.1). Most of the impact factors are $\geq$ 3.0 (7 in 10), and half of them have H-index $\geq$ 25. "International Journal of Medicinal Mushrooms" is the most productive journals with 203 publications related to *Ganoderma lucidum* research. But it's TC and H-index is not the highest, with 2,751 and 20, respectively. "Carbohydrate Polymers" ranks in the second place, and has the highest TC, H-index, and IF of 4120, 36, and 7.182, respectively.

Besides, we analyzed the keywords plus of 3,286 publications. The word cloud is shown in

Table 2.1 Top 10 most productive sources during 1936–2019

Rank	Source	TP	TC	H_index	IF (2019)
1	International Journal of Medicinal Mushrooms	203	2751	20	1.423 (2018)
2	Carbohydrate Polymers	84	4120	36	7.182
3	International Journal of Biological Macromolecules	79	2010	26	5.162
4	Phytochemistry	59	3189	29	3.044
5	Molecules	49	858	17	3.267
6	Journal of Agricultural and Food Chemistry	43	1815	27	4.192
7	Journal of Ethnopharmacology	43	1707	26	3.69
8	Food Chemistry	40	2066	25	6.306
9	Plos One	37	863	19	2.74
10	Process Biochemistry	36	1296	20	2.952

Fig. 2.4. The word minimum frequency is chosen as 5. "*Ganoderma lucidum*" was recorded 844 times in 2013, and "Polysaccharides" had 334 records in 2014. Besides, the records of "lucidum," "in-vitro," "growth," "cells," "expression," and "acid" are more than 200 between 2012 and 2015, indicating that this period is a rapid-growth stage in *G. lucidum* research.

2.3.2 Thematic Analysis

Thematic analysis is used to define the conceptual structure of the topic. The thematic map exploits the Keywords Plus field. Those keywords are associated with Thomson Reuters editorial experts supported by a semi-automated algorithm. They review all references' titles and highlight additional relevant but overlooked keywords that the authors did not list. Different

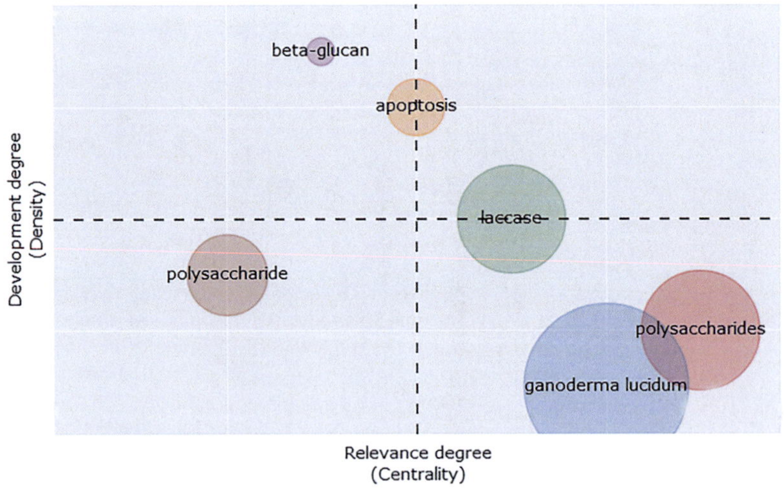

Fig. 2.4 Thematic map. The x-axis represents centrality, and the y-axis represents density

from the authors' keywords, the Keywords Plus field is normalized. Keywords Plus terms can differentiate content with greater depth and variety.

Applying a clustering algorithm on the keyword network makes it possible to highlight a given domain's different themes. Each cluster can be presented on a particular plot. Centrality measures the importance of the subject in the entire research field. In contrast, density measures the theme's development.

Thematic mapping allows visualization of six different types of subjects, as shown in Fig. 2.4. Each bubble represents a network cluster. The bubble name is the word belonging to the cluster, with the higher occurrence value. The bubble size is proportional to the cluster word occurrences. The bubble position is set according to the cluster centrality and density.

The upper-right quadrant shows the "motor themes." They are characterized by both high centrality and density. Regarding the upper-left quadrant, it shows high-density themes. Still, unimportant external links are of only limited importance for the field (low centrality). The main theme is on "beta-glucan." This theme is connected with different concepts, such as "antitumor," "basidiomycetes," and "grifola frondosa." In the lower-left quadrant are the emerging or declining themes. We then analyzed the frequency of the keyword using wordcloud (Fig. 2.5), the theme of "polysaccharide" is emerging. Besides, "antioxidant activity," "antitumor activity," and "structure" are also involved in this theme. Finally, the lower-right quadrant shows the themes that are basic and transversal. These themes concern general topics transversal to the different research areas of the field. In this area, the appearing themes are "*Ganoderma lucidum*" and "polysaccharides." Besides, "laccase" is between motor theme and basic theme, And "submerged culture," "mushrooms," and "response surface methodology" also in this cluster. This shows that the research work on glucan, antitumor effects of Lingzhi are in foucs and it is likely that there will be literatures related to these topics in the future. Active ingredients, such as laccase, polysaccharides, and triterpenoids, have been the research hotspot recently.

2.3.3 Co-Word Analysis

2.3.3.1 Keyword Co-Occurrence

The keywords co-occurrence network is shown in Fig. 2.6a. Among them, words such as Ganoderma-lucidum, polysaccharides, in vitro, growth, lucidum have higher centrality in the whole network, appear more frequently and have a more substantial influence. The above keywords are connected with other keywords, such as antitumor, expressions, fruiting bodies, apoptosis, and purification. These results reflect the wide variety of topics of *G. lucidum* that are currently under study.

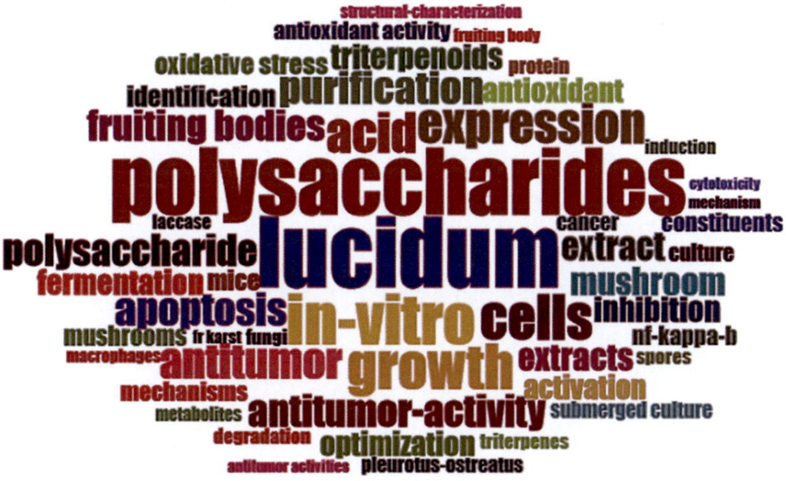

Fig. 2.5 The WordCloud of the keywords in *G. lucidum* research from 1936 to 2019

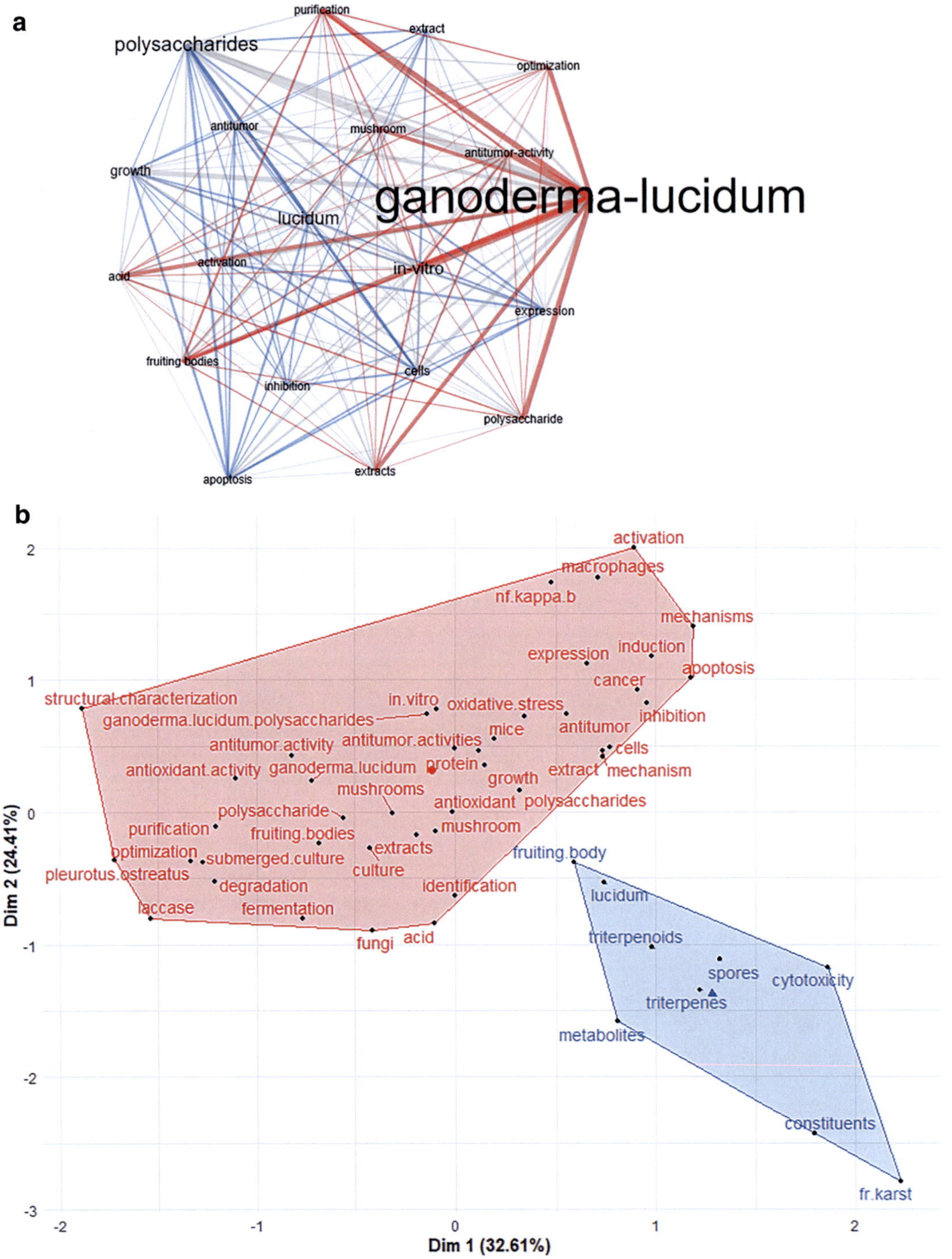

Fig. 2.6 Conceptual structure. **a** Keyword co-occurrence network. **b** Multiple correspondence analysis

2.3.3.2 Correspondence Analysis

The conceptual structure–function in Bibliometrix uses the Multivariate Correspondence Analysis (MCA) to draw the research area's conceptual structure. The k-means clustering method is used to cluster the literature. As shown in Fig. 2.6b, each color represents a cluster of a word. Cluster I in red has the keywords such as structural characterization, optimization, laccase, mechanisms, activation, etc. Cluster II in blue has the keywords such as metabolites, constituents, cytotoxicity, etc.

2.3.4 Co-Citation Analysis

2.3.4.1 Papers Co-Citation Analysis

In the co-citation network (Fig. 2.7a), 20 articles are divided into two clusters with red and blue. The most influential paper is Paterson et al. (2006) with the title of "Ganoderma—A therapeutic fungal biofactory." This review collates the publications detailing activities and compounds by representative species whilst considering the most valid claims of effectiveness.

2.3.4.2 Sources Co-Citation Analysis

In the sources co-citation network (Fig. 2.7b), 20 sources are divided into 3 clusters. Phytochemistry, life science, Journal of Ethnopharmacology, Journal of Natural Products, Phyototherapy Research, Planta Medica and Chemical & Pharmaceutical Bulletin form cluster I in red. Applied Microbiology and Biotechnology, PLOS one, Biochemical and Biophysical Research Communication, Journal of Biological Chemistry and Proceeding of the National Academy of Sciences of the United States of America form cluster II in blue. Journal of Agricultural and Food Chemistry, Food Chemistry, Carbohydrate Polymers, International Journal of Medicinal Mushrooms, Internal Journal of Biological Macromolecules, Food and Chemical Toxicology and International Immunopharmacology form cluster III in green. Phytochemistry, Journal of Ethnopharmacology, and Journal of Agricultural and Food Chemistry correlate with other sources.

2.3.5 Collaboration Analysis

2.3.5.1 Author Collaboration Network

In total, 9,339 authors published a total of 3,286 papers by 2019. In the author collaboration network (Fig. 2.8a), 50 authors are listed, and 18 clusters are formed. Zhao mw, Ren a, and Shi l have more cooperative relationships. Besides, authors Zhang js, Yang y, Tang qj, Liu yf, and Zhou s also have a special collaborative relationship.

2.3.5.2 Institution Collaboration Network

A total of 3,286 papers were published from 2,406 institutions by 2019. In the institutions' collaboration analysis (Fig. 2.8b), National Taiwan University, National Yang-Ming University, Academia Sinica, China Medical University, Taipei Medical University, and National Chung Hsing University have many contacts in *G. lucidum* research. Besides, Shanghai Jiao Tong University and East China University of Science and Technology, Kunming Institute of Botany, and University of Chinese Academy of Sciences also have a cooperative relationship.

2.3.5.3 Country Collaboration Network

Academic cooperation in different countries/regions is vital. Scholars can communicate with one another, exchange novel opinions, and determine ideal solutions through international collaboration. Figure 2.8c shows a clear cooperative relationship between countries/regions. Each color represents a different country/region, and the thickness of the connecting lines denotes the strength of cooperation. China takes first place for *G. lucidum* related international collaboration, especially with the USA and Japan. China is the country with the most abundant *G. lucidum* resources and has established cooperative relations with the United States and Japan, which are technologically advanced at the time of the collaborations.

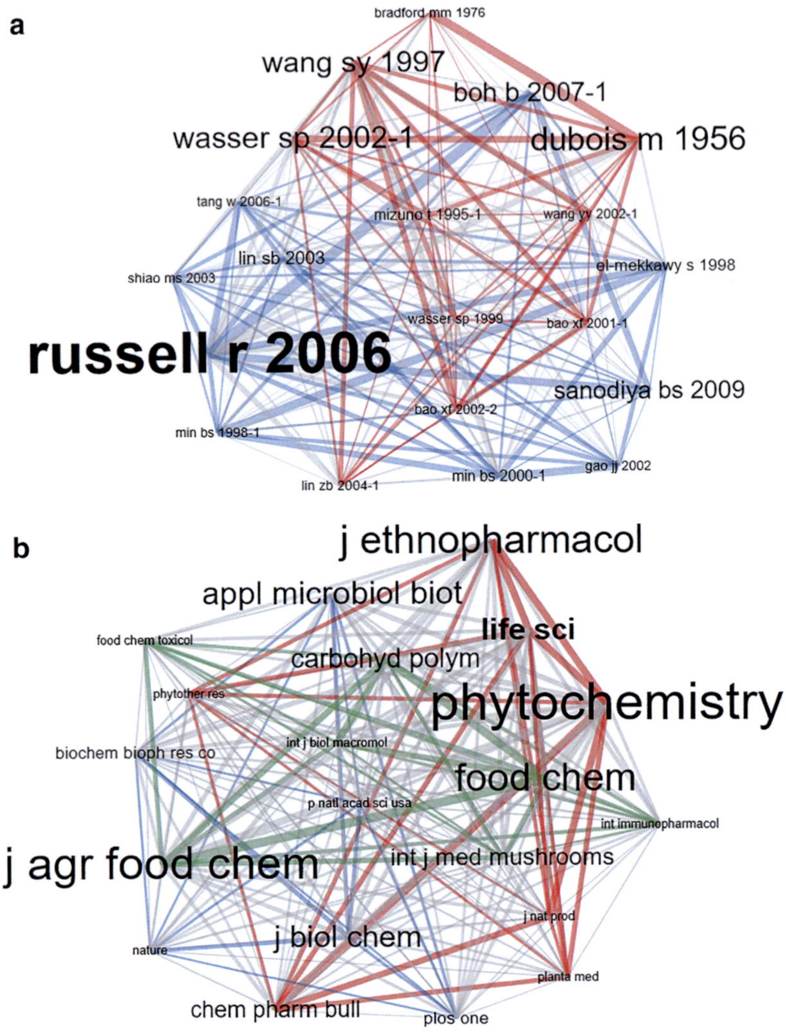

Fig. 2.7 Intellectual structure. **a** Papers co-citation network. **b** Sources co-citation network

2.3.6 Keyword Detection and Analysis

Detection of keywords can be used to identify *G. lucidum* research trends (Chen et al. 2014). Burst detection is a method of detecting dramatic changes in events, measured by a burst intensity parameter, measured by a score calculated by Kleinberg J algorithm (Kleinberg 2003). After importing data into Citespace, eight keywords bursts continued into 2017 and 2019 (Fig. 2.9). The keyword with the largest burst strength score of 9.37 was "lingzhi," followed by "structural characterization" with 8.9. The keywords of "reishi medicinal mushroom" and "meroterpenoid" ended in 2019. Therefore, future research directions may aim to determine the structural feature and meroterpenoid of Lingzhi.

2.3.7 Most Highly Cited Papers Analysis

We analyze the most highly cited articles to define the research hotspots (Table 2.2). The most highly cited paper is a review titled "Antioxidative peptides from food proteins: A review," published in Peptides in 2010 (Sarmadi and Ismail 2010). It has been cited 765 times ever since. This paper reviews bioactive peptides

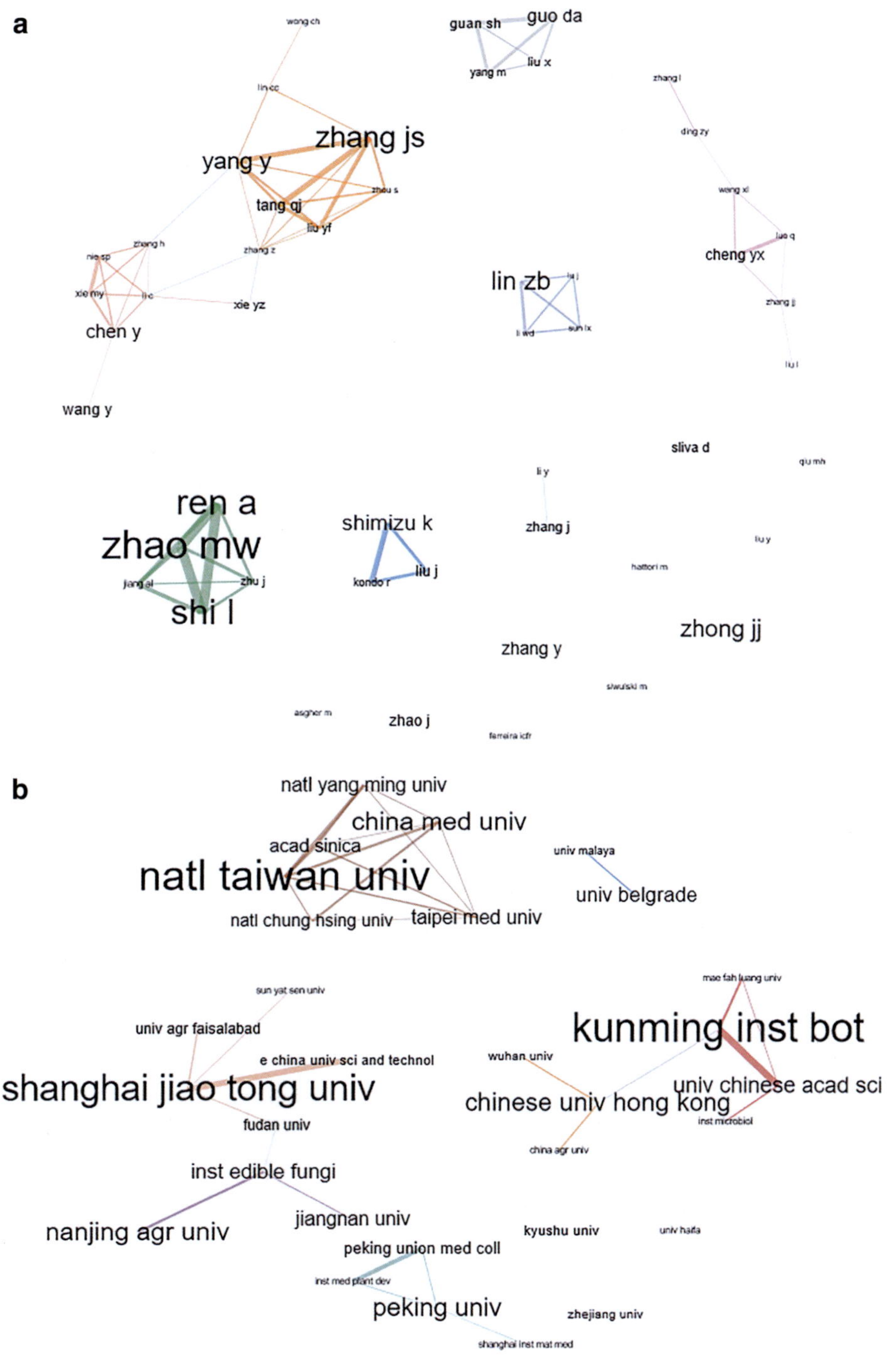

Fig. 2.8 Collaboration network. **a** Authors collaboration network. **b** Institutions collaboration network. **c** Countries collaboration network

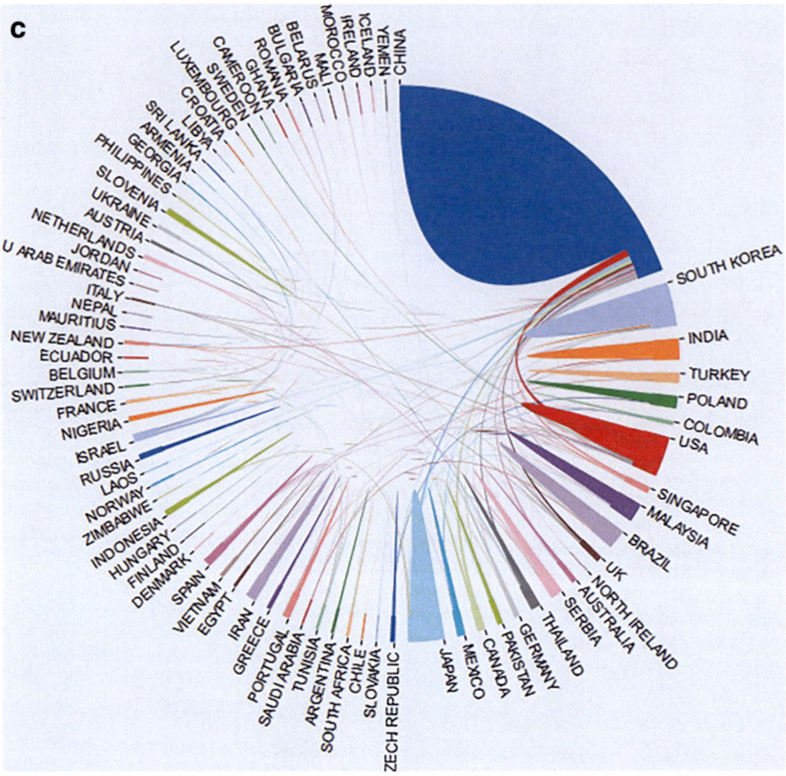

Fig. 2.8 (continued)

Fig. 2.9 Results of keyword burst continuing to 2017 and 2019

from food sources concerning their antioxidant activities. The properties of active antioxidant peptides, enzymatic production, antioxidant capacity evaluation methods, bioavailability, and safety are also reviewed. Similarly, Mau et al. (2002) studied the antioxidant properties of methanolic extracts from the medicinal mushrooms in Taiwan, namely, *G. lucidum* (Ling-chih), *G. tsugae* (Sung-shan-ling-chih), and *Coriolus versicolor* (Yun-chih). Overall, the antioxidant capacity, reducing capacity, chelation scavenging capacity, and total phenolic content were found higher in *G. lucidum* and *G. tsugae*.

The pharmacologic action of *G. lucidum* is the mostly studied topic studied. For instance, Wang et al. (1997) isolated polysaccharides (PS) from fresh seeds of *G. lucidum* (PS-G) for enhancing cytokine production by human monocyte-

Table 2.2 Summary of the most highly cited articles

Rank	Title	TC	PY	Summary
1	Antioxidative peptides from food proteins: A review	765	2010	This paper reviews bioactive peptides from food sources concerning their antioxidant activities. Then, it reviewed the specific characteristics of antioxidative bioactive peptides, enzymatic production, and methods to evaluate antioxidant capacity and bioavailability. Lastly, safety concerns of peptides were discussed
2	*Ganoderma*—A therapeutic fungal biofactory	549	2006	This review collates the publications detailing activities and compounds by representative species whilst considering the most valid claims of effectiveness
3	The pharmacological potential of mushrooms	521	2005	This review describes pharmacologically active compounds from mushrooms.
4	Medicinal Mushroom Science: History, Current Status, Future Trends, and Unsolved Problems	476	2010	This review summarized the data on mushroom polysaccharides for approximately 700 species of higher Hetero- and Homobasidiomycetes. It discussed the chemical structure of polysaccharides. Furthermore, their connection to antitumor activity was investigated. These include possible ways of chemical modification, experimental testing, and clinical use of antitumor or immunostimulating polysaccharides.
5	Purification, composition analysis and antioxidant activity of a polysaccharide from the fruiting bodies of *Ganoderma atrum*	413	2008	A water-soluble protein-bound polysaccharide was extracted from the fruiting bodies of *Ganoderma atrum* and isolated by gel-filtration chromatography. Its primary structural features and molecular weight were characterized by infrared spectrometry, gas chromatography, size exclusion chromatography, amino acid analyzer, and high-performance liquid chromatography (HPLC). The results suggest the polysaccharides' strong DPPH free radical and superoxide anion radical scavenging activities. The purified polysaccharides could potentially be used as natural antioxidants.
6	The antitumor effect of *Ganoderma lucidum* is mediated by cytokines released from activated macrophages and T lymphocytes	371	1997	The present study was to ascertain the immunomodulating and antitumor effects of *Ganoderma lucidum.*
7	*Ganoderma lucidum* reduces obesity in mice by modulating the composition of the gut microbiota	369	2015	Here, the authors show that a water extract of *Ganoderma lucidum* mycelium (WEGL) reduces body weight, inflammation, and insulin resistance in mice fed a high-fat diet (HFD). They further show that high molecular weight polysaccharides (>300 kDa) isolated from the WEGL extract produce similar anti-obesity and microbiota-modulating effects. Our results indicate that

(continued)

Table 2.2 (continued)

Rank	Title	TC	PY	Summary
				G. lucidum and its high molecular weight polysaccharides may be used as prebiotic agents to prevent gut dysbiosis and obesity-related metabolic disorders in obese individuals.
8	The Role of Culinary-Medicinal Mushrooms on Human Welfare with a Pyramid Model for Human Health	295	2012	This review presents a pyramid model for mushroom uses (industries), like food, dietary supplements (tonic), and medicine. A regular intake of mushrooms can make us healthier, fitter, and happier and help us live longer. The sense of purpose and vision for the mushroom industries is also briefly discussed.
9	Antioxidative and immunomodulating activities of polysaccharide extracts of the medicinal mushrooms Agaricus bisporus, Agaricus brasiliensis, *Ganoderma lucidum*, and Phellinus linteus	232	2011	Partially purified polysaccharides were obtained from four medicinal mushroom species, Agaricus bisporus, Agaricus brasiliensis, Phellinus linteus, *Ganoderma lucidum*, by hot water extraction, followed by ethanol precipitation. EC50 values of the DPPH scavenging activity of the polysaccharides from *G. lucidum* spores and *P. linteus* fruiting bodies were found to be particularly low, i.e., EC50 < 0.1 mg/ml. EC50 values of the antioxidant activity were 7.07 mg/ml for *G. lucidum*, EC50 values of ferrous ions' chelating activity ranged from 0.59 mg/ml for *G. lucidum*
10	Genome sequence of the model medicinal mushroom *Ganoderma lucidum*	231	2012	Here we report its 43.3-Mb genome, encoding 16,113 predicted genes. Twenty-four physical CYP gene clusters are identified. Moreover, 78 CYP genes are coexpressed with lanosterol synthase, and 16 of these show high similarity to fungal CYPs that specifically hydroxylate testosterone, suggesting their possible roles in triterpenoid biosynthesis
11	Antitumor and immunoregulatory activities of *Ganoderma lucidum* and its possible mechanisms	229	2004	This review summarized *G. lucidum* modulate many components of the immune system. The water extract and the polysaccharides fraction of *G. lucidum* exhibited a significant antitumor effect in several tumor-bearing animals. Besides, the alcohol extract or the triterpene fraction of *G. lucidum* possessed an antitumor effect, which seemed to be related to the cytotoxic activity against tumor cells directly. A preliminary study indicated that the antiangiogenic effect might be involved the antitumor activity of *G. lucidum*

(continued)

Table 2.2 (continued)

Rank	Title	TC	PY	Summary
12	Antioxidant properties of several medicinal mushrooms	215	2002	Three species of medicinal mushrooms are commercially available in Taiwan, namely, *Ganoderma lucidum* (Ling-chih), *Ganoderma tsugae* (Sung-shan-ling-chih), and *Coriolus versicolor* (Yun-chih). Methanolic extracts were prepared from these medicinal mushrooms and their antioxidant properties studied. Overall, *G. lucidum* and *G. tsugae* were higher in antioxidant activity, reducing power, scavenging and chelating abilities, and total phenol content
13	Structural features of immunologically active polysaccharides from *Ganoderma lucidum*	196	2002	Three polysaccharides, two heteroglycans (PL-1 and PL-4) and one glucan (PL-3) were extracted from the fruit bodies, followed by anion-exchange and gel-filtration chromatography. Their structural features were elucidated by glycosyl residue and glycosyl linkage composition analyses, partial acid hydrolysis, acetolysis, periodate oxidation, 1D and 2D NMR spectroscopy, and ESI–MS experiments.
14	Triterpenes from the spores of *Ganoderma lucidum* and their inhibitory activity against HIV-1 protease	189	1998	Two new lanostane-type triterpenes, lucidumol A and ganoderic acid beta, were isolated from the spores of *Ganoderma lucidum*, together with a new natural one and seven that were known. Of the compound isolated, ganoderic acid beta, (24S)-lanosta-7,9(11)-diene-3 beta,24,25-triol (called lucidumol B), ganodermanondiol, ganodermanontriol, and ganolucidic acid A showed significant anti-human immunodeficiency virus (anti-HIV)-1 protease activity.
15	Studies on the immuno-modulating and antitumor activities of *Ganoderma lucidum* (Reishi) polysaccharides: functional and proteomic analyses of a fucose-containing glycoprotein fraction responsible for the activities	187	2002	A fucose-containing glycoprotein fraction that stimulates spleen cell proliferation and cytokine expression has been identified from the water-soluble extract of *Ganoderma lucidum*. Proteomic analysis of mouse spleen cells treated with this glycoprotein fraction showed up to a 50% change of the proteome. Further studies indicate a polysaccharide fraction is responsible for stimulating the expression of cytokines.
16	Triterpene-enriched extracts from *Ganoderma lucidum* inhibit the growth of hepatoma cells via suppressing protein kinase C, activating mitogen-activated protein kinases and G2-phase cell cycle arrest	182	2003	In this report, we studied the anticancer mechanism of triterpene-enriched extracts from *G. lucidum*. Our findings suggest that the triterpenes contained in *G. lucidum* are potential anticancer agents.

(continued)

Table 2.2 (continued)

Rank	Title	TC	PY	Summary
17	Effects of Ganopoly (R) (A *Ganoderma lucidum* polysaccharide extract) on the immune functions in advanced-stage cancer patients	178	2003b	This study aimed to investigate the effects of Ganopoly(R), the polysaccharides fractions extracted from *G. lucidum*, on advanced-stage cancer patients' immune function. Thirty-four advanced-stage cancer patients were entered into this study and treated with 1800 mg Ganopoly(R), three times daily orally before meals for 12 weeks. The present study indicates that Ganopoly enhanced the immune responses in patients with advanced-stage cancer. Clinical evaluations of response and toxicity are ongoing.
18	Fed-batch fermentation of *Ganoderma lucidum* for hyperproduction of polysaccharide and ganoderic acid	173	2002	A process was developed for simultaneous efficient production of ganoderic acid (GA) and polysaccharides by fed-batch fermentation. Sucrose as a carbon source was suitable for the extracellular polysaccharide (EPS) production, although the cells could not grow well. Lactose was beneficial for the cell growth and production of GA and intracellular polysaccharide (IPS).
19	Extraction, purification, characterization, and antitumor activity of polysaccharides from *Ganoderma lucidum*	172	2010	Ultrasonic-aid extraction (UAE) was applied to extract polysaccharides from *Ganoderma lucidum*. Then the crude polysaccharides were purified by filtration, DEAE cellulose-52 chromatography, and Sephadex G-100 size-exclusion chromatography in that order.
20	Structural and immunological studies of a major polysaccharide from spores of *Ganoderma lucidum* (Fr.) Karst	172	2001	A polysaccharide isolated from *Ganoderma lucidum* spores was found to be a complex glucan. Its conformational in aqueous solution was analyzed. And the immunological activities of the native and degraded glucans were also investigated.

macrophages and T lymphocytes. The results had shown that the levels of interleukin (IL)-1 beta, tumor necrosis factor (TNF)-alpha, and IL-6 in macrophage cultures treated with PS-G (100 mu g/ml) were 5.1-, 9.8-, and 29-fold higher, respectively, than those of untreated controls. Lin and Zhang (2004) reviewed that the aqueous extract and polysaccharide fraction of *G. lucidum* exhibited significant antitumor effects in several tumor-bearing animals, mainly through its immune-enhancing activity. Recent studies have also shown that the alcoholic extract or triterpene parts of *G. lucidum* have antitumor effects, directly related to their cytotoxic activity against tumor cells. Preliminary studies suggest that the antiangiogenic effect of *G. lucidum* may be associated with the antitumor activity. Gao et al. (2003b) investigated the effect of Ganopoly(R), a polysaccharide fraction of *Ganoderma lucidum*, on immune function in patients with advanced cancer. Thirty-four patients with advanced cancer participated in this study and received 1800 mg Ganopoly(R) orally three times daily before meals for 12 weeks. Immune function was assessed in 30 patients. This study demonstrates that Ganopoly enhances the immune response in patients with advanced cancer. Clinical evaluation of efficacy and toxicity is ongoing. Lin et al.

(2003) prepared the triterpene-rich component WEES—G6 from mycelium of *G. lucidum*. They found that WEES-G6 inhibited the growth of human carcinoma Huh-7 cells, but not normal human hepatocytes. Their findings suggest that the triterpenes contained in *G. lucidum* are potential anticancer agents. Wang et al. (2002) identified a fucose-containing glycoprotein fraction that stimulates spleen cell proliferation and cytokine expression from the water-soluble extract of *G. lucidum*. The active ingredients of *G. lucidum* include polysaccharides and triterpenes. They have attracted scholars' attention at home and abroad.

Besides, research on *G. lucidum* extract methods is common. Firstly, Min et al. (1998) isolated two new lanolin-type triterpenes and a new natural triterpene from spores. Secondly, Bao et al. (2002) identified three polysaccharides, two heteroglycans, and one glucan from the fruit bodies. Thirdly, Bao et al. (2001) found a complex glucan from *G. lucidum* spores. Forthly, Tang and Zhing (2002) developed a process for the simultaneous and efficient production of bioactive ganoderic acid (GA) and polysaccharides using supplement fermentation. Fifthly, Zhao et al. (2010) applied ultrasonic-aid extraction (UAE) to extract polysaccharides from *G. lucidum*. Then the crude polysaccharides were purified by filtration, DEAE cellulose-52 chromatography, and Sephadex G-100 size-exclusion chromatography in that order. Lastly, Kozarski et al. (2011) extraction of partially purified polysaccharides from three medicinal mushrooms, *Agaricus bisporus, Agaricus brasiliensis*, and *G. lucidum*, by hot water extraction and ethanol precipitation. More and more advanced extraction methods have been proposed, conducive to the full utilization of active ingredients of *G. lucidum*.

Most recently, genetic studies of *G. luciudm* become a hot area. Chen et al. (2012) reported a 43.3-Mb genome of *G. lucidum* encoding 16,113 predicted genes, which was obtained using next-generation sequencing and optical mapping methods. The genome also encodes the most abundant set of wood-degrading enzymes of all sequenced Streptomyces. In total, the authors identified 24 physical CYP gene clusters. Moreover, 78 CYP genes are coexpressed with lanosterol synthase. 16 of them showed high similarity to hydroxylate testosterone, suggesting their possible triterpenoid biosynthesis roles. Therefore, the genomic information of *G. lucidum* has been improved, allowing researchers to have a better understanding of *G. lucidum* at the molecular level.

2.4 Conclusion and Discussion

We have conducted a comprehensive bibliometric analysis on "*Ganoderma lucidum*" based on 3286 publications collected from the WOS core collection database from 1936 to 2019. To sum up, the number of publications on *G. lucidum* has been growing rapidly. Geographically, China is the most active region for *G. lucidum* research. The journal "International Journal of Medicinal Mushrooms" is the most productive source. The keyword analysis identified the following most frequently used keywords "polysaccharides," "in-vitro," "expression," and "antitumor-activity". Based on the most cited papers, we found a large number of papers focused on the pharmacologic actions and composition of the active components of *G. lucidum*, followed by papers studying the genetics of *G. lucidum*. China has a long history of using *G. lucidum* for health-promotion. It will likely continue to play a leading role in this area. Increasing cooperation among countries, organizations cooperation, and international exchange will be the inevitable trend. In the past, studies have focused on the pharmacological effects and active ingredients of *G. lucidum*. We think genetic engineering studies of *G. lucidum* will be the focus of further research.

References

Aria M, Cuccurullo C (2018) bibliometrix: an r-tool for comprehensive science mapping analysis. J Informet 11(4):959–975

Bao X, Liu C, Fang J, Li X (2001) Structural and immunological studies of a major polysaccharide from spores of *Ganoderma lucidum* (Fr.) Karst. Carbohydr Res 332 (1):67–74

Bao XF, Wang XS, Dong Q, Fang JN, Li XY (2002) Structural features of immunologically active polysaccharides from *Ganoderma lucidum*. Ganoderma Lucidum Phytochem 59(2):175–181

Barbero-Sierra C, Marques MJ, Ruiz-Pérez M, Escadafal R, Exbrayat W (2015) How is desertification research addressed in spain? land versus soil approaches. Land Degrad Dev 26(5):423–432

Batra P (2019) Bioactive metabolites of *Ganoderma lucidum*: factors, mechanism and broad spectrum therapeutic potential. J Herb Med 17-18(8):100268

Chang CJ, Lin CS, Lu CC, Martel J, Ko YF, Ojcius DM, Tseng SF, Wu TR, Chen YYM, Young JD, Lai HC (2015) *Ganoderma lucidum* reduces obesity in mice by modulating the composition of the gut microbiota. Nat Commun 6:17. https://doi.org/10.1038/ncomms8489

Chen S, Xu J, Liu C, Zhu Y, Nelson D, Zhou S, Li C, Wang L, Guo X, Sun Y, Luo H, Li Y, Song J, Henrissat B, Levasseur A, Qian J, Li J, Luo X, Shi L, Sun C (2012) Genome sequence of the model medicinal mushroom *Ganoderma lucidum*. Nat Commun 3:913. https://doi.org/10.1038/ncomms1923

Chen C, Dubin R, Kim MC (2014) Emerging trends and new developments in regenerative medicine: a scientometric update (2000–2014). Expert Opin Biol Ther 14(9):1295

Franz G (1989) Polysaccharides in pharmacy: current applications and future concepts. Planta Med 55 (06):493–497

Furusawa E, Chou SC, Furusawa S, Hirazumi A, Dang Y (1992) Antitumour activity of *Ganoderma lucidum*, an edible mushroom, on intraperitoneally implanted Lewis lung carcinoma in synergenic mice. Phytother Res 6(6):300–304

Gao Y, Zhou SW, Huang M, Dai X (2003b) Effects of ganopoly (a *Ganoderma lucidum* polysaccharide extract) on the immune functions in advanced-stage cancer patients. Immunol Invest 32(3):201–215

Gao Y, Zhou S, Huang M, Xu A (2003a) Antibacterial and antiviral value of the genus *Ganoderma* P. Karst. Species (Aphyllophoromycetideae): a review. 5(3):12. https://doi.org/10.1615/InterJMedicMush.v5.i3.20

Gao C, Sun M, Geng Y, Wu R, Chen W (2016) A bibliometric analysis based review on wind power price. Applied Energy 182:602-612

Ge Y, Guo P, Xu X, Chen G, Zhang X, Shu H, Zhang B, Luo Z, Chang C, Fu Q (2017) Selective analysis of aristolochic acid I in herbal medicines by dummy molecularly imprinted solid-phase extraction and HPLC. J Sep Sci 40(13):2791–2799

Heleno SA, Barros L, Martins A, Queiroz MJRP, Santos-Buelga C, Ferreira ICFR (2012) Fruiting body, spores and in vitro produced mycelium of *Ganoderma lucidum* from Northeast Portugal: a comparative study of the antioxidant potential of phenolic and polysaccharidic extracts. Food Res Int 46(1):135-140

Karsten P (1881) Enumeratio boletinearum et polyporearum fennicarum, systemate novo dispositarum. Rev Mycol 3:1-19

Kleinberg J (2003) Bursty and Hierarchical Structure in Streams. Data Min Knowl Disc 7(4):373–397

Kozarski M, Klaus A, Niksic M, Jakovljevic D, Helsper J, Van Griensven L (2011) Antioxidative and immunomodulating activities of polysaccharide extracts of the medicinal mushrooms *Agaricus bisporus*, *Agaricus brasiliensis*, *Ganoderma lucidum* and *Phellinus linteus*. Food Chem 129(4):1667-1675

Li K, Gao P, Xiang P, Zhang X, Cui X, Ma LQ (2016) Molecular mechanisms of PFOA-induced toxicity in animals and humans: Implications for health risks. Environ Int 99:43–54

Li M, Porter AL, Wang ZL (2017) Evolutionary trend analysis of nanogenerator research based on a novel perspective of phased bibliographic coupling. Nano Energy 34:93–102

Lin Z-B, Zhang H-N (2004) Anti-tumor and immunoregulatory activities of *Ganoderma lucidum* and its possible mechanisms. Acta Pharmacol Sin 25 (11):1387–1395

Lin SB, Li CH, Lee SS, Kan LS (2003) Triterpene-enriched extracts from *Ganoderma lucidum* inhibit growth of hepatoma cells via suppressing protein kinase C, activating mitogen-activated protein kinases and G2-phase cell cycle arrest. Life Sci 72(21):2381–2390

Liu L, Xu FR, Wang YZ (2020) Traditional uses, chemical diversity and biological activities of Panax L. (Araliaceae): A review. J Ethnopharmacol 263:41. https://doi.org/10.1016/j.jep.2020.112792

Lyu H-N, Liu H-W, Keller NP, Yin W-B (2020) Harnessing diverse transcriptional regulators for natural product discovery in fungi. Nat Prod Rep 37 (1):6–16. https://doi.org/10.1039/C8NP00027A

Mau J-L, Lin H-C, Chen C-C (2002) Antioxidant properties of several medicinal mushrooms. J Agric Food Chem 35(6):519-526

Min B-S, Nakamura N, Miyashiro H, Bae K-W, Hattori M (1998) Triterpenes from the spores of *Ganoderma lucidum* and their inhibitory activity against HIV-1 protease. Chem Pharm Bull 46(10):1607–1612. https://doi.org/10.1248/cpb.46.1607

Nishita M, Park S-Y, Nishio T, Kamizaki K, Wang Z, Tamada K, Takumi T, Hashimoto R, Otani H, Pazour GJ, Hsu VW, Minami Y (2017) Ror2 signaling regulates Golgi structure and transport through IFT20 for tumor invasiveness. Sci Rep 7(1):1. https://doi.org/10.1038/s41598-016-0028-x

Paterson R et al. (2006) *Ganoderma*—a therapeutic fungal biofactory. Phytochemistry 67(18):1985–2001

Pritchard A (1969) Statistical bibliography or bibliometrics? J Doc 25(4):348–349

Sarmadi BH, Ismail A (2010) Antioxidative peptides from food proteins: a review. Peptides 31(10):1949–1956

Shaoping N, Hui Z, Wenjuan Li, Mingyong, (2013) Current development of polysaccharides from Ganoderma: isolation, structure and bioactivities. Bioact Carbohydr Diet Fibre 1(1):10–20

Shiao MS, Lee KR, Lin LJ, Wang CT (1994) Natural products and biological activities of the Chinese medicinal fungus *Ganoderma lucidum*. Acs Symposium Series 204:342–354

Soccol CR, Bissoqui LY, Rodrigues C, Rubel R, Sella SRBR, Leifa F, de Souza Vandenberghe LP, Soccol VT (2016) Pharmacological properties of biocompounds from spores of the lingzhi or reishi medicinal mushroom *Ganoderma lucidum* (Agaricomycetes): a review. 18(9):757–767. https://doi.org/10.1615/IntJMedMushrooms.v18.i9.10

Sun YF, Costa-Rezende DH, Xing JH, Zhou JL, Zhang B, Gibertoni TB, Gates G, Glen M, Dai YC, Cui BK (2020) Multi-gene phylogeny and taxonomy of Amauroderma s. lat. (Ganodermataceae). Persoonia —Molecular Phylogeny and Evolution of Fungi 44 (8):206-239

Tan W, Lu J, Huang M, Li Y, Chen M, Wu G, Gong J, Zhong Z, Xu Z, Dang Y, Guo J, Chen X, Wang Y (2011) Anti-cancer natural products isolated from chinese medicinal herbs. Chin Med 6(1):27. https://doi.org/10.1186/1749-8546-6-27

Tang YJ, Zhong JJ (2002) Fed-batch fermentation of *Ganoderma lucidum* for hyperproduction of polysaccharide and ganoderic acid. Enzyme Microb Technol 31(1–2):20–28

Torres L, Abraham EM, Rubio C, Barbero-Sierra C, Ruiz-Perez M (2015) Desertification research in Argentina. Land Degrad Dev 26(5):433–440

Wang S-Y, Hsu M-L, Hsu H-C, Lee S-S, Shiao M-S (1997) The antitumor effect of *Ganoderma lucidum* is mediated by cytokines released from activated macrophages and T lymphocytes. Int J Cancer 70(6):699-705

Wang Y-Y, Khoo K-H, Chen S-T, Lin CC (2002) Studies on the immuno-Modulating and antitumor activities of *Ganoderma lucidum* (Reishi) polysaccharides: functional and proteomic analyses of a fucose-Containing glycoprotein fraction responsible for the activities. Bioorg Med Chem 10(4):1057-1062

Wanqi X, Hyung-Kyoon C, Linfang H (2017) State of Panax ginseng research: a global analysis. Molecules 22(9):1518

Xu W, Zou Z, Pei J, Huang L (2018) Longitudinal trend of global artemisinin research in chemistry subject areas (1983–2017). Bioorg Med Chem 26(20):5379-5387

Yu D, Xu Z, Pedrycz W, Wang W (2017) Information sciences 1968–2016: A retrospective analysis with text mining and bibliometric. Inf Sci 418–419:619–634

Zhang S, Liu Xi, Huibin M, Guozhu C, John, (2017) Groundwater remediation from the past to the future: a bibliometric analysis. Water Res J Int Water Assoc 119:114–125

Zhao L, Dong Y, Chen G, Hu Q (2010) Extraction, purification, characterization and antitumor activity of polysaccharides from *Ganoderma lucidum*. Carbohydr Polym 80(3):783-789

Zhao C, Zhang C, Xing Z, Ahmad Z, Li JS, Chang MW (2018) Pharmacological effects of natural Ganoderma and its extracts on neurological diseases: a comprehensive review. Int J Biol Macromol 121:1160-1178

Zhou L-W, Nakasone KK, Burdsall HH, Ginns J, Vlasák J, Miettinen O, Spirin V, Niemelä T, Yuan H-S, He S-H, Cui B-K, Xing J-H, Dai Y-C (2016) Polypore diversity in North America with an annotated checklist. Mycol Prog 15(7):771–790. https://doi.org/10.1007/s11557-016-1207-7

The Nuclear Genome of Lingzhi Mushroom

3

Jingting Liu, Jingling Li, Mei Jiang, and Chang Liu

Abstract

Abstract By November 2020, sixteen whole-genome sequencing projects for *Ganoderma* species have been registered in the international databases. Three of them have been assembled to high quality. This chapter provides an overview of these genome sequencing projects. To support the utilization of genomic information of *Ganoderma lucidum*, we constructed the *G. lucidum* Database (GaLuDB). This database's contents include the genome assembly, predicted gene models, and their corresponding gene function annotations. Besides, GaLuDB also contains transcriptomic data for *G. lucidum* obtained from three different developmental stages or those under various conditions. To facilitate the browsing and querying of these data, GaLuDB provides three groups of utilities: GBrowse, Tools, and Wiki pages. GBrowse allows users to navigate the genome to view the gene structure, their annotation, and the expression levels. The tools can be divided into four sub-categories: (1) database searching tools based on sequence similarity; (2) retrieval of annotations and sequences giving a list of gene IDs; (3) curation of gene models; and (4) identification of gene clusters enriched with genes belonging to particular functional groups. Lastly, a wiki site has been set up for the *G. lucidum* research community members to share and exchange information. GaLuDB can be accessed publicly at http://www.1kmpg.cn/galu.

J. Liu · J. Li · M. Jiang · C. Liu (✉)
Institute of Medicinal Plant Development, Chinese Academy of Medical Sciences & Peking Union Medical College, No. 151 Malianwa North Road, Haidian, Beijing 100193, China
e-mail: cliu6688@yahoo.com

3.1 Introduction

3.1.1 Medical and Economic Significance of *Ganoderma* species

Ganoderma P. Karst. refers to a fungal genus including species such as *Ganoderma lucidum* Karst and *Ganoderma sinense*. All of them have distinctive double-wall basidiospores and decorative endospores. Dictionary of Fungi contains 80 *Ganoderma* species. In contrast, 427 name records are recognized by Index Fungorum (http://www.indexfungorum.org/). *Ganoderma* genus is of great importance from three aspects: as plant pathogens (Paterson 2006), as bio-bags of ligninolytic enzymes (Hushiarian et al. 2013), and as therapeutic fungal bio-factories (Zhou et al. 2013). For instance, *G. lucidum* Karst. and *G. sinense* has been used as high-value traditional medicines in East Asian countries, including China, Korea, and Japan. Besides, *Ganoderma* as plant pathogens has been reported

C. Liu (ed.), *The Lingzhi Mushroom Genome*, Compendium of Plant Genomes,
https://doi.org/10.1007/978-3-030-75710-6_3

in many countries in North America, Africa, and South Asia, including Indonesia, Pakistan, India, and Malaysia. Lastly, *Ganoderma* species have been reported for their potential uses for bioenergy production in North America and East Asia. It should be pointed out that the *Ganoderma* genus is a fungus group with a highly diverse genetic makeup. *Ganoderma* species reported from different areas might represent genetically different strains having different biochemical properties (Flood et al. 2000).

As medicinal fungi, *Ganoderma* species can produce many bioactive components, including polysaccharides, triterpenoids, alcohols, and phenols, oligosaccharides, peptides and proteins. These chemical compounds have been reported to prevent and treat various conditions as "arthritis, nephritis, hepatopathy, insomnia, chronic hepatitis, diabetes, hypertension, neurasthenia, anorexia, bronchitis, gastric ulcers, asthma, hyperlipidemia, atherosclerosis and leucopenia" (Rai et al. 2005; Xu et al. 2011). In particular, the immuno-modulation and anticancer properties of *G. lucidum* have been studied extensively recently (see Chap. 10). Besides, biological activities, chemical components, and mechanisms of action have been studied in-depth. Readers interested in that subject are advised to check several excellent reviews published recently (Boh 2013; Jin et al. 2012; Xu et al. 2012a; Habijanic et al. 2013; Lee et al. 2012; Popovic et al. 2013; Soares et al. 2013; Wu et al. 2013a).

As a plant pathogen, *Ganoderma* species grows on various dying or dead shrubs and trees in the north of Australia, including economically important broad-leaved trees and ornamental palms, perennial crop plants, and several other woody species, whose seeds, wood, gum, fragrances, bioactive compounds,etc., are valuable products (Hennesey and Daly 2007). The annual economic loss caused by this pathogen has been estimated to be as high as half-billion US$ (Hushiarian et al. 2013). Five pathogens are reported the most. The first one, *G. zonatum* Murrill, is a well-known pathogen for mature palms (US Forest Service 2011). The second one, *G. applanatum* (Pers. ex Wallr.) Pat., can kill aspen grown in the western USA (Elliot and Broschat 2000). The third one, *G. boninense* Pat., causes major loss of oil palms in Indonesia and Malaysia. The fourth one, *G. oerstedii* (Fr.) Murrill, was reported as a parasite for trees in Mexico (Mendoza et al. 2011). The fifth one, *G. applanatum*, and complex species of *G. lucidum* are widespread pathogens in India (Sankaran et al. 2005). Similarly, infections of *G. lucidum* and *G. applanatum* have been recorded in Pakistan (Nasir 2005). It is well-known that *Ganoderma* species infect oil palm trees and cause basal stem rots in many tropical African countries (Miller et al. 2000). Unfortunately, measures to prevent and cure the disease caused by *Ganoderma* species have yet to be developed.

Ganoderma species can degrade lignocellulose, which is the molecular basis for their being plant pathogens (Paterson 2007). In nature, *Ganoderma* species are saprotrophic plants, which play important roles in nutrient mobilization through the enzymatic decomposition of deadwood (Irianto et al. 2006; Rees et al. 2012; Clinton et al. 2009). This capability has been applied in bioremediation, wastewater treatment, and bioenergy production (Coelho-Moreira et al. 2013). Furthermore, several results showed that *Ganoderma* species had strong enzymatic degradation of lignocellulose (Adaskaveg et al. 1990; Martínez et al. 2011; Son et al. 2010; Andrade et al. 2012). Screening for *Ganoderma* species with strong activity of lignin modifying enzyme (LME), and tolerance to extreme conditions are the current research hotspot.

3.1.2 Advantages of Whole-Genome Sequencing (WGS) Projects

There are several difficulties in studying *Ganoderma species* Firstly, the identity of the *Ganoderma* species remains controversial. Historically, common names are used for various *Ganoderma* species. For instance, the *Ganoderma* species used as traditional medicines are called "Lingzhi" in China, "Youngzhi" in Korea, and "Munnertake", "Sachitake", and "Reishi" in

Japan. It is unknown if these common names refer to the same fungal species. It should be pointed out that both *G. lucidum* and *G. Sinensis* are specified as the authentic sources for Lingzhi in Chinese pharmacopeia (Chinese Pharmacopoeia Commission 2010). Even if a taxonomy classification was provided for a particular species, there is no way to verify that the source materials were genetically equivalent. It would not be easy to compare the chemical composition and therapeutic effects described in different studies if the study materials' taxonomy classifications are not determined. Likewise, one difficulty in controlling the infections caused by *Ganoderma* species is the uncertainty of the disease-causing species' identity, leading to the difficulty in developing effective tools for infection detection early in the field (Hushiarian et al. 2013; Flood et al. 2000; Paterson 2007; Naher et al. 2013). The gene-level markers, for the most time, are not sufficient for the identification of *Ganoderma* species (Smith and Sivavithamparam 2000; Moncalvo and Buchanan 2008; Glen et al. 2009). Furthermore, it is unknown about the scales of the intra-specific and inter-specific variations of *Ganoderma* species. (Pilotti 2005; Pilotti et al. 2003; Rakib et al. 2014). Lastly, it is unclear whether these molecular markers are related to specific host ranges (Nusaibah et al. 2011).

The second difficulty is the lack of information about the enzymatic genes and regulatory genes implicated in the biosynthesis of secondary metabolites, ligninolytic activities, and pathogenesis. This hinders the efforts attempting to screen for or breed *Ganoderma* species or strains with favorable traits. Without precise genomic information, it is impossible to utilize the metabolic engineering approach for the large-scale production of ligninolytic enzymes and active compounds for industrial and medical applications. Similarly, it is not easy to develop specific drugs controlling pathogenic *Ganoderma* species.

Through genome annotation, WGS projects can lead to molecular marker development, gene discovery, and pathway identification. Firstly, the complete genome contains all the information for the development of molecular markers suitable for various purposes. Secondly, critical genes responsible for the biosynthesis of chemical compounds can be identified and used to reconstruct the entire biosynthetic pathways. Thirdly, comparative genome analysis of the genomes will help to elucidate the evolutionary history of the genomes and key genes (Andersen et al. 2013; Brakhage and Schroeckh 2011; Floudas et al. 2012). This information is invaluable for the selection of species with favorable traits. On the other hand, this information can help develop anti-fungi strategies or develop better strategies against pathogens.

3.2 The current status of *Ganoderma* Nuclear Genome Whole-Genome Sequencing

By November 2020, sixteen whole-genome sequencing projects for *Ganoderma* species have been registered in the Bioproject section of GenBank. Detailed information is listed in Table 3.1. Five of these projects were carried out by researchers from various Chinese research institutes, including Huazhong University of Science and Technology, China Medical University, and The Institute of Edible Fungi. Researcher from the Institute of Medicinal Plant Development (IMPLAD) and The National Yang-Ming University registered two projects each. Research groups around the world conducted the remaining nine projects. Broad institute registered the largest numbers of WGS projects for *Ganoderma* species. Interestingly, six projects sequenced the genomes of *G. lucidum*; the most studied *Ganoderma* species. Notably, four assembled genomes have been published, two for *G. lucidum*, one for *G. sinense*, and one for an unidentified *Ganoderma* species (Binder et al. 2013; Chen et al. 2012; Liu et al. 2012; Zhu et al. 2015).

The genome assembly from IMPLAD used the next-generation sequencing technology combined with optical mapping technology. The entire genome is 43.3 Mb long, encoding 16,113 genes. Furthermore, the assembly has the largest

Table 3.1. *Ganoderma* WGS projects listed in public databases

Strain	Status	Accession	Organization*	References
G. tornatum strain NPG1	In progress	PRJNA182009	BROAD	NA
G. zonatum strain POR69	In progress	PRJNA182007	BROAD	NA
G. miniatocinctum strain 337,035	In progress	PRJNA182006	BROAD	NA
G. boninense strain PER71	In progress	PRJNA182005	BROAD	NA
G. lucidum strain BCRC 37,177	Whole-genome shotgun sequences available	PRJDA61381	GLRC/NYMU	Huang et al. (2013), Kües et al. (2015)
G. lucidum strain BCRC 37,180	In progress	PRJDA61379	GLRC/NYMU	NA
G. lucidum strain Xiangnong No.1	High quality genome assembly available	PRJNA77007	HNAU/HUST	Liu et al. (2012)
G. lucidum strain G.260125–1	High quality genome assembly available	PRJNA71455	IMPLAD	Chen et al. (2012)
G. sinense strain gasi0214-1	High quality genome assembly	PRJNA42807	IMPLAD	Zhu et al., (2015)
G. lucidum strain Hu-nongke no. 1	In progress	PRJNA42873	IEF	NA
Ganoderma sp. strain 10,5977 SS1	High quality genome assembly available	PRJNA68313	JGI	Binder et al. (2013)
G. boninense strain NJ3	In progress	PRJNA287769	CIRAD	NA
G. lucidum strain BCRC 36,111	In progress	PRJNA564693	CMU	NA
Ganoderma sp. strain BRIUMSc	In progress	PRJNA553124	MSU	NA
G. bominense strain G3	In progress	PRJNA421251	PST	NA
G. tsugae strain s90	In progress	PRJNA445345	BNTRI	NA

*BROAD: BROAD Institute of Massachusetts Institute of Technology (MIT) and Harvard, Boston, USA; IMPLAD: Institute of Medicinal Plant Development, Chinese Academy of Medical Sciences, Beijing, China; IEF: Institute of Edible Fungi, Shanghai Academy of Agricultural Sciences, Shanghai, China; JGI: DOE Joint Genome Institute, Walnut Creek, California USA; GLRC/NYMU: *Ganoderma lucidum* Research Consortium, National Yang-Ming University, Taiwan; HNAU: Hunan Agriculture University; HUST: HuaZhong University of Science and Technology, Wuhan, China; CIRAD: Recherche Agronomique pour le développement; CMU: China Medical University; MSU: Malaysia Sabah University; PST: PT SMART Tbk; BNTRI: Biotechnology and Nuclear Technology Research Institute; NA: Not Available.

number of wood degradation enzymes among the basidiomycete genomes. Besides, the genome contains 24 physical CYP gene clusters (Chen et al. 2012). The forming of gene clusters might allow better-coordinated expression regulation of genes involved in a particular process.

Using Illumina sequencing technology alone, the HNAU (Hunan Agricultural University) group sequenced *G. lucidum* strain Xiangnong No.1. The assembly is 39.9 Mb, encoding 12,080 genes. About 83% of the genes have homologs in international databases. The HNAU group annotated their *G. lucidum* assembly and then annotated the genome. They also analyzed the genes implicated in the biosynthesis of the various major active components in detail, particularly for the ganoderic acids (Liu et al. 2012).

The JGI genome assembly was sequenced using combined ABI3730, 454-Titanium, and Illumina GAII sequencing platforms. The genome is 39.52 Mb long, encoding 12,910 genes. The blended sequencing reads were assembled using a JGI's custome assembly pipeline. The JGI annotation pipeline was used to annotate the assembly. Three datasets with various numbers of genes were constructed for species from Polyporales (Binder et al. 2013) to conduct phylogenomic analysis.

The summary of *Ganoderma* species genome assemblies is described in Table 3.2. These ten genome assemblies have sizes ranging from 39.5 to 79.2 Mb. Among the four genome assemblies for *G. lucidum*, the quality of the assembly from IMPLAD appears to be the highest. Taking advantage of the optical mapping technology, they constructed a physical map consisting of thirteen chromosomes. Eight-two scaffolds were mapped to thirteen chromosomes. The raw data for the high throughput sequencing of *Ganoderma* species are shown in Table 3.3. The raw data sizes range from 368 to 1517 Mb.

The genes cloned and characterized from the *Ganoderma* species genome are listed in Table 3.4. A total of 35 genes were cloned from *Ganoderma* species. Among them, 28 genes were clone from *G. lucidum*. *Ganoderma* triterpenes, particularly the ganoderic acids, are *G. lucidum*'s main effective components. In 2008, a farnesyl-diphosphate synthase gene (GLFPS), was isolated from *G. lucidum* (Ding et al. 2008). Moreover, an isopentenyl diphosphate isomerase (IDI) gene, GLIDI, was isolated from *G. lucidum*, potentially playing an important role in triterpene biosynthesis (Wu et al. 2013b). In 2013, a hydroxymethylglutaryl-CoA synthase gene, which is potentially involved in the ganoderic acid biosynthesis, was cloned from *G. lucidum* (Ren et al. 2013). Further expression analysis revealed that the overexpression of this gene enhanced ganoderic acid content in *G. lucidum*. Among these cloned genes in *G. lucidum*, the laccase gene was cloned twice in 2008 and 2014. These two laccase genes were all cloned from the white-rot fungus *G. lucidum*. Both of them were expressed in *Pichia pastoris* and had shown high laccase activity (Joo et al. 2008; Xu et al. 2014). They proved that laccase genes have antioxidative activities and can protect the cellular proteins from degradation.

3.3 Genome Resources for *Ganoderma* Species

Although public databases like GenBank are the most commonly used general-purpose genome sequence database, taxonomic-specific databases are also needed to provide detailed information about a species. There are two website resources available for *Ganoderma* species. One is the JGI portal (Grigoriev et al. 2009). The JGI portal provides a variety of datasets and tools for the JGI's studies, including (1) a genome browser for browsing the genome; (2) gene annotations; and (3) a download page. The data are well-organized and easy to search and download. Another website resource provides data generated from the WGS project led by IMPLAD (Chen et al. 2012). Details are described in the following text.

3.3.1 Background

For a long time, *G. lucidum* has been called "the mushroom of immortality" and "the symbol of Traditional Chinese Medicine". It thus represents one of the most well-known medicinal macrofungi in the world. It is included in the American herbal pharmacopeia and therapeutic compendium because of its various pharmacological benefits (Sanodiya et al. 2009). Presently, more than 400 different compounds have been found in *G. lucidum* (Boh et al. 2007). Tri-terpenoids and polysaccharides represent two major categories of pharmacologically active components that might contribute to multiple therapeutic activities, including antitumor, antihypertensive, antivirus, and immunomodulatory activities (Boh

Table 3.2. Comparison of sequenced *Ganoderma* genomes

Organism name	*G. tsugae*	*G. boninense*	*G. boninense*	*G. lucidum*	*G. lucidum*	*G. lucidum*	*G. lucidum*	*G. sinense*	*G. multipileum*	*Ganoderma sp.*
Strain	s90	G3	NJ3	G.260125–1	BCRC 36,111	Xiangnong No.1	BCRC 37,177	ZZ0214-1	BCRC 37,180	BRIUMSc
BioSample	SAMN08792530	SAMN08135413	SAMN03786845	SAMN02981363	SAMN12720022	SAMN00754288	SAMD00036554	SAMN02225658	SAMD00036548	SAMN12229584
BioProject	PRJNA445345	PRJNA421251	PRJNA287769	PRJNA71455	PRJNA564693	PRJNA77007	PRJDA61381	PRJNA42807	PRJDA61379	PRJNA553124
Assembly	GCA_003057275.1	GCA_002900995.2	GCA_001855635.1	GCA_000271565.1	GCA_012655175.1	GCA_000262775.1	GCA_000338035.1	GCA_002760635.1	GCA_000338015.1	GCA_008694245.1
Level	Scaffold	Contig	Contig	Scaffold	Contig	Scaffold	Contig	Scaffold	Contig	Scaffold
Size (Mb)	45.5036	79.1885	60.3258	43.2926	48.9134	39.9458	44.0804	48.9557	46.3836	52.2843
GC%	/	55.9	55.9	56.1	55.1	55.3	55.5	55.6	55.3	55.6
WGS	PYSH01	PJEW02	LFMK01	AGAX01	JAAIFM01	AHGX01	BACH01	AYKW01	BACB01	VJXU01
Scaffolds	6,638	495	18,903	82	173	634	3,275	69	6,173	12,158
Scaffold N50	11,802	/	/	1,387,710	/	322,991	/	2,256,307	/	6,788
Scaffold L50	886	/	/	11	/	34	/	8	/	1,869
Contigs	6,742	495	18,903	194	173	1,708	3,275	196	6,173	14,519
Contig N50	11,659	272,644	6,116	649,708	1,281,108	80,796	63,041	753,893	50,471	6,197
Contig L50	897	83	2,724	19	10	108	208	21	250	2,053
Release Date	2018-04-18T00:00:00Z	2018-01-26T00:00:00Z	2016-10-28T00:00:00Z	2012-06-21T00:00:00Z	2020-04-22T00:00:00Z	2012-05-14T00:00:00Z	2013-02-01T00:00:00Z	2017-11-07T00:00:00Z	2013-02-01T00:00:00Z	2019-09-25T00:00:00Z
GenBank FTP	ftp://ftp.ncbi.nlm.nih.gov/genomes/all/GCA/003/057/275/GCA_003057275.1_ASM305727v1	ftp://ftp.ncbi.nlm.nih.gov/genomes/all/GCA/002/900/995/GCA_002900995.2_ASM290099v2	ftp://ftp.ncbi.nlm.nih.gov/genomes/all/GCA/001/855/635/GCA_001855635.1_ASM185563v1	ftp://ftp.ncbi.nlm.nih.gov/genomes/all/GCA/000/271/565/GCA_000271565.1_GanLuc1.0	ftp://ftp.ncbi.nlm.nih.gov/genomes/all/GCA/012/655/175/GCA_012655175.1_ASM1265517v1	ftp://ftp.ncbi.nlm.nih.gov/genomes/all/GCA/000/262/775/GCA_000262775.1_GanLuc1.0	ftp://ftp.ncbi.nlm.nih.gov/genomes/all/GCA/000/338/035/GCA_000338035.1_Glu37177_1.0	ftp://ftp.ncbi.nlm.nih.gov/genomes/all/GCA/002/760/635/GCA_002760635.1_GanSi1.6	ftp://ftp.ncbi.nlm.nih.gov/genomes/all/GCA/000/338/015/GCA_000338015.1_Glu37180 1.0	ftp://ftp.ncbi.nlm.nih.gov/genomes/all/GCA/008/694/245/GCA_008694245.1_ASM869424v1
References	Not available	Ultomo et al. (2018)	Merclere et al. (2015)	Chen et al. (2012)	Chen et al. (2012)	Liu et al. (2012)	Liu et al. (2012)	Zhu et al. (2015)	Zhu et al. (2015)[1]	Voo et al. (2020)

Table 3.3. List of High throughput sequencing raw data in the public databases

Scientific name	Size (MB)	Experiment ID	SequencingPlatform	BioProject ID	BioSample ID	TaxID	Center name	Download path	ReleaseDate	References
Ganoderma sp. BRIUMSc	3835	SRX6934198	ILLUMINA	PRJNA553124	SAMN12229584	1,072,361	MSU	https://sra-download.ncbi.nlm.nih.gov/traces/sra59/SRR/009975/SRR10214518	2019/10/2 11:03	Voo et al. (2020)
G. boninense	95	SRX5162931	PACBIO_SMRT	PRJNA421251	SAMN08135413	34,458	PST	https://sra-download.ncbi.nlm.nih.gov/traces/sra69/SRR/008156/SRR8351970	2018/12/19 10:00	Ultomo et al. (2018)
G. lucidum	0.19	SRX3191046	LS454	PRJNA350580	SAMN05942096	5315	IEF	https://sra-downloadb.be-md.ncbi.nlm.nih.gov/sos2/sra-pub-run-11/SRR6043982/SRR6043982.1	2017/9/16 5:16	NA[a]
G. sinense ZZ0214-1	7698	SRX319002	ILLUMINA	PRJNA42807	SAMN02225658	1,077,348	IMPLAD	https://sra-downloadb.be-md.ncbi.nlm.nih.gov/sos1/sra-pub-run-5/SRR927414/SRR927414.1	2014/7/3 21:29	Zhu et al. (2015)
G. sinense ZZ0214-1	7033	SRX319001	ILLUMINA	PRJNA42807	SAMN02225658	1,077,348	IMPLAD	https://sra-downloadb.be-md.ncbi.nlm.nih.gov/sos1/sra-pub-run-5/SRR927413/SRR927413.1	2014/7/3 21:29	NA
G. sinense ZZ0214-1	513	SRX319000	LS454	PRJNA42807	SAMN02225658	1,077,348	IMPLAD	https://sra-downloadb.be-md.ncbi.nlm.nih.gov/sos1/sra-pub-run-5/SRR927412/SRR927412.1	2014/7/3 21:29	NA
G. sinense ZZ0214-1	368	SRX318999	LS454	PRJNA42807	SAMN02225658	1,077,348	IMPLAD	https://sra-downloadb.be-md.ncbi.nlm.nih.gov/sos1/sra-pub-run-5/SRR927411/SRR927411.1	2014/7/3 21:29	NA
G. sinense ZZ0214-1	980	SRX318998	LS454	PRJNA42807	SAMN02225658	1,077,348	IMPLAD	https://sra-downloadb.be-md.ncbi.nlm.nih.gov/sos1/sra-pub-run-5/SRR927409/SRR927409.1	2014/7/3 21:29	NA
G. sinense ZZ0214-1	745	SRX318998	LS454	PRJNA42807	SAMN02225658	1,077,348	IMPLAD	https://sra-downloadb.be-md.ncbi.nlm.nih.gov/sos1/sra-pub-run-5/SRR927410/SRR927410.1	2014/7/3 21:29	NA
G. sinense ZZ0214-1	1959	SRX318997	LS454	PRJNA42807	SAMN02225658	1,077,348	IMPLAD	https://sra-downloadb.be-md.ncbi.nlm.nih.gov/sos2/sra-pub-run-7/SRR927381/SRR927381.1	2014/7/3 21:29	NA
Ganoderma sp. 10,5977 SS1	3333	SRX1955656	ILLUMINA	PRJNA68313	SAMN00630372	767,862	JGI	https://sra-downloadb.be-md.ncbi.nlm.nih.gov/sos1/sra-pub-run-1/SRR3927440/SRR3927440.1	2016/7/17 3:11	Syed et al. (2013)
Ganoderma sp. 10,5977 SS1	3299	SRX1955655	ILLUMINA	PRJNA68313	SAMN00630372	767,862	JGI	https://sra-downloadb.be-md.ncbi.nlm.nih.gov/sos1/sra-pub-run-1/SRR3927439/SRR3927439.1	2016/7/17 3:11	NA
Ganoderma sp. 10,5977 SS1	575	SRX079502	LS454	PRJNA68313	SAMN00630372	767,862	JGI	https://sra-downloadb.be-md.ncbi.nlm.nih.gov/sos1/sra-pub-run-1/SRR292169/SRR292169.4	2011/6/24 11:59	NA

(continued)

Table 3.3. (continued)

Scientific name	Size (MB)	Experiment ID	SequencingPlatform	BioProject ID	BioSample ID	TaxID	Center name	Download path	ReleaseDate	References
Ganoderma sp. 10,5977 SS1	135	SRX079501	LS454	PRJNA68313	SAMN00630372	767,862	JGI	https://sra-downloadb.be-md.ncbi.nlm.nih.gov/sos1/sra-pub-run-1/SRR292168/SRR292168.4	2011/6/24 11:57	NA
Ganoderma sp. 10,5977 SS1	91	SRX079500	LS454	PRJNA68313	SAMN00630372	767,862	JGI	https://sra-downloadb.be-md.ncbi.nlm.nih.gov/sos1/sra-pub-run-1/SRR292167/SRR292167.4	2011/6/24 11:57	NA
Ganoderma sp. 10,5977 SS1	92	SRX079499	LS454	PRJNA68313	SAMN00630372	767,862	JGI	https://sra-downloadb.be-md.ncbi.nlm.nih.gov/sos1/sra-pub-run-1/SRR292166/SRR292166.4	2011/6/24 11:57	NA
Ganoderma sp. 10,5977 SS1	219	SRX079498	LS454	PRJNA68313	SAMN00630372	767,862	JGI	https://sra-downloadb.be-md.ncbi.nlm.nih.gov/sos1/sra-pub-run-1/SRR292165/SRR292165.3	2011/6/24 11:41	NA
Ganoderma sp. 10,5977 SS1	813	SRX079497	LS454	PRJNA68313	SAMN00630372	767,862	JGI	https://sra-downloadb.be-md.ncbi.nlm.nih.gov/sos1/sra-pub-run-1/SRR292164/SRR292164.3	2011/6/24 11:59	NA
Ganoderma sp. 10,5977 SS1	782	SRX079496	LS454	PRJNA68313	SAMN00630372	767,862	JGI	https://sra-downloadb.be-md.ncbi.nlm.nih.gov/sos1/sra-pub-run-1/SRR292163/SRR292163.3	2011/6/24 11:59	NA
Ganoderma sp. 10,5977 SS1	1005	SRX079495	LS454	PRJNA68313	SAMN00630372	767,862	JGI	https://sra-downloadb.be-md.ncbi.nlm.nih.gov/sos1/sra-pub-run-1/SRR292162/SRR292162.3	2011/6/24 11:59	NA
Ganoderma sp. 10,5977 SS1	418	SRX079494	LS454	PRJNA68313	SAMN00630372	767,862	JGI	https://sra-downloadb.be-md.ncbi.nlm.nih.gov/sos1/sra-pub-run-1/SRR292161/SRR292161.3	2011/6/24 11:43	NA
Ganoderma sp. 10,5977 SS1	450	SRX079493	LS454	PRJNA68313	SAMN00630372	767,862	JGI	https://sra-downloadb.be-md.ncbi.nlm.nih.gov/sos1/sra-pub-run-1/SRR292160/SRR292160.3	2011/6/24 11:43	NA
Ganoderma sp. 10,5977 SS1	533	SRX079492	LS454	PRJNA68313	SAMN00630372	767,862	JGI	https://sra-downloadb.be-md.ncbi.nlm.nih.gov/sos1/sra-pub-run-1/SRR292159/SRR292159.3	2011/6/24 11:42	NA
Ganoderma sp. 10,5977 SS1	35	SRX079491	LS454	PRJNA68313	SAMN00630372	767,862	JGI	https://sra-downloadb.be-md.ncbi.nlm.nih.gov/sos1/sra-pub-run-1/SRR292158/SRR292158.4	2011/6/24 11:41	NA
Ganoderma sp. 10,5977 SS1	41	SRX079490	LS454	PRJNA68313	SAMN00630372	767,862	JGI	https://sra-downloadb.be-md.ncbi.nlm.nih.gov/sos1/sra-pub-run-1/SRR292157/SRR292157.4	2011/6/24 11:57	NA
	4	SRX079489	LS454	PRJNA68313	SAMN00630372	767,862	JGI		2011/6/24 11:41	NA

(continued)

Table 3.3. (continued)

Scientific name	Size (MB)	Experiment ID	SequencingPlatform	BioProject ID	BioSample ID	TaxID	Center name	Download path	ReleaseDate	References
Ganoderma sp. 10,5977 SS1								https://sra-downloadb.be-md.ncbi.nlm.nih.gov/sos1/sra-pub-run-1/SRR292156/SRR292156.4		
Ganoderma sp. 10,5977 SS1	41	SRX079488	LS454	PRJNA68313	SAMN00630372	767,862	JGI	https://sra-downloadb.be-md.ncbi.nlm.nih.gov/sos1/sra-pub-run-1/SRR292155/SRR292155.4	2011/6/24 11:39	NA
Ganoderma sp. 10,5977 SS1	4245	SRX027726	ILLUMINA	PRJNA68313	SAMN00114960	767,862	JGI	https://sra-downloadb.be-md.ncbi.nlm.nih.gov/sos1/sra-pub-run-2/SRR067807/SRR067807.1	2010/10/6 0:00	NA
G. lucidum G.260125–1	4548	SRX105334	ILLUMINA	PRJNA71455	SAMN00769599	1,077,286	IMPLAD	https://sra-downloadb.be-md.ncbi.nlm.nih.gov/sos1/sra-pub-run-5/SRR364077/SRR364077.3	2012/6/30 0:00	Li et al. (2014)
G. lucidum G.260125–1	2546	SRX105333	ILLUMINA	PRJNA71455	SAMN00769599	1,077,28686	IMPLAD	https://sra-downloadb.be-md.ncbi.nlm.nih.gov/sos1/sra-pub-run-5/SRR364076/SRR364076.3	2012/6/30 0:00	NA
G. lucidum G.260125–1	4270	SRX105332	ILLUMINA	PRJNA71455	SAMN00769599	1,077,286	IMPLAD	https://sra-downloadb.be-md.ncbi.nlm.nih.gov/sos1/sra-pub-run-5/SRR364075/SRR364075.3	2012/6/30 0:00	NA
G. lucidum G.260125–1	488	SRX089233	LS454	PRJNA71455	SAMN00769599	1,077,286	IMPLAD	https://sra-downloadb.be-md.ncbi.nlm.nih.gov/sos1/sra-pub-run-1/SRR327450/SRR327450.3	2012/6/30 0:00	NA
G. lucidum G.260125–1	627	SRX089231	LS454	PRJNA71455	SAMN00769599	1,077,286	IMPLAD	https://sra-downloadb.be-md.ncbi.nlm.nih.gov/sos1/sra-pub-run-1/SRR327448/SRR327448.3	2012/6/30 0:00	NA
G. lucidum G.260125–1	358	SRX089228	LS454	PRJNA71455	SAMN00769599	1,077,286	IMPLAD	https://sra-downloadb.be-md.ncbi.nlm.nih.gov/sos1/sra-pub-run-1/SRR327444/SRR327444.3	2012/6/30 0:00	NA
G. lucidum G.260125–1	410	SRX089226	LS454	PRJNA71455	SAMN00769599	1,077,286	IMPLAD	https://sra-downloadb.be-md.ncbi.nlm.nih.gov/sos1/sra-pub-run-1/SRR327441/SRR327441.3	2012/6/30 0:00	NA
G. lucidum G.260125–1	368	SRX089221	LS454	PRJNA71455	SAMN00769599	1,077,286	IMPLAD	https://sra-downloadb.be-md.ncbi.nlm.nih.gov/sos1/sra-pub-run-1/SRR327438/SRR327438.3	2012/6/30 0:00	NA
G. lucidum G.260125–1	844	SRX089205	LS454	PRJNA71455	SAMN00769599	1,077,286	IMPLAD	https://sra-downloadb.be-md.ncbi.nlm.nih.gov/sos1/sra-pub-run-1/SRR327428/SRR327428.3	2012/6/30 0:00	NA
G. lucidum G.260125–1	732	SRX089204	LS454	PRJNA71455	SAMN00769599	1,077,286	IMPLAD		2012/6/30 0:00	NA

(continued)

Table 3.3. (continued)

Scientific name	Size (MB)	Experiment ID	SequencingPlatform	BioProject ID	BioSample ID	TaxID	Center name	Download path	ReleaseDate	References
								https://sra-downloadb.be-md.ncbi.nlm.nih.gov/sos1/sra-pub-run-1/SRR327427/SRR327427.3		
G. lucidum G.260125–1	573	SRX089203	LS454	PRJNA71455	SAMN00769599	1,077,286	IMPLAD	https://sra-downloadb.be-md.ncbi.nlm.nih.gov/sos1/sra-pub-run-1/SRR327423/SRR327423.3	2012/6/30 0:00	NA
G. lucidum G.260125–1	705	SRX089202	LS454	PRJNA71455	SAMN00769599	1,077,286	IMPLAD	https://sra-downloadb.be-md.ncbi.nlm.nih.gov/sos1/sra-pub-run-1/SRR327430/SRR327430.3	2012/6/30 0:00	NA
G. lucidum	4827	DRX080875	ILLUMINA	PRJDB5525	SAMD00074308	5315	NITE	https://sra-downloadb.be-md.ncbi.nlm.nih.gov/sos1/sra-pub-run-2/DRR087046/DRR087046.1	2017/3/22 8:22	NA
G. lucidum	5335	DRX080874	ILLUMINA	PRJDB5525	SAMD00074309	5315	NITE	https://sra-downloadb.be-md.ncbi.nlm.nih.gov/sos1/sra-pub-run-2/DRR087045/DRR087045.1	2017/3/22 8:22	NA
G. lucidum	851	SRX105750	ILLUMINA	PRJNA77007	SAMN00754288	5315	HUST	https://sra-downloadb.be-md.ncbi.nlm.nih.gov/sos1/sra-pub-run-5/SRR364503/SRR364503.3	2012/5/14 14:52	Liu et al. (2012)
G. lucidum	1517	SRX105750	ILLUMINA	PRJNA77007	SAMN00754288	5315	HUST	https://sra-downloadb.be-md.ncbi.nlm.nih.gov/sos1/sra-pub-run-2/SRR365034/SRR365034.2	2012/5/14 14:52	NA

[a]*NA* Not available

et al. 2007; Shiao 2003). *G. lucidum* is also a white-rot basidiomycete that can break down both cellulose and lignin effectively. *Lastly, G. lucidum* is also a rich source of biomass utilization enzymes for fiber bleaching and organo-pollutant degradation (Ko et al. 2001). *G. lucidum* has been studied actively to develop new drugs and utilize its lignin-degradation enzymatic arsenal.

To realize the full potential of *G. lucidum* as a source of pharmacologically active components and industrial enzymes, we have completed the sequencing of the *G. lucidum* genome using combined approaches of optical mapping, Roche's 454, Illumina's Solexa, and Sanger sequencing technologies. Detailed analyses of the genome have been reported (Chen et al. 2012). To our knowledge, two other research groups have also sequenced different strains of *G. lucidum* simultaneously (Liu et al. 2012) (http://genome.jgi-psf.org/), demonstrating wide interest in *G. lucidum* biology by understanding its genetic makeup. Although we have made our sequences available in GenBank, many analysis results cannot be delivered through GenBank because of its complex data format. Further, we have developed several tools during the analysis of the *G. lucidum* genome, which may be of interest to *G. lucidum* research communities. To allow users access to the extensive set of data and analytic tools, we have constructed GaLuDB or the *Ganoderma* lucidum DataBase. GaLuDB is a valuable resource for research community members interested in understanding the biosynthesis of natural compounds for this, arguably, the most pharmacologically diverse macrofungus. GaLuDB is publicly available at http://www.1kmpg.cn/galu/.

3.3.2 Construction and content

Computation and Database Design

GaLuDB was developed using the Perl Catalyst Framework (5.16) and MySQL (5.7) as the database management system. The GaLuDB application is deployed on an apache server (2.2.14). The web server is run on a Centos 7 operating system. The contents of GaLuDB include an extensive set of data and utilities, which are discussed in detail below.

Content of Data

GaLuDB provides two categories of data: the original experimental data and secondary analysis data derived from the experimental data. The experimental data include the assembled genome, 454 transcriptomic data, and RNA-Seq data generated from different developmental stages. This same set of data can also be downloaded from GenBank (PRJNA71455). The analysis data include gene annotations, 2D structures of the chemical compounds, and the list and members of gene clusters (Table 3.5). The assembled genome contains 13 pseudo-chromosomes constructed using optical mapping technologies. Eighty-two scaffolds were mapped to these 13 chromosomes. *G. lucidum* genes are annotated by searching the following databases: Nt (4,103a), Nr (10,921), Swissprot (6,369), KOGs (6,141), COGs (4,934), Pfam (8,682), InterPro (7,768), and GO (5,598). There are approximately 16,113 gene models in GFF3 format, and 16495 transcripts and predicted protein-coding sequences in gene models. Antismash predicted Twenty-four gene clusters involved in secondary metabolites. One hundred fifty-six structures of Chemical compounds were found in *G. lucidum*.

The genome assembly includes pseudo chromosome sequences, the optical map, the final genome scaffolds, and the contigs. Data were generated from the genome of the haploid *G. lucidum* strain 260125-1 using Roche 454 GS FLX and Illumina platforms. The combination of CABOG (version 6.1) (Miller et al. 2008) and SSPACE (version 1.1) assemblers (Pirovano et al. 2011) was used to assemble the genome. The Roche 454 sequencing data were used for the primary assembly, and the Illumina data were used for scaffolding, gap filling, and error correction. The final assembled sequence consisted of 194 contigs, spanning 43.3 Mb, organized into 82 scaffolds.

The gene models included predicted proteins, coding sequences, transcripts, and gene sequences. Gene models were predicted using MAKER (version 2.10), an automatic annotation pipeline

Table 3.4. Genes cloned and characterized from *Ganoderma* sp.

Accession number	Organism name	Strain/isolate	Gene name	Product	References
KP410689.1	*G. lucidum*	isolate XQ9	LAEA	Global regulation factor	Baby et al. (2015)
HQ377510.1	*G. lucidum*	NA	GMD	GDP-D-mannose 4,6-dehydratase	Chiu et al. (2008)
HQ377511.1	*G. lucidum*	NA	GFS	GDP-L-fucose synthetase	Chiu et al. (2008)
HQ377513.1/HQ377514.1	*G. lucidum*	NA	GPPS	FPP synthase-like geranyl pyrophosphate synthase	Chiu et al. (2008)
HQ377515.1/HQ377516.1	*G. lucidum*	NA	IPI	Isopentenyl pyrophosphate isomerase	Chiu et al. (2008)
EU399544.1/EU399545.1	*G. lucidum*	NA	NA[a]	Farnesyl-diphosphate synthase	Ding et al. (2008)
HQ596496.1/HQ596497.1	*G. lucidum*	NA	AACT	acetyl-CoA acetyltransferase	Fang et al. (2013)
AH015702.2	*G. lucidum*	NA	GPD	Glyceraldehyde-3-phosphate dehydrogenase	Fei et al. (2006)
DQ404343.1	*G. lucidum*	NA	GPD	Glyceraldehyde-3-phosphate dehydrogenase	Fei et al. (2006)
EU526904.1	*G. lucidum*	NA	MAP	Manganese peroxidase	Huang et al. (2009)
MG212669.1	*G. tsugae*	isolate Gps1	NA	Laccase	Jin et al. (2018)
AF185275.2	*G. lucidum*	strain 7071-9	NA	Laccase	Joo et al. (2008)
LN623998.1	*G. applanatum*	strain CBS 250.61	MNP1	Manganese peroxidase	Lanfermann et al. 2015
LN623999.1	*G. applanatum*	strain CBS 250.61	MNP2	Manganese peroxidase	Lanfermann et al. (2015)
MF668107.1	*G. lucidum*	isolate A	PLD	Phospholipase D	Liu et al. (2017)
AB035734.1	*G. applanatum*	NA	EA.MNP1	Manganese peroxidase	Maeda et al. (2001)
JQ406674.1/JQ406675.1	*G. lucidum*	NA	URA3	Orotidine 5'-monophosphate decarboxylase enzyme	Mu et al. (2012)
U56403.1	*G. Microsporum*	NA	NA	Manganese-superoxide dismutase	Pan et al. (1997)
JN391468.1/JN391469.1	*G. lucidum*	NA	HMGS	Hydroxymethylglutaryl-CoA synthase	Ren et al. (2013)
GQ169528.1/GQ169529.1	*G. lucidum*	NA	OSC	Lanosterol synthase	Shang et al. (2010)
EU263989.1	*G. lucidum*	NA	NA	3-hydroxy-3-methylglutaryl-coenzyme A reductase	Shang et al. (2008)
HQ596494.1/HQ596495.1	*G. lucidum*	NA	MVD	Mevalonate pyrophosphate decarboxylase	Shi et al. (2012)
MH919404.1	*G. boninense*	NA	NEP	Necrosis and ethylene inducing protein	Teh et al. (2019)
MF581928.1	*G. lucidum*	strain ACCC53264	SOD	Superoxide dismutase	Wu et al. (2017)
JX524564.1/JX524565.1	*G. lucidum*	NA	IDI	Isopentenyl diphosphate isomerase	Wu et al. (2013a, b)
ACA48488.1	*G. formosanum*	NA	MAP	Manganese peroxidase	Xu et al. (2017)
KF233761.1	*G. lucidum*	isolate 260125-1	RPB1	RNA polymerase subunit 1	Xu et al. (2014)
KF233762.1	*G. lucidum*	isolate 260125-1	RPB2	RNA polymerase subunit 2	Xu et al. (2014)
KF233763.1	*G. lucidum*	isolate 260125-1	APT	Adenine phosphoribosyl transferase	Xu et al. (2014)

(continued)

Table 3.4. (continued)

Accession number	Organism name	Strain/isolate	Gene name	Product	References
KF233764.1	*G. lucidum*	isolate 260125-1	PP2A	Protein phosphatase 2	Xu et al. (2014)
KF233765.1	*G. lucidum*	isolate 260125-1	BTU	Beta-tubulin	Xu et al. (2014)
KF233768.1	*G. lucidum*	isolate 260125-1	UBQ	Polyubiquitin	Xu et al. (2014)
KF384100.1	*G. weberianum*	strain TZC-1	LAC1	Laccase	Xu et al. (2014)
JN377411.1	*G. lucidum*	strain	NA	CGMCC 5.616 succinate dehydrogenase iron-sulfur protein	Xu et al. (2012a)
GQ293361.1	*G. lucidum*	NA	NA	G-protein complex beta subunit	Xu et al. (2012b)
GQ293362.1	*G. lucidum*	NA	NA	Mitogen-activated protein kinase	Xu et al. (2012a)
MH431572.1	*G. lucidum*	NA	CYP512U6	Cytochrome P450	Yang et al. (2018)
MH431573.1	*G. lucidum*	NA	GLCPR	Cytochrome P450 reductase	Yang et al. (2018)
AY485825.1/FJ473385.2	*G. lucidum*	strain TR6	LACCASE GENE	Laccase	You et al. (2014)
KY211742.1	*G. lucidum*	strain CGMCC 5.616	NA	Squalene epoxidase	Zhang et al. (2017)
KX452685.1	*G. lucidum*	strain CGMCC 5.616	NA	Lanosterol synthase	Zhang et al. (2017)
MG428710.1	*G. lucidum*	NA	SWI6 GENE		Zhang et al. (2018)
DQ494674.1/DQ494675.1	*G. lucidum*	NA	SQS	Squalene synthase	Zhao et al. (2007)
KJ155729.1	*G. lucidum*	NA	NA	Squalene synthase	Zhou et al. (2014)
KJ591029.1	*G. lucidum*	isolate CGMCC 5.616	NA	Phosphoglucomutase	Zhou et al. (2014)
KM260167.1	*G. lucidum*	strain CGMCC 5.616	NA	UDP-glucose pyrophosphorylase	Zhou et al. (2014)
HM569745.1	*Ganoderma sp.*	NA	NA	En3 laccase	Zhuo et al. (2011)
EU680479.1	*G. lucidum*	NA	NA	Immunomodulatory protein 8	Ding (2006)
HQ709106.1	*G. lucidum*	NA	NA	Sterol 14alpha-demethylase, promoter region	Fang (2011)
JN167598.1	*G. applanatum*	NA	FIP	Fungal immunomodulatory protein	Lin et al. (2016)
AY987805.1	*G. japonicum*	NA	FIP-GJA	Immunomodulatory protein	Liu et al. (2006)
MG866077.1	*G. pseudoferreum*	strain GP-020	GPFTR1	Iron permease	Sun et al. (2019)

[a]*NA* Not available

(Cantarel et al. 2019). Three types of prediction methods were used, namely, ab initio methods, homology-based methods, and Expressed Sequence Tag (EST)-based methods, in which the predictor was seeded with 454 contigs derived from mycelia, primordia, and fruiting bodies, respectively.

All predicted gene models were annotated functionally using BLAST (version 2.2.17) against a series of protein and nucleotide databases, including the NCBI nucleotide (Nt), the non-redundant set (Nr), the UniProt/Swissprot, Clusters of Orthologous Groups (COGs) of proteins, eukaryotic orthologous groups (KOGs), and the Kyoto Encyclopedia of Genes and Genomes (KEGG) protein databases (Kanehisa et al. 2004), with a significance threshold E-value of 1e-5. The gene models were also annotated using InterProScan (version 4.6) (Hunter et al. 2011) and were assigned Gene Ontology (GO) terms.

GaLuDB also contains RNA-Seq data, which are used to understand the expression regulations of various genes across the developmental stages of *G. lucidum*. RNA-Seq analysis was performed based on the protocol recommended by the manufacturer (Illumina, USA). Reads from different phases were mapped on the whole-genome assembly using BLAT (version 0.33). Abundances were reported as normalized fragments per kb of transcript per million mapped reads (FPKM).

The data described above can all be downloaded directly. Besides, users can browse two types of secondary analysis data, namely, the gene clusters and the chemical compounds identified from *G. lucidum*, both related to secondary metabolites and the genes responsible for their biosynthesis. Recent studies have found that genes involved in the secondary metabolisms are organized in clusters frequently (Brakhage and Schroeckh 2011; Osbourn 2010). These clusters of genes can then be regulated, particularly in response to environmental stimuli (Bok et al. 2006). To identify gene clusters in *G. lucidum*, we used the software tool Antismash (Medema et al. 2011) to identify a total of 24 clusters that were potentially involved in the production of secondary metabolites in *G. lucidum*. A putative cluster containing two NRPS-like proteins is shown in Fig. 3.1a. Domain structures of these two proteins are shown in Fig. 3.1b. From this page, users can browse the gene clusters. They can then use other tools provided in GaLuDB to further characterize the genes inside and around the gene clusters.

In the chemical data section, *G. lucidum* produces large varieties of chemical compounds. To allow chemists to identify potentially active compounds and modify them for better therapeutic efficacy, we have compiled a list of compounds reported for *G. lucidum*. Further, 2D and 3D structures of the compounds can be viewed. The structure of a *G. lucidum* compound, Ganoderic acid A, is shown in Fig. 3.1C as an example. Files for the compound structures can be downloaded for downstream analyses.

Utilities

To facilitate the usage of GaLuDB, we provide three groups of utilities: (1) Genome browser, (2) Tools, and (3) Wiki page (Table 3.6). Their functions are described in detail below.

Genome Browser

GBrowse is a combination of database and interactive web pages for manipulating and displaying annotations on genomes (Donlin 2009; Stein et al. 2002). It is one of the most popular components of the Generic Model Organism Database (GMOD) (O'Connor et al. 2008). We have installed the *G. lucidum* genome browser to perform many functions, such as simultaneous bird's eye and detailed views of the genome; scrolling, zooming, and centering on particular regions; information search by annotation ID, name, or comment; viewing gene expression levels (Fig. 3.2A); and dumping DNA sequences or gene models in the generic feature format (GFF), among others.

Tools

Tools can be divided further into four groups based on their functions, such as (1) BLAST search; (2) retrieval of sequences and annotations for a list of gene IDs; (3) curation of gene models; and (4) identification of gene clusters. An overview of the tools is shown in Table 3.6.

Table 3.5. An overview of data included in GaLuDB

Genome assembly
194 assembled contigs
82 assembled scaffolds
13 pseudo-chromosomes identified by optical mapping
13 pseudo-chromosome integrating the scaffolds and the optical map
Gene models
16,113 gene models in GFF3 format
16,495 gene sequences
16,495 transcript sequences
16,495 predicted protein-coding sequences
Gene functional annotations
G. lucidum genes are annotated by searching the following databases: Nt (4,103[a]), Nr (10,921), Swissprot (6,369), KOGs (6,141), COGs (4,934), Pfam (8,682), InterPro (7,768), and GO (5,598)
Transcriptomic data
12,420 assembled 454 contigs from mycelia 14,431 assembled 454 contigs from primordial 12,881 assembled 454 contigs from fruiting bodies
Gene clusters 24 gene clusters involved in secondary metabolites predicted by Antismash
Chemical compounds 156 structures for compounds of *G. lucidum*

[a]Numbers in the parentheses are numbers of *G. lucidum* genes annotated by corresponding databases

Database search tools allow users to use BLAST and pfamscan to search against *G. lucidum* scaffold, mRNA, protein, and Coding DNA Sequence (CDS) databases. It allows the user to search the databases for genes that are critical in fungal secondary metabolisms, such as Carbohydrate-Active Enzyme (CAZy) families (Cantarel et al. 2009), Cytochrome P450 (Nelson 2011), and Fungal Oxidative Lignin enzyme (FOLy) families (Levasseur et al. 2008). Also, pfamscan, a utility from the Pfam database (Paterson 2007), has been integrated into this module to annotate proteins of interest with protein family domains (Fig. 3.2B, Step 1). Users can retrieve sequences from GaLuDB or GenBank by clicking on identifiers in the BLAST results. Besides, links are included to view the BLAST hits on the genome browsers.

The second category of tools supports the retrieval of information related to genes. Using "Find Gene Annotations", users can retrieve gene annotations using a list of internal *G. lucidum* gene IDs (Fig. 3.2B, Step 2) or using a keyword of interest. Information on the hits from Interpro (Hunter et al. 2011), KOGs (Koonin et al. 2004), COGs (Tatusov et al. 2003), KEGG (Kanehisa et al. 2004), GO (Harris et al. 2004), Pfam (Finn et al. 2010), Nt, Nr, and Swissprot databases are retrieved (Fig. 3.2B). After retrieving the IDs, users can then use "fetch sequences" to retrieve the sequences of genes, mRNA, and proteins, given a list of *G. lucidum* sequence IDs. The corresponding type of sequences for the given IDs is returned in FASTA format. Using "Fetch IDs of Adjacent Genes", users can retrieve a given number of genes that are adjacent to a given gene ID or a list of all genes within a certain number of bases from the query gene.

One of the important roles of any genome database is maintaining a single reference gene set, mapping gene features to new assemblies, and making these data easily accessible to the research community. A crucial step in fulfilling this goal is to accumulate and curate data generated by this community and then allow the gene models to be uploaded back into the central database. The third category of tools supports the curation of gene models. First, users can use "Fetch a Gene Model" to retrieve a gene model in GFF3 format (Fig. 3.2C, Step 1) and then edit the gene model using a third-party gene model editor, such as Apollo (Fig. 3.2C, Step 2) (Lewis et al. 2002). The tool can also be used to export the corresponding protein sequence. Performing gene model editing based on reference sequences is the most efficient. As a result, we provide a tool, "Get a Sequence Alignment," which users can use to retrieve homologous sequences and build a multiple sequence alignment (Fig. 3.2C, Step 3) using the protein sequence exported from Step 2. The input is a general FASTA format protein sequence. After the user has entered a query sequence, the program then does a BLAST to search against the backend database and retrieves the number of top hits (as specified in the "No. of Top Hits") and the corresponding sequences. Finally, the program returns a multiple sequence alignment that contains the query and its best hit sequences. The multiple sequence files can be used to curate the gene models (Fig. 3.2C, Step 4). We support four databases for four major gene families: CAZy, FOLy, CYP450, and Transporter genes. After curation, users can use "Upload a Gene Model" to upload the gene models (Fig. 3.2C, Step 5). To date, *G. lucidum* research communities have annotated over 1600 gene models (Chen et al. 2012).

As described above, genes involved in similar pathways are frequently clustered together in fungal genomes. As a result, genes within the same clusters are possibly involved in the same or related pathways. Identification clusters of genes can thus help to predict the functions of these clustered genes based on those genes with known functions in the clusters. The fourth category of tools is developed to identify and visualize gene clusters. The "Find Gene Clusters" tool is based on the Clusters of Adjacent and Similarly Expressed Genes (CASEG) algorithm (Liu et al. 2005). Briefly, the tool takes a list of

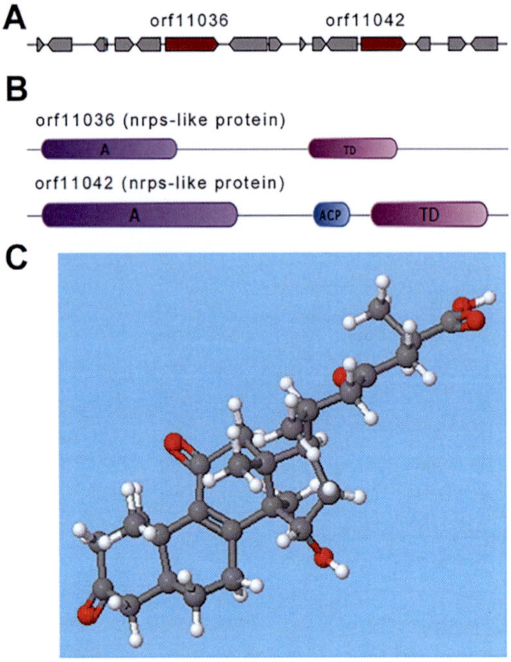

Fig. 3.1 Examples are showing the retrieval of genomic, genetic, and chemical information from *G. lucidum*. **A** a cluster containing two NRPS-like proteins on the *G. lucidum* genome. **B** a domain organization of the two NRPS-like proteins, orf11036 and orf11042. They belong to an NRPS-like protein family and have corresponding signature domains, such as Adenylation Domain (A), Terminal Domain (TD), and Acyl Carrier Protein (ACP). **C** a three-dimensional structure of a secondary metabolic compound

G. lucidum gene IDs as the input and processes them in five steps. First, a sliding window, with a size specified by the user, is used to scan through the chromosomes. Second, the number of genes on the input list in each window is counted. Third, hypergeometric statistics (p-value) is calculated to measure the enrichment of the genes on the input list in each window. Fourth, sliding windows with a significant p-value are retained. Finally, adjacent significant sliding windows are combined, and hypergeometric statistics are recalculated. These final combined sliding windows with p-values greater than the cutoff contain significantly enrich genes compared with those on the input list. Once these clusters of genes are identified, users can use "Map Genes

Table 3.6. An overview of the utility tools implemented in GaLuDB

Categories	Modules	Descriptions
Genome	GBrowse	Allows users to browse genome to view information, such as genome characteristics, structure of gene model, and gene expression levels
Tools	Blast+pfamscan	Allows users to search sequences based on BLAST and Hidden Markov Model (HMM). The query sequences can be searched again the *G. lucidum* genome, mRNA, CDS, and protein sequence databases using BLASTn, BLASTp, BLASTx, tBLASTn, tBLASTx, and pfamscan
	Fetch gene annotations	Allows users to obtain annotation results for the *G. lucidum*-predicted genes based on the hits from Nr, Nt, KEGG, Swissprot, KOGs, COGs, Pfam, InterPro, and GO
	Fetch sequences	Allows users to fetch *G. lucidum* sequences for genome, mRNA, CDS, and proteins
	Fetch IDs of adjacent genes	Allows users to fetch IDs of genes adjacent to a gene of interest. Additional parameters include numbers of genes or ranges on the upstream and the downstream of a query gene
	Fetch gene models	Allows users to fetch gene models in GFF3 format for manual curation
	Get a sequence alignment	Allows users to retrieve homologous sequences for their query and build multiple sequence alignment for gene structure curation. At present, this tool supports annotating homologs of CAZy, FOLy, CYP450, and transporter genes only
	Upload a gene model	Allows users to upload curated gene models in GFF3 format
	Find gene clusters	Allows users to search for gene clusters based on CASEG identification algorithm (Adaskaveg et al. 1990), given a list of genes
	Map genes to chromosomes	Allows users to view the locations of genes on chromosomes. The genes are colored according to their functional categories provided from the input
Wiki	Wiki	Allows users to add, modify, or delete *G. lucidum*-related information via a web browser

to Chromosomes' to find the relative location of this list of genes.

A scenario that identifies CYP450 gene clusters is shown in Fig. 3.3. First, the IDs of all *G. lucidum* CYP450 genes (Fig. 3.3A, Response) are retrieved using "Fetch Gene Annotations' with the keyword "CYP450' (Fig. 3.3A, Request). The list of *G. lucidum* CYP450 gene IDs (Fig. 3.3b, Request) is used as input for the tool "Find Gene Clusters.' The output is the input gene identifiers (IDs) along with their cluster numbers if found (Fig. 3.3B, Response). Finally, the list of CYP450 gene IDs, along with their cluster assignment, is used as the input (Fig. 3.3C, Request) for the tool "Map Genes to Chromosomes" to generate an image map (Fig. 3.3C, Response). The map can be used for close examination or edited further for publication.

Wiki Pages

A wiki is a website allowing users to add, modify, or delete its content via a web browser using a simplified markup language or a rich-text editor. We set up a Wiki page for GaLuDB to support both knowledge management and note-taking. Wikis can be used to post topics on "Discussion Board" pages, create or modify annotation information data on genes of interest, disseminate news to the research community, and share experimental methods and protocols. GaLuDB Wiki allows subscribed researchers to add or edit its content. This is particularly suited for the *G. lucidum* genome-sequencing project, in which the assembly and the annotations are updated regularly. Subscription is obtained after identity verification. The entered information does not need to be error-proof or in its final revision stages to be posted. Setting up this way

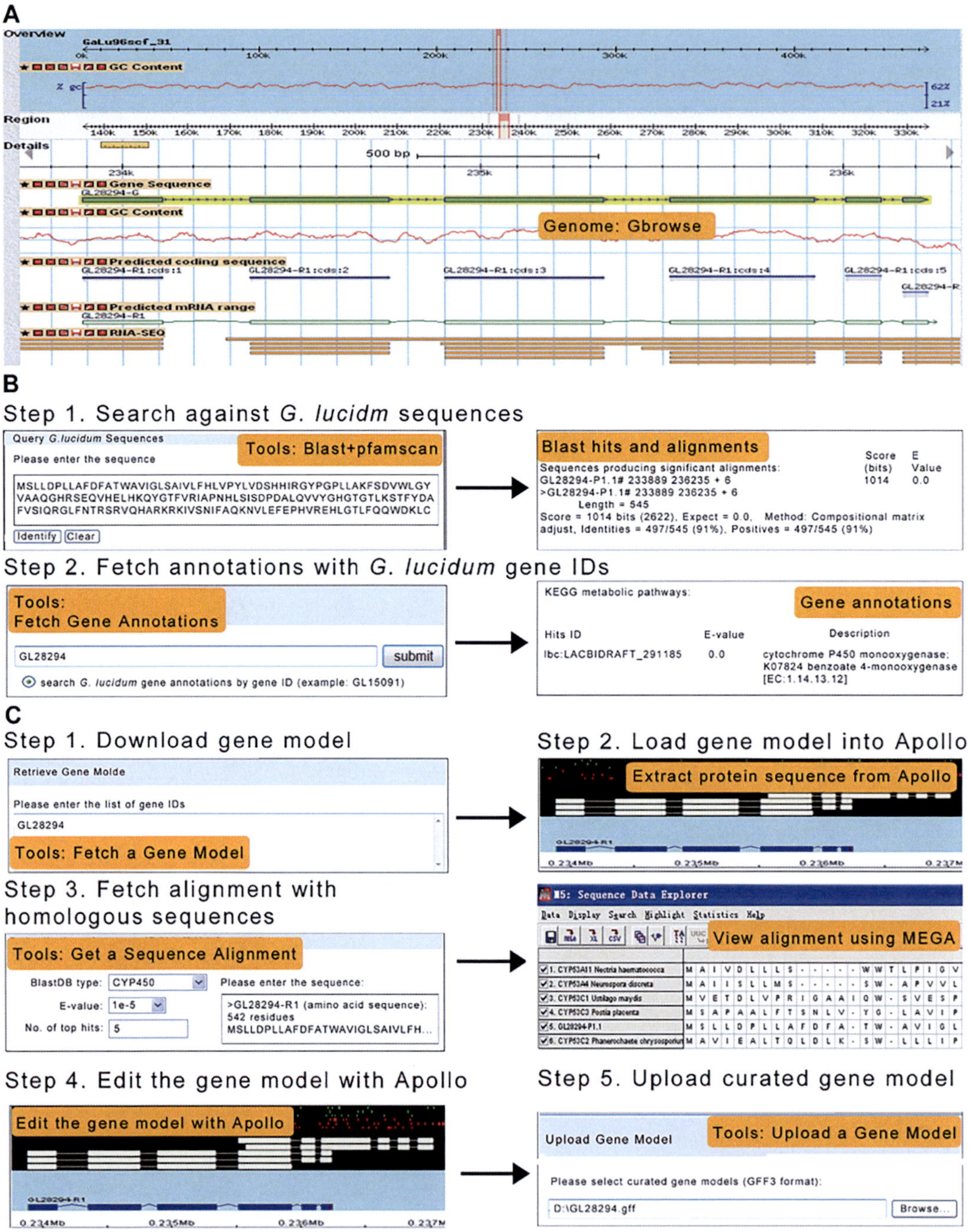

Fig. 3.2 Three main functions of GaLuDB. **A** a genome browser to view, extract, and analyze genome organizations using GBrowse. In this figure, information, including gene structure, predicted CDS, predicted mRNA range, and mapped RNA-Seq data, is displayed. **B** a set of tools supporting several analysis steps of genes: (1) search against *G. lucidum* sequences and (2) fetch annotations with *G. lucidum* gene IDs. **C** a set of tools supporting four steps in the curation of gene models: (1) download gene model, (2) load gene model into Apollo, (3) fetch alignment with homologous sequences, (4) edit gene model with Apollo, and (5) upload curated gene model

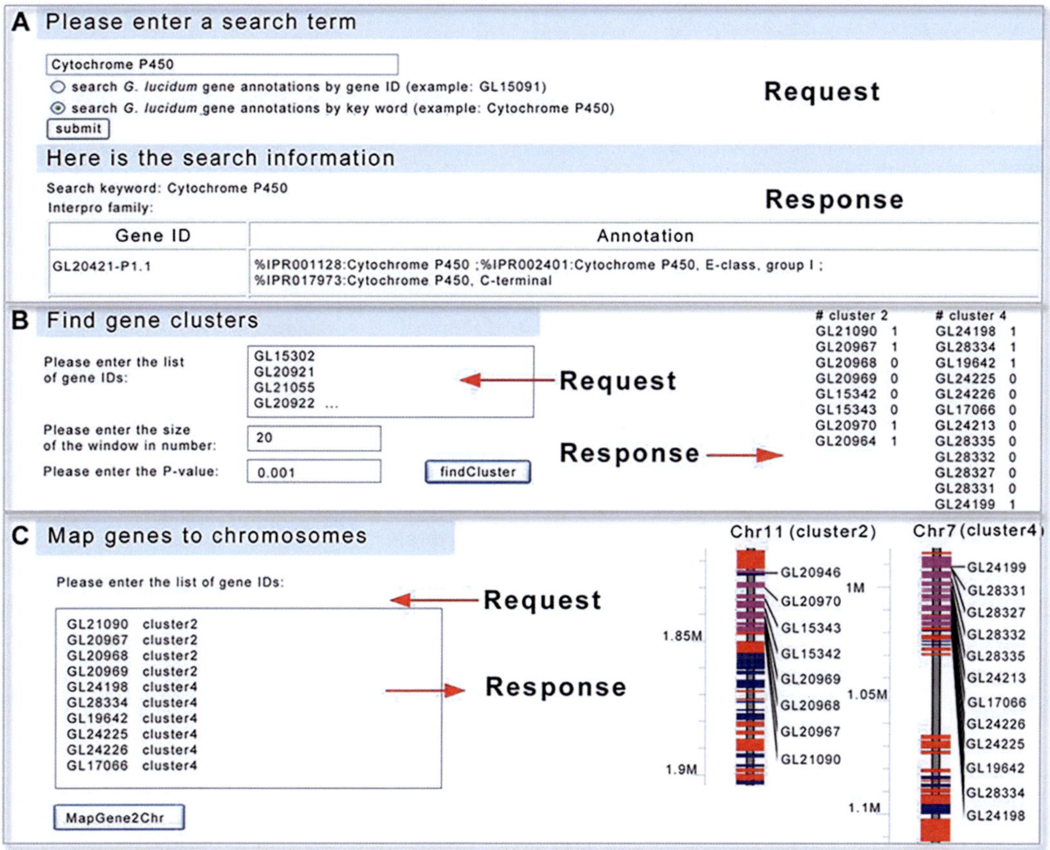

Fig. 3.3 An example scenario to identify CYP450 gene clusters on *G. lucidum* genome. The process includes three steps, namely, **A** retrieving the IDs of all *G. lucidum* CYP450 genes; **B** identifying gene clusters that contain significantly enriched *G. lucidum* CYP450 genes; and **C** mapping and visualizing these gene clusters. The blue and the red rectangles represent the forward strand and reverse strand, respectively. The length of the genes is proportional to the height of the rectangles

allows the author to dynamically update all contents, allowing the community to keep the information as current as possible. GaLuDB Wiki is a valuable resource for the *Ganoderma* community to exchange and distribute information and ideas related to *Ganoderma* species.

3.4 Discussion

With the rapid development of whole-genome sequencing technologies and bioinformatics tools, sequencing the complete fungi genomes with a size of around 50 Mb become relatively straightforward. This chapter summarized the progress made in the WGS projects of *G. lucidum*. In particular, it described the GaluDB, a genetic portal for *Ganoderma* species.

Most WGS projects focused on *G. lucidum*. Two groups have published studies on the complete genomes of *G. lucidum* (Chen et al. 2012; Liu et al. 2012). Another group has also made its data publicly available (http://genome.jgi-psf.org/). Raw data and general genome annotations from both studies have been submitted to GenBank. However, as a general-purpose sequence repository, GenBank cannot present additional data and provide tools to support in-depth analysis of *G. lucidum* on time. For example, like any genome-sequencing project, the sequence assembly of the genome continues to be improved, which leads to the continuous improvement of genome annotations. Various experiments will be conducted to validate the predicted sequences, which will lead to the correction of possible errors. Finally, a forum supporting the discussion of any related topics would further enhance information flow and improve the design of new experiments. Consequently, the *G. lucidum*-specific database still needs to meet the requirements from members of the *G. lucidum* community. GaLuDB is constructed to serve this purpose.

The most valuable output of the WGS project is the identification of genes involved in various biological processes. For instance, 24 gene clusters found in the *G. lucidum* strain genome G.260125-1 may be responsible for the biosynthesis of active compounds. The clustering of these genes allows for better-coordinated expression regulation. More studies can be used to compare genes related to lignin degradation and plant disease by getting more sequences from strains with different phenotypes, which may lead to the identification of genes related to these processes.

However, the current research still has some limitations, such as limited sample size, inaccurate species sampling, unconfirmed species characteristics, and unsatisfactory sequencing quality. First, there is a lack of consensus on the specific characteristics of *the G. lucidum* complex. Many *G. lucium* specimens might originate from a different species (Moncalvo et al. 1995). Recently, Wang et al. analyzed the *G. lucidum* specimens collected in Asia and found that they can be divided into two clades. Additional tests suggest that the tropical collections represent *G. multipileum*. The characteristics of the other clade suggest their belonging to *G. sichuanense* (Wang et al. 2009, 2012). Later, Cao et al. found that the *G. sichuanense* holotype was different from those described above and speculated that the unknown clade is a new species, *G. lingzhi* (Cao et al. 2012). These studies have raised the alert that how many papers on *G. lucidum* are actually from different species. Fortunately, with the accumulation of additional complete nuclear and mitochondrial genomes of *Ganoderma* species, we will be more confident with the taxonomy classification system of *Ganoderma* species.

There are a few plans for the improvement of GaLuDB. *G. lucidum* is a heterogeneous complex species, and there are significant variations among different populations and individuals. Based on our preliminary comparison of the genomes for *G. lucidum*, the strains sequenced by the various research groups described above differs significantly. As a result, extensive evaluations of the intra-specific and inter-specific variations are needed. In the meantime, information regarding multiple strains should be collected and organized. The information should include the phenotypic characteristics, distribution, growth, chemical compositions, and pharmacological effects. By incorporating this wealth of knowledge, GaLuDB will become a valuable resource for selecting and molecular breeding strains with favorable traits.

Another plan is to organize genome information further using a well-developed GMOD database schema, such as Chado or BioSQL (Zhou et al. 2006; Reese et al. 2010). When the community is better developed, and the curators have adopted a standard curation procedure, we will connect the Chado database directly with the client application, like Apollo, to give curators more freedom in editing gene models.

3.5 Conclusions

In summary, the whole genome sequence contains the ultimate genetic information for an organism. Several WGS projects have been carried out for *G. lucidum*. GaluDB was originally developed as a companion to our genome analysis paper. It serves as a portal for researchers to query, retrieve, analyze, and integrate genomic, genetic, and chemical information from *G. lucidum*. It will significantly promote research *G. lucidum* biology and the development of *G.*

lucidum-based medical or industrial products. And it is on its way to becoming the portal for the genetic analysis of all *Ganoderma* species.

References

Adaskaveg JE, Gilbertson RL, Blanchette RA (1990) Comparative studies of delignification caused by *Ganoderma* species. Appl Environ Microbiol 56:1932–1943

Andersen MR, Nielsen JB, Klitgaard A, Petersen LM, Zachariasen M, Hansen TJ, Blicher LH, Gotfredsen CH, Larsen TO, Nielsen KF, Mortensen UH (2013) Accurate prediction of secondary metabolite gene clusters in filamentous fungi. Proc Natl Acad Sci USA 110:E99–E107

Baby S, Johnson AJ, Govindan B (2015) Secondary metabolites from *Ganoderma*. Phytochemistry 114:66–101

Binder M, Justo A, Riley R, Salamov A, Lopez-Giraldez F, Sjokvist E, Copeland A, Foster B, Sun H, Larsson E, Larsson KH, Townsen J, Grigoriev IV, Hibbett DS (2013) Phylogenetic and phylogenomic overview of the Polyporales. Mycologia 105:1350–1373

Boh B (2013) *Ganoderma lucidum*: a potential for biotechnological production of anti-cancer and immunomodulatory drugs. Recent Pat Anticancer Drug Discov 8:255–287

Boh B, Berovic M, Zhang J, Zhi-Bin L (2007) *Ganoderma lucidum* and its pharmaceutically active compounds. Biotechnol Annu Rev 13:265–301

Bok JW, Noordermeer D, Kale SP, Keller NP (2006) Secondary metabolic gene cluster silencing in *Aspergillus nidulans*. Mol Microbiol 61(6):1636–1645

Brakhage AA, Schroeckh V (2011) Fungal secondary metabolites—strategies to activate silent gene clusters. Fungal Genet Biol 48:15–22

Cantarel BL, Coutinho PM, Rancurel C, Bernard T, Lombard V, Henrissat B (2009) The Carbohydrate-Active EnZymes database (CAZy): an expert resource for Glycogenomics. Nucl Acids Res 37(Database issue):D233–238

Cantarel BL, Korf I, Robb SMC, Parra G, Ross E, Moore B, Holt C, Sánchez Alvarado A, Yandell M (2019) MAKER: An easy-to-use annotation pipeline designed for emerging model organism genomes. Genome Res 18(1):188–196

Cao Y, Wu SH, Dai YC (2012) Species clarification of the prize medicinal *Ganoderma* mushroom "Lingzhi." Fungal Divers 56:49–62

Chen S, Xu J, Liu C, Zhu Y, Nelson DR, Zhou S, Li C, Wang L, Guo X, Sun Y, Luo H, Li Y, Song J, Henrissat B, Levasseur A, Qian J, Li J, Luo X, Shi L, He L, Xiang L, Xu X, Niu Y, Li Q, Han MV, Yan H, Zhang J, Chen H, Lv A, Wang Z, Liu M, Schwartz DC, Sun C (2012) Genome sequence of the model medicinal mushroom *Ganoderma lucidum*. Nat Commun 3:913

Chinese Pharmacopoeia Commission (2020) Pharmacopoeia of the People's Republic of China. Chinese Medical Science, Beijing, China

Chiu H-T, Wu CC, Deng S-J, Huang P-N (2008) BIOL 68-Identification, functional expression and kinetic characterization of two GDP-L-fucose biosynthetic enzymes from *Ganoderma lucidum*. Abstr Pap Am Chem Soc 235

Clinton PW, Buchanan PK, Wilkie JP, Smaill SJ, Kimberley MO (2009) Decomposition of Nothofagus wood in vitro and nutrient mobilization by fungi. Can J Forest Res 39:2193–2202

da Coelho-Moreira JS, Bracht A, de Souza AC, Oliveira RF, de Sa-Nakanishi AB, de Souza CG, Peralta RM (2013) Degradation of diuron by *Phanerochaete chrysosporium*: role of ligninolytic enzymes and cytochrome P450. Biomed Res Int 251354

de Andrade FA, Calonego FW, Severo ETD, Furtado EL (2012) Selection of fungi for accelerated decay in stumps of *Eucalyptus* spp. Biores Technol 110:456–461

Ding YX (2006) Cloning and prokaryotic expression of LZ8 gene from *Ganoderma lucidum* and detection of LZ-8 protein by Western blot. Mycosystema 25 (4):587–591

Ding YX, Ou-Yang X, Shang CH, Ren A, Shi L, Li YX, Zhao MW (2008) Molecular cloning, characterization, and differential expression of a farnesyl-diphosphate synthase gene from the basidiomycetous fungus *Ganoderma lucidum*. Biosci Biotechnol Bioche 72 (6):1571–1579

Donlin MJ (2009) Using the generic genome browser (GBrowse). Current protocols in bioinformatics/editorial board, Andreas D Baxevanis Chapter 9:Unit 9 9

Elliot ML, Broschat TK (2000) Ganoderma butt rot in plants. Fact sheet PP-54. The University of Florida, The Institute of Food and Agriculture Sciences (IFAS), Gainsville, FL

Fang X (2011) Cloning of 14 α—demethylase gene from *Ganoderma lucidum* sterol and effect of overexpression on triterpene synthesis. Doctoral thesis, Nanjing Agricultural University

Fang X, Shi L, Ren A, Jiang A-L, Wu F-L, Zhao M-W (2013) The cloning, characterization and functional analysis of a gene encoding an acetyl-CoA acetyltransferase involved in triterpene biosynthesis in *Ganoderma lucidum*. Mycoscience 54(2):100–105

Fei X, Zhao MW, Li YX (2006) Cloning and sequence analysis of a glyceraldehyde-3-phosphate dehydrogenase gene from *Ganoderma lucidum*. J Microbiol 44 (5):515–522

Finn RD, Mistry J, Tate J, Coggill P, Heger A, Pollington JE, Gavin OL, Gunasekaran P, Ceric G,

Forslund K et al (2010) The Pfam protein families database. Nucl Acids Res 38(Database issue):D211–222

Flood J, Bridge PD, Holderness M (eds) (2000) *Ganoderma* diseases of perennial crops. CABI Publishing Wallingford, UK

Floudas D, Binder M, Riley R, Barry K, Blanchette RA, Henrissat B, Martínez AT, Otillar R, Spatafora JW, Yadav JS, Aerts A, Benoit I, Boyd A, Carlson A, Copeland A, Coutinho PM, de Vries RP, Ferreira P, Findley K, Foster B, Gaskell J, Glotzer D, Górecki P, Heitman J, Hesse C, Hori C, Igarashi K, Jurgens LA, Kallen N, Kersten P, Kohler A, Kües U, Kumar TK, Kuo A, LaButti K, Larrondo LF, Lindquist E, Ling A, Lombard V, Lucas S, Lundell T, Martin R, MyLaughlin DJ, Morgenstern I, Morin E, Murat C, Nagy LG, Nolan M, Ohm RA, Patyshakuliyeva A, Rokas A, Ruiz-Dueñas FJ, Sabat G, Salamov A, Samejima M, Schmutz J, Slot JC, John FS, Stenlid J, Sun H, Sun S, Syed K, Tsang A, Wiebenga A, Young D, Pisabarro A, Eastwood D, Martin F, Cullen D, Grigoriev IV, Hibbett D (2012) The paleozoic origin of enzymatic lignin decomposition reconstructed from 31 fungal genomes. Science 336:1715–1719

Glen M, Bougher NL, Francis AA, Nigg SQ, Lee SS, Irianto R, Barry KM, Beadle CL, Mohammed CL (2009) *Ganoderma* and *Amauroderma* species associated with root-rot disease of *Acacia mangium* plantation trees in Indonesia and Malaysia. Australian Plant Path 38:345–356

Grigoriev IV, Nordberg H, Shabalov I, Aerts A, Cantor M, Goodstein D, Kuo A, Minovitsky S, Nikitin R, Ohm RA, Otillar R, Poliakov A, Ratnere I, Riley R, Smirnova T, Rokhsar D, Dubchak I (2009) The genome portal of the Department of Energy Joint Genome Institute. Nucleic Acids Res 40:D26-32

Habijanic J, Berovic M, Boh B, Wraber B, Petravic-Tominac V (2013) Production of biomass and polysaccharides of Lingzhi or Reishi medicinal mushroom, *Ganoderma lucidum* (W.Curt. :Fr.) P. Karst. (higher Basidiomycetes), by submerged cultivation. Int J Med Mushrooms 15:81–90

Harris MA, Clark J, Ireland A, Lomax J, Ashburner M, Foulger R, Eilbeck K, Lewis S, Marshall B, Mungall C et al (2004) The gene ontology (GO) database and informatics resource. Nucl Acids Res 32(Database issue):D258–261

Hennesey C, Daly A (2007) *Ganoderma* diseases. Agnote no 167. NSW Government, Orange, NSW, Australia

Huang ST, Tzean SS, Tsai BY, Hsieh HJ (2009) Cloning and heterologous expression of a novel ligninolytic peroxidase gene from poroid brown-rot fungus *Antrodia cinnamomea*. Microbiology (reading) 155(Pt 2):424–433

Huang YH, Wu HY, Wu KM, Liu TT, Liou RF, Tsai SF, Shiao MS, Ho LT, Tzean SS, Yang UC (2013) Generation and analysis of the expressed sequence tags from the Mycelium of *Ganoderma lucidum*. Plos One 8(5):e61127

Hunter S, Jones P, Mitchell A, Apweiler R, Attwood TK, Bateman A, Bernard T, Binns D, Bork P, Burge S et al (2012) New developments in the family and domain prediction database. Nucl Acids Res 40(Database issue):D306–312

Hushiarian R, Yusof NA, Dutse SW (2013) Detection and control of *Ganoderma boninense*: strategies and perspectives. Springerplus 2:555

Irianto RSB, Barry K, Hidayati N, Ito S, Fiani A, Rimbawanto A, Mohammed C (2006) Incidence and spatial analysis of root rot of *Acacia mangium* in Indonesia. J Trop Forest Sci 18:157–165

Jin X, Ruiz Beguerie J, Sze DM, Chan GC (2012) *Ganoderma lucidum* (Reishi mushroom) for cancer treatment. Cochrane Database Syst Rev 6:CD007731. https://doi.org/10.1002/14651858.CD14007731.pub14651852

Jin W, Li J, Feng H, You S, Zhang L, Norvienyeku J, Hu K, Sun S, Wang Z (2018) Importance of a Laccase Gene (Lcc1) in the development of *Ganoderma tsugae*. Int J Mol Sci 19(2):471

Joo SS, Ryu IW, Park JK, Yoo YM, Lee DH, Hwang KW, Choi HT, Lim CJ, Lee DI, Kim K (2008) Molecular cloning and expression of a laccase from *Ganoderma lucidum*, and its antioxidative properties. Mol Cells 25(1):112–118

Kanehisa M, Goto S, Kawashima S, Okuno Y, Hattori M (2004) The KEGG resource for deciphering the genome. Nucl Acids Res 32(Database issue):D277–280

Ko EM, Leem YE, Choi HT (2001) Purification and characterization of laccase isozymes from the white-rot basidiomycete *Ganoderma lucidum*. Appl Microbiol Biotechnol 57(1–2):98–102

Koonin EV, Fedorova ND, Jackson JD, Jacobs AR, Krylov DM, Makarova KS, Mazumder R, Mekhedov SL, Nikolskaya AN, Rao BS et al (2004) A comprehensive evolutionary classification of proteins encoded in complete eukaryotic genomes. Genome Biol 5(2):R7

Kües U, Nelson DR, Liu C, Yu GJ, Zhang JH, Li J, Wang XC, Sun H (2015) Genome analysis of medicinal *Ganoderma* spp. with plantpathogenic and saprotrophic life-styles. Phytochemistry 114:18–37

Lanfermann I, Linke D, Nimtz M, Berger RG (2015) Manganese peroxidases from *Ganoderma applanatum* degrade β-carotene under alkaline conditions. Appl Biochem Biotechnol 175(8):3800–3812

Lee KH, Morris-Natschke SL, Yang X, Huang R, Zhou T, Wu SF, Shi Q, Itokawa H (2012) Recent progress of research on medicinal mushrooms, foods, and other herbal products used in traditional Chinese medicine. J Tradit Complement Med 2:84–95

Levasseur A, Piumi F, Coutinho PM, Rancurel C, Asther M, Delattre M, Henrissat B, Pontarotti P, Record E (2008) FOLy: an integrated database for the classification and functional annotation of fungal oxidoreductases potentially involved in the degradation of lignin and related aromatic compounds. Fungal Gene Biol FG & B 45(5):638–645

Lewis SE, Searle SM, Harris N, Gibson M, Lyer V, Richter J, Wiel C, Bayraktaroglir L, Birney E, Crosby MA et al (2002) Apollo: a sequence annotation editor. Genome Biol 3(12):RESEARCH0082

Li J, Wu B, Xu J, Liu C (2014) Genome-wide identification and characterization of long intergenic noncoding RNAs in *Ganoderma lucidum*. PLoS ONE 9 (6):e99442

Lin JW, Duan ZW, Guan SY et al (2016) Cloning, bioinformatics analysis and construction of eukaryotic expression vector of immunoregulatory protein gene from *Ganoderma lucidum*. J Shenyang Agricult Univ 180(1):1–7

Liu C, Ghosh S, Searls DB, Saunders AM, Cossman J, Roses AD (2005) Clusters of adjacent and similarly expressed genes across normal human tissues complicate comparative transcriptomic discovery. OMICS 9 (4):351–363

Liu Y, Guo LQ, Wang HY et al (2006) Cloning and analysis of immunoregulatory protein gene from *Ganoderma lucidum*. J Trop Crops 27(1):54–58

Liu D, Gong J, Dai W, Kang X, Huang Z, Zhang H-M, Liu W, Liu L, Ma J, Xia Z (2012) The genome of *Ganoderma lucidum* provide insights into triterpene biosynthesis and wood degradation. PLoS ONE 7: e36146

Liu YN, Lu XX, Chen D, Lu YP, Ren A, Shi L, Zhu J, Jiang AL, Yu HS, Zhao MW (2017) Phospholipase D and phosphatidic acid mediate heat stress-induced secondary metabolism in *Ganoderma lucidum*. Environ Microbiol 19(11):4657–4669

Martínez AT, Rencoret J, Nieto L, Jiménet-Barbero J, Gutiérrez A, del Río JC (2011) Selective lignin and polysaccharide removal in natural fungal decay of wood as evidenced by in situ structural analyses. Env Microbiol 13:96–107

Medema MH, Blin K, Cimermancic P, de Jager V, Zakrzewski P, Fischbach MA, Weber T, Takano E, Breitling R: antiSMASH, (2011) rapid identification, annotation and analysis of secondary metabolite biosynthesis gene clusters in bacterial and fungal genome sequences. Nucl Acids Res 39(suppl 2): W339–W346

Mendoza G, Guzman G, Ramirez-Guillen F, Luna M, Trigos A (2011) *Ganoderma oerstedii* (Fr.) Murrill (higher Basidiomycetes), a tree parasite species in Mexico: taxonomic description, rDNA study, and review of its medical applications. Int J Med Mushrooms 13:545–552

Mercière M, Laybats A, Carasco-Lacombe C et al (2015) Identification and development of new polymorphic microsatellite markers using genome assembly for *Ganoderma boninense*, causal agent of oil palm basal stem rot disease. Mycol Progress 14:103

Miller RNG, Holderness M, Bridge PD (2000) Molecular and morphological characterization of *Ganoderma* in oil-palm plantings. In: Flood J, Bridge PD, Holderness M (eds) *Ganoderma* diseases of perennial crops. CABI Publishing, Wallingford, UK, pp 159–182

Miller JR, Delcher AL, Koren S, Venter E, Walenz BP, Brownley A, Johnson J, Li K, Mobarry C, Sutton G (2008) Aggressive assembly of pyrosequencing reads with mates. Bioinformatics 24(24):2818–2824

Moncalvo JM, Buchanan PK (2008) Molecular evidence for long distance dispersal across the Southern Hemisphere in the *Ganoderma applanatum*-australe species complex (Basidiomycota). Mycol Res 112:425–436

Moncalvo JM, Wang HF, Hseu RS (1995) Gene phylogeny of the *Ganoderma lucidum* complex based on ribosomal DNA sequences. Comparison with traditional taxonomic characters. Myco Res 99:1489–1499

Mu D, Shi L, Ren A, Li M, Wu F, Jiang A, Zhao M (2012) The development and application of a multiple gene co-silencing system using endogenous URA3 as a reporter gene in *Ganoderma lucidum*. PLoS ONE 7 (8):e43737

Naher L, Yusuf UK, Ismail A, Tan SG, Monal MMA (2013) Ecological status of *Ganoderma* and basal stem rot disease of oil palms (Elaeis guineensis Jacq.). AJCS 7:1723–1727

Nasir N (2005) Diseases caused by *Ganoderma* spp. on perennial crops in Pakistan. Mycopathologia 159:119–121

Nelson DR (2011) Progress in tracing the evolutionary paths of cytochrome P450. Biochem Biophys Acta 1814(1):14–18

Nusaibah SA, Latiffah Z, Hassaan AR (2011) ITS-PCR-RFLP analysis of *Ganoderma* sp. infecting industrial crop. Pertanika J Trop Agric Sci 34:83–91

O'Connor BD, Day A, Cain S, Arnaiz O, Sperling L, Stein LD (2008) GMODWeb: a web framework for the generic model organism database. Genome Biol 9 (6):R102

Osbourn A (2010) Secondary metabolic gene clusters: evolutionary toolkits for chemical innovation. Trend Gene TIG 26(10):449–457

Pan SM, Ye JS, Hseu RS (1997) Purification and characterization of manganese superoxide dismutase from *Ganoderma microsporum*. Biochem Mol Biol Int 42(5):1035–1043

Paterson RRM (2006) *Ganoderma*—a therapeutic fungal biofactory. Phytochemistry 67:1985–2001

Paterson RRM (2007) *Ganoderma* disease of oil palm—a white rot perspective necessary for integrated control. Crop Protect 26:1369–1376

Pilotti CA (2005) Stem rots of oil palm caused by *Ganoderma boninense*: pathogen biology and epidemiology. Mycopathologia 159:129–137

Pilotti CA, Sanderson FR, Aitken EAB (2003) Genetic structure of a population of *Ganoderma boninense* on oil palm. Plant Pathol 52:455–463

Pirovano W, Boetzer M, Henkel CV, Jansen HJ, Butler D (2011) Scaffolding pre-assembled contigs using SSPACE. Bioinformatics 27(4):578–579

Popovic V, Zivkovic J, Davidovic S, Stevanovic M, Stojkovic D (2013) Mycotherapy of cancer: an update on cytotoxic and antitumor activities of mushrooms,

bioactive principles and molecular mechanisms of their action. Curr Top Med Chem 13:2791–2806
Rai M, Tidke G, Wasser SP (2005) Therapeutic potential of mushrooms. Nat Prod Rad 4:246–257
Rakib MRM, Bong C-FJ, Khairulmazmi A, Idris AS (2014) Genetic and morphological diversity of *Ganoderma* species isolated from infected oil palms (Elaeis guineensis). Int J Agricult Biol 16:691–699
Rees RW, Flood J, Hasan Y, Wills MA, Cooper RM (2012) *Ganoderma boninense* basidiospores in oil palm plantations: evaluation of their possible role in stem roots of Elaeis guineensis. Plant Pathol 61:567–578
Reese JT, Childers CP, Sundaram JP, Dickens CM, Childs KL, Vile DC, Elsik CG (2010) Bovine Genome Database: supporting community annotation and analysis of the *Bos taurus* genome. BMC Genomics 11:645
Ren A, Ouyang X, Shi L, Jiang AL, Mu DS, Li MJ, Han Q, Zhao MW (2013) Molecular characterization and expression analysis of GlHMGS, a gene encoding hydroxymethylglutaryl-CoA synthase from *Ganoderma lucidum* (Ling-Zhi) in ganoderic acid biosynthesis pathway. World J Microbiol Biotechnol 29 (3):523–531
Sankaran KV, Bridge PD, Gokulapalan C (2005) *Ganoderma* diseases of perennial crops in India—an overview. Mycopathologia 159:143–152
Sanodiya BS, Thakur GS, Baghel RK, Prasad GB, Bisen PS (2009) *Ganoderma lucidum*: a potent pharmacological macrofungus. Curr Pharm Biotechnol 10(8):717–742
Shang CH, Zhu F, Li N, Yang XO, Shi L, Zhao MW, Li YX (2008) Cloning and characterization of a gene encoding HMG-CoA reductase from *Ganoderma lucidum* and its functional identification in yeast. Biosci Biotechnol Biochem 72(5):1333–1339
Shang CH, Shi L, Ren A, Qin L, Zhao MW (2010) Molecular cloning, characterization, and differential expression of a lanosterol synthase gene from *Ganoderma lucidum*. Biosci Biotechnol Biochem 74 (5):974–978
Shi L, Qin L, Xu Y, Ren A, Fang X, Mu D, Tan Q, Zhao M (2012) Molecular cloning, characterization, and function analysis of a mevalonate pyrophosphate decarboxylase gene from *Ganoderma lucidum*. Mol Biol Re 39(5):6149–6159
Shiao MS (2003) Natural products of the medicinal fungus *Ganoderma lucidum*: occurrence, biological activities, and pharmacological functions. Chem Rec 3 (3):172–180
Smith BJ, Sivavithamparam K (2000) Internal transcribed spacer ribosomal DNA sequence of five species of *Ganoderma* from Australia. Mycol Res 104:943–951
Soares AA, de Sa-Nakanishi AB, Bracht A, da Costa SM, Koehnlein EA, de Souza CG, Peralta RM (2013) Hepatoprotective effects of mushrooms. Molecules 18:7609–7630
Son E, Au-Yeung TT, Yang CYH, Breui C (2010) Diversity and decay ability of basidiomycetes isolated from lodgepole pines killed by the mountain pine beetle. Can J Microbiol 57:33–41
Stein LD, Mungall C, Shu S, Caudy M, Mangone M, Day A, Nickerson E, Stajich JE, Harris TW, Arva A et al (2002) The generic genome browser: a building block for a model organism system database. Genome Res 12(10):1599–1610
Sun QQ, He QG, Qi PF et al (2019) Cloning, subcellular localization and expression analysis of gpftr1 gene from *Rhizoctonia gelatinosa*. Genomics Appl Biol 38 (8):244–251
Syed K, Nelson DR, Riley R, Yadav JS (2013) Genome-wide annotation and comparative genomics of cytochrome P450 monooxygenases (P450s) in the polypore species Bjerkandera adusta. *Ganoderma* Sp. Phlebia Brevispora. Mycologia 105(6):1445–1455
Tatusov RL, Fedorova ND, Jackson JD, Jacobs AR, Kiryutin B, Koonin EV, Krylov DM, Mazumder R, Mekhedov SL, Nikolskaya AN, a, (2003) The COG database: an updated version includes eukaryotes. BMC Bioinform 4:41
Teh C-Y, Pang C-L, Tor X-Y, Ho P-Y, Lim Y-Y, Namasivayam P, Ho C-L (2019) Molecular cloning and functional analysis of a necrosis and ethylene inducing protein (NEP) from *Ganoderma boninense*. Physiol Mol Plant Pathol 106:42–48
US Forest Service (2011) White mottled rot.. Root rot that topples live aspen. US Department of Agriculture, Washington
Utomo C, Tanjung ZA, Aditama R et al (2018) Draft genome sequence of the phytopathogenic fungus *Ganoderma boninense*, the causal agent of basal stem rot disease on oil palm. Genome Announc 6(17): e00122-e218
Voo CLY, Yeo DET, Chong KP, Rodrigues KF (2020) Draft genome sequence of a phytopathogenic *Ganoderma* sp. strain that causes basal stem rot disease on oil palm in Sabah, Malaysia. Microbiol Resour Announc 9(1):e01240–19
Wang DM, Wu SH, Su CH, Peng JT, Shi YH, Chen LC (2009) Ganoderma multipileum, the correct name for '*G. lucidum*' in tropical Asia. Bot Stud 50:451–458
Wang XC, Xi RJ, Li Y, Wang DM, Yao YJ (2012) The species identity of the widely cultivated *Ganoderma*, '*G. lucidum*' (Ling-Zhi), in China. PLoS ONE 7: e40857
Wu GS, Guo JJ, Bao JL, Li XW, Chen XP, Lu JJ, Wang YT (2013a) Anti-cancer properties of triterpenoids isolated from *Ganoderma lucidum*—a review. Expert Opin Investig Drugs 22:981–992
Wu FL, Shi L, Yao J, Ren A, Zhou C, Mu DS, Zhao MW (2013b) The cloning, characterization, and functional analysis of a gene encoding an isopentenyl diphosphate isomerase involved in triterpene biosynthesis in the Lingzhi or Reishi medicinal mushroom *Ganoderma lucidum* (higher Basidiomycetes). Int J Med Mushrooms 15(3):223–232
Wu CG, Tian JL, Liu R, Cao PF, Zhang TJ, Ren A, Shi L, Zhao MW (2017) Ornithine decarboxylase-mediated production of putrescine influences ganoderic acid

biosynthesis by regulating reactive oxygen species in *Ganoderma lucidum*. Appl Environ Microbiol 83(20): e01289-e1317

Xu Z, Chen X, Zhong Z, Chen L, Wang Y (2011) *Ganoderma lucidum* polysaccharides: immunomodulation and potential antitumor activities. Am J Chin Med 39:15–27

Xu T, Beelman RB, Lambert JD (2012a) The cancer-preventive effects of edible mushrooms. Anticancer Agents Med Chem 12:1255–2126

Xu JW, Xu YN, Zhong JJ (2012b) Enhancement of ganoderic acid accumulation by overexpression of an N-terminally truncated 3-hydroxy-3-methylglutaryl coenzyme A reductase gene in the basidiomycete *Ganoderma lucidum*. Appl Environ Microbiol 78 (22):7968–7976

Xu J, Xu Z, Zhu Y, Luo H, Qian J, Ji A, Hu Y, Sun W, Wang B, Song J, Sun C, Chen S (2014) Identification and evaluation of reference genes for qRT-PCR normalization in *Ganoderma lucidum*. Curr Microbiol 68(1):120–126

Xu H, Guo MY, Gao YH, Bai XH, Zhou XW (2017) Expression and characteristics of manganese peroxidase from *Ganoderma lucidum* in *Pichia pastoris* and its application in the degradation of four dyes and phenol. BMC Biotechnol 17(1):19

Yang C, Li W, Li C, Zhou Z, Xiao Y, Yan X (2018) Metabolism of ganoderic acids by a Ganoderma lucidum cytochrome P450 and the 3-keto sterol reductase ERG27 from yeast. Phytochemistry 155:83–92

You LF, Liu ZM, Lin JF, Guo LQ, Huang XL, Yang HX (2014) Molecular cloning of a laccase gene from *Ganoderma lucidum* and heterologous expression in Pichia pastoris. J Basic Microbiol 54(Suppl 1):S134–S141

Zhang DH, Jiang LX, Li N, Yu X, Zhao P, Li T, Xu JW (2017) Overexpression of the squalene epoxidase gene alone and in combination with the 3-hydroxy-3-methylglutaryl coenzyme a gene increases Ganoderic acid production in *Ganoderma lingzhi*. J Agric Food Chem 65(23):4683–4690

Zhang G, Ren A, Shi L, Zhu J, Jiang A, Shi D, Zhao M (2018) Functional analysis of an APSES transcription factor (GlSwi6) involved in fungal growth, fruiting body development and ganoderic-acid biosynthesis in *Ganoderma lucidum*. Microbiol Res 207:280–288

Zhao MW, Liang WQ, Zhang DB, Wang N, Wang CG, Pan YJ (2007) Cloning and characterization of squalene synthase (SQS) gene from *Ganoderma lucidum*. J Microbiol Biotechnol 17(7):1106–1112

Zhou P, Emmert D, Zhang P (2006) Using Chado to store genome annotation data. Curr Protocol Bioinform 9 (9):6

Zhou XW, Cong WR, Su KQ, Zhang YM (2013) Ligninolytic enzymes from *Ganoderma* spp: current status and potential applications. Crit Rev Microbiol 39:416–426

Zhou J-S, Ji S-L, Ren M-F, He Y-L, Jing X-R, Xu J-W (2014) Enhanced accumulation of individual ganoderic acids in a submerged culture of *Ganoderma lucidum* by the overexpression of squalene synthase gene. Biochem Eng J 90:178–183

Zhu Y, Xu J, Sun C, Zhou S, Xu H, Nelson DR, Qian J, Song J, Luo H, Xiang L, Li Y, Xu Z, Ji A, Wang L, Lu S, Hayward A, Sun W, Li X, Schwartz DC, Wang Y, Chen S (2015) Chromosome-level genome map provides insights into diverse defense mechanisms in the medicinal fungus *Ganoderma sinense*. Sci Rep 5:11087

Zhuo R, Ma L, Fan F, Gong Y, Wan X, Jiang M, Zhang X, Yang Y (2011) Decolorization of different dyes by a newly isolated white-rot fungi strain *Ganoderma* sp. En3 and cloning and functional analysis of its laccase gene. J Hazard Mater 192 (2):855–73

4 Lingzhi Mitochondrial Genome

Xin-Cun Wang

Abstract

The mitochondrial genome is the second genome for eukaryotes. It has essential functions in many cellular processes, including producing energy, signaling, aging, and programmed cell death. Up to now, ten mitochondrial genomes from seven *Ganoderma* species had been released on the public database. The mitochondrial genomes of each species were described and compared here, and some intra-specific differences were also issued. The mitochondrial phylogenomic tree of Polyporales, which *Ganoderma* is belonged to in taxonomy, was constructed, and the potential DNA barcodes from the mitochondrial genome for this genus were discussed.

4.1 General Introduction

4.1.1 Concept

Mitochondria: the mitochondrion (plural mitochondria) is a highly specialized organelle with double membranes, existing in the cytoplasm of almost all eukaryotic cells (cells with clearly defined nuclei). As the cell's powerhouse, it produces large quantities of energy stored in adenosine triphosphate (ATP) through oxidative phosphorylation (Chinnery and Hudson 2013; van der Giezen and Tovar 2005). The number of mitochondria in a cell can vary significantly by organism and tissue, and each mitochondrion contains multiple mitochondrial DNA molecules (Basse 2010).

Mitochondrial genome (Mitogenome): mitochondria have their DNA, known as mitochondrial DNA or mtDNA. mtDNA was first discovered in animal cells (Nass and Nass 1963). mtDNA consists of a heavy (H) and a light (L) strand. The heavy strand/light strand terminology dates back to the 1970s and refers to the relative CsCl banding of the two single strands of the mitochondrial DNA. These strands have a G/C asymmetry leading to the sense strand being "light" and the reverse complement being "heavy" in their relative molecular weights. The sense strand (L) encodes 13 of the more than 90 subunits of the electron-transfer chain and 22 tRNAs and 2 rRNAs in humans (*Homo sapiens*). Humans' mitochondrial genome is a closed circular duplex DNA of estimated 16,500 nucleotide pairs (Brown et al. 1979). Then, the first complete sequence of a mitochondrial genome was reported from humans, and its exact size of 16.569 base pairs (bp) was presented (Anderson et al. 1981). In fungi, the first complete mitochondrial genome was sequenced from the yeast

X.-C. Wang (✉)
State Key Laboratory of Mycology, Institute of Microbiology, Chinese Academy of Sciences, Beijing 100101, People's Republic of China
e-mail: wangxc@im.ac.cn

C. Liu (ed.), *The Lingzhi Mushroom Genome*, Compendium of Plant Genomes,
https://doi.org/10.1007/978-3-030-75710-6_4

Saccharomyces cerevisiae in 1998, with the size of 85,779 bp (Foury et al. 1998).

4.1.2 Principle

4.1.2.1 Mitochondrial Origin

The current theory as to the origin of eukaryotic cells is endosymbiosis. Based on the endosymbiotic theory, mitochondria are the descendants of once free-living α-proteobacteria. The proteobacteria became endosymbiotic organelles of eukaryotes two billion years ago (Andersson et al. 1998; Gray et al. 1999). It is believed that mitochondria and chloroplasts began as prokaryotic organisms engulfed by the larger cells, either as food or parasites. That is why they are enclosed in double membranes. At some point, the relationship became mutually beneficial, and the mitochondria and chloroplasts evolved into permanent members in the cells as cellular machinery.

4.1.2.2 Mitochondrial Structure

Mitochondria are typically oval, spherical, or rod-shaped. Its size ranges from 0.5 to 1.0 μm in diameter. Double membranes enclose them. The membranes are lipid bilayers with proteins embedded within the layers. The region between the two membranes is the intermembrane space. The inner membrane is folded to form cristae. This folding increases the surface area of the membrane and maximizes cellular respiration output. Inside the inner membrane is the mitochondrial matrix, and within the matrix, there are ribosomes, enzymes, and mitochondrial DNA. The mitochondrion can reproduce and synthesize proteins independently. It contains the enzymes necessary for transcription and the transfer of RNAs and ribosomes required for translation and protein formation. Recent research revealed that, instead of small, oblong, and independent entities in the eukaryotic cytoplasm, mitochondria form dynamic networks of long tubular structures undergoing fusions and separations during the cell cycle or after changes of growth conditions (Fig. 4.1; Dujon 2020; Lackner 2014).

4.1.2.3 Mitochondrial Genetics

Mitochondrial DNA (mtDNA) is typically a small circular double-stranded DNA molecule that encodes many proteins and RNA involved primarily in cellular respiration. In some fungi and protists, mtDNA can be linear (van de Vossenberg et al. 2018). Mitochondrial DNA is well conserved within taxa.

Unlike nuclear DNA passed on from both parents, mitochondrial DNA is generally uniparentally inherited (with some notable exceptions). In animals, mtDNA is passed on maternally through the egg. In plants, mtDNA may be passed on maternally, paternally, or biparentally. There is also evidence for paternal leakage of mtDNA, where the offspring inherits most of their mtDNA from their mother and receives a small amount from their father. Uniparental inheritance of mitochondria is a common phenomenon in sexual eukaryotes. It leads to little opportunity for genetic recombination between different lineages of mitochondria.

4.1.2.4 Mitochondrial Evolution

Researches on mitochondrion evolution began with animal mitochondrial DNA. There is a widely accepted generalization concerning molecular evolution rates: the more important the function of a gene or protein, the more slowly it undergoes an evolutionary change in the primary structure. As mitochondria have essential cellular functions, the life of animals is crucially dependent on these functions. Mitochondrial evolution would be expected to be highly constrained. Besides, animal mitochondria genome is small in size and relatively uniform in structure among vertebrate and invertebrate animals. The implication is strong that this genome was reduced at an early stage of animal evolution to the minimum size compatible with function.

However, evidence from restriction endonuclease cleavage maps implies that mtDNA evolves 5–10 times faster than single-copy nuclear DNA (Brown et al. 1979). The mean substitution rate in the three studied mitochondrial tRNA genes is at least 100 times higher than that in nuclear tRNA genes (Brown et al. 1982). The base substitution

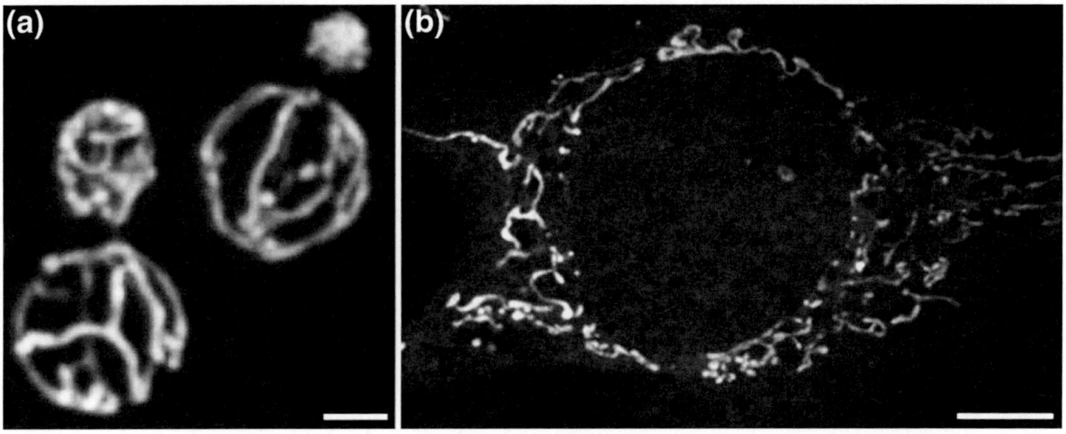

Fig. 4.1 Networks of mitochondria. **a** Mitochondria were forming connected, tubular networks. **b** Mitochondria were creating well-distributed tubular networks. Reprinted from Fig. 1 of Lackner (2014) with permissions

rate of mammalian mtDNA has been estimated to be 0.5–1.0% per lineage per 10^6 years (Brown 1980). For this reason, mitochondrial DNA is commonly used to study evolutionary relationships and population genetics in animals. Plant mtDNA evolves reasonably slowly and is less commonly used in evolutionary studies.

4.1.2.5 Mitochondrial Function

Mitochondria host the tricarboxylic acid (TCA) cycle and oxidative phosphorylation, breaking down sugars and fats into energy through aerobic respiration (cellular respiration). This metabolic process creates ATP, the energy source of a cell, through a series of steps that require oxygen. Beyond their well-known function as energy-producing entities, mitochondria are involved in a multitude of cellular processes, including the production of iron-sulfur clusters, calcium homeostasis, calcium signaling, haem synthesis, steroid synthesis, cellular differentiation, aging, programmed cell death (apoptosis), as well as maintaining the control of the cell cycle and cell growth (Malina et al. 2018).

The role of mitochondria in disease has been expanded beyond the respiratory chain, as defects in additional mitochondrial functions and behaviors have been linked to cancer, metabolic disorders, and neurodegenerative diseases, such as Alzheimer's, Parkinson's, and Huntington's disease. Mutations in mitochondrial DNA can result in many human genetic disorders. Examples include diabetes, heart disease, myoclonic epilepsy, and Kearns-Sayre neuromuscular syndrome (Friedman and Nunnari 2014).

4.1.3 Research Methods For Mitochondrial Genome Analysis

4.1.3.1 Molecular Experiments

DNA Extraction

Genomic DNA could be extracted from fruiting bodies using Tiangen DNA Extraction Kit (DP305) or a fungal DNA kit (Cat. #D3390–00, Omega Bio-Tek, Norcross, GA, USA) according to the manufacturer's instructions.

Next-Generation Sequencing

Whole genomic sequencing was conducted on an Illumina HiSeq 2500 Platform (Illumina, San Diego, CA, USA). Two 300 bp genomic DNA libraries were constructed for two *Ganoderma* species. For *G. meredithiae*, the produced paired-end reads were 100 bp long, while the paired-end reads of *G. applanatum* were 125 bp long.

4.1.3.2 Data Analyses

Mitochondrial Genome Assembly and Annotation

De novo assembly of the mitogenomes was performed using ABySS-pe version 1.5.2 (Simpson et al. 2009) for *G. meredithiae*, CLC Genomics Workbench Version 7.5.1 (CLC Bio, Aarhus, Denmark) for *G. applanatum*, and SPAdes 3.9.0 (Bankevich et al. 2012) for the five ones published in 2019. The assembled scaffolds were compared to the reference sequence using BLASTN version 2.2.23 (Altschul et al. 1990). The screened scaffolds were compared with the reference using Gepard version 1.30 (Krumsiek et al. 2007) and then assembled into a circular mitogenome sequence by manual editing. The coverage was examined using Tablet version 1.14.10.20 (Milne et al. 2013).

The mitogenome was annotated using MFannot (Lang et al. 2014) or MITOS (Bernt et al. 2013). The Sequence Manipulation Suite Version 2 (SMS2, http://www.bioinformatics.org/sms2/) was used to identify the hypothetical proteins. tRNA genes were initially identified using tRNAscan-SE (Lowe and Chan 2016). DNASTAR Lasergene (http://www.dnastar.com/) was used to analyze the base composition of the mitogenomes. The synonymous substitution rate (Ks) and the nonsynonymous substitution rate (Ka) for all PCGs in each of the mitogenomes were calculated using DnaSP (Rozas et al. 2017). Mitochondrial genome maps were generated by OGDRAW (Lohse et al. 2007).

Phylogenomic Analysis

The concatenated amino acid sequences were extracted from the GenBank files of the mitogenomes of Polyporales. They were then aligned using MAFFT v7.221 (Katoh et al. 2002) or MUSCLE version 3.6 (Edgar 2004). The neighbor-Joining analysis was conducted using MEGA version 6.06 (Tamura et al. 2013). Bootstrap values were calculated from 1,000 replicates.

4.2 Progress of Lingzhi Mitochondrial Genome

Up to date (October 2020), ten mitochondrial genomes of *Ganoderma* species were released from the NCBI database, seven of them recognized by the RefSeq database. Their information is summarized in Table 4.1. All of them were submitted by Chinese researchers.

4.2.1 Structure of Lingzhi Mitochondrial Genome

As listed in Table 4.1, three mitochondrial genomes of '*Ganoderma lucidum* (Curtis) P. Karst.' was released online. The identity of this species is *G. sichuanense* J.D. Zhao and XQ. Zhang, as detailed elucidated in the previous taxonomic researches (Wang et al. 2012; Thawthong et al. 2017; Yao et al. 2020, 2013).

Strain CGMCC 5.26 was isolated from the fruiting body collected in China and then deposited to China General Microbiological Culture Collection Center (CGMCC, Beijing, China) in 1963 for medicinal usage. Its mitogenome (HF570115) was first reported in this genus *Ganoderma* (Li et al. 2013). This mitogenome is a typical circular DNA molecule of 60,630 bp with 26.67% GC content. About 62.69% of the mitochondrial genome contains 50 genes encoding two ribosomal RNAs, 26 transfer RNAs, one ribosomal protein gene *rps3*, 14 genes involved in respiratory chain complexes, four ORFs in the intron of other genes (ip1–4), and two ORFs in the intergenic regions (*orf1* and *orf2*). These genes were in the same orientation, except those genes encoding for *trnW-CCA* and three ORFs. In total, thirteen introns were found in the protein-coding genes *cox1*, *cox2*, *cox3*, *nad4*, *nad5*, and *rnl* and *rns* genes. All introns were the group I introns, except for the single type II intron *cox1 i6*.

The other released mitogenome (KC763799 = NC_021750) of the same strain was submitted by Qian, but it was not published. It has only five base pairs more than the one

Table 4.1. Information on the released *Ganoderma* mitochondrial genomes

Taxon	Strain	Accession	Length	GC content (%)	References	Submitted date
G. applanatum	CGMCC 5.249	KR109212 NC_027188	119,803 bp	26.66	Wang et al. (2016a)	2015.04.13
G. calidophilum	s136	MH252535 NC_037938	124,588 bp	25.43	Li et al. (2019)	2018.04.20
G. leucocontextum	s116	MH252534 NC_037937	88,194 bp	27.08	Li et al. (2019)	2018.04.20
G. lucidum	CGMCC 5.26	HF570115	60,630 bp	26.67	Li et al. (2013)	2012.12.18
G. lucidum	CGMCC 5.26	KC763799 NC_021750	60,635 bp	26.67	Qian unpublished	2013.05.10
G. lucidum	s26	MH252532	57,932 bp	26.75	Li et al. (2019)	2018.04.20
G. meredithiae	CGMCC 5.766	KP410262 NC_026782	78,447 bp	26.14	Wang et al. (2016b)	2015.01.09
G. sinense	CGMCC 5.69	KF673550 NC_022933	86,451 bp	26.78	Qian unpublished	2013.09.14
G. sp.	s8	MH252531	57,232 bp	26.46	Li et al. (2019)	2018.04.20
G. tsugae	s90	MH252533 NC_037936	92,511 bp	26.67	Li et al. (2019)	2018.04.20

(HF570115) reported by Li et al. (Li et al. 2013), while the GC content is almost the same. The gene order is the same between them, as listed in Table 4.2. But the introns reported in *rnl* and *rns* of HF570115 were not identified, so only 11 introns exist in the mitogenome KC763799. An ideogram of the mitogenome organization and gene classification is shown in Fig. 4.2. The two mitogenomes were from the same strain and assembled from the same raw data. Thus, the differences do not reflect the genetic variation of the same species. These differences need to be further validated using Sanger sequencing.

The more recently released mitogenome (MH252532) of *G. lucidum* s26 was sequenced from the strain isolated from a fruiting body collected from Tai'an City, Shandong Province, China. A comparison between this one and KC763799 was described (Li et al. 2019). The nucleotide sequence similarity between the two mitogenomes was 95.18%. Furthermore, most protein-coding genes and tRNA genes were conserved between them. *Ganoderma* CGMCC 5.26 has longer intergenic regions than *G. lucidum* s26 does. In addition, single nucleotide polymorphisms were frequently observed in the intergenic region of the two mitogenomes. Besides, *G. lucidum* s26 had one more intron (12 introns in total, one more in *rns*) than *G. lucidum* CGMCC 5.26, indicating that the mitochondrial genomes of *G. lucidum* from different origin have been differentiated.

4.2.2 Evolution of Mitochondrial Genome in Genus *Ganoderma*

Besides the three mitogenomes of *G. lucidum*, there are still seven ones released in this genus (Table 4.1). A mount of inter-specific variance has been observed (Table 4.2).

4.2.2.1 Mitochondrial Genome of *Ganoderma meredithiae*

The strain CGMCC 5.766 equals CBS 271.88. It is the holotype culture isolated from the holotype

Table 4.2 Gene contents in *Ganoderma* mitochondrial genomes

Gene order[#]	*G. lucidum*	*G. meredithiae*	*G. applanatum*	*G. sinense*	*G. tsugae*	*G. calidophilum*	*G. leucocontextum*	Gene order 2[#]
1	*rnl*	*rnl* (1i[*])	*rnl* (6i)	*rnl* (3i)	*rnl* (1i)	*rnl* (4i)	*rnl* (2i)	1
2	trnS-GCU	trnS-GCU						
3	atp6	atp6	atp6	atp6	atp6	atp6	atp6	3
2'					trnS-GCU			
37'		trnR-UCG						
4	trnM-CAU	trnM-CAU	trnM-CAU	trnM-CAU	trnM-CAU	trnM-CAU	*cox1* (9i)	15
5	trnH-GUG	trnH-GUG	trnH-GUG	trnH-GUG	trnH-GUG	trnH-GUG	*nad4* (1i)	14
6	trnS-UGA	trnS-UGA	trnS-UGA	trnS-UGA	trnS-UGA	trnS-UGA	rps3	35
7	trnM-CAU	trnM-CAU	trnM-CAU	trnM-CAU	trnM-CAU	trnM-CAU	nad6	34
8	trnY-GUA	trnY-GUA	trnY-GUA	trnY-GUA	trnY-GUA	trnY-GUA	atp8	33
9	trnK-UUU	trnK-UUU	trnK-UUU	trnK-UUU	trnK-UUU	trnK-UUU	trnS-GCU	32
10	trnQ-UUG	trnQ-UUG	trnQ-UUG	trnQ-UUG	trnQ-UUG	trnQ-UUG	nad2	31
11	trnT-UGU	trnT-UGU	trnT-UGU	trnT-UGU	trnT-UGU	trnT-UGU	*nad3* (1i)	30
12	trnF-GAA	trnF-GAA	trnF-GAA	trnF-GAA	trnF-GAA	trnF-GAA	trnG-UCC	29
13	trnA-UGC	trnA-UGC	trnA-UGC	trnA-UGC	trnA-UGC	trnA-UGC	trnW-CCA	28
14	*nad4* (1i)	*nad4* (1i)	*nad4* (2i)	*nad4* (1i)	*nad4* (1i)	*nad4* (1i)	trnD-GUC	27
15	*cox1* (6i)	*cox1* (10i)	*cox1* (14i)	*cox1* (12i)	*cox1* (11i)	*cox1* (14i)	atp9	26
16	trnE-UUC	trnE-UUC	trnE-UUC	trnE-UUC	trnE-UUC	trnE-UUC		
17	trnM-CAU	trnM-CAU	trnM-CAU	trnM-CAU	trnM-CAU	trnM-CAU	trnR-UCU	24
18	trnL-UAA	trnL-UAA	trnL-UAA	trnL-UAA	trnL-UAA	trnL-UAA	nad1	23
19	trnN-GUU	trnN-GUU	trnN-GUU	trnN-GUU	trnN-GUU	trnN-GUU	trnI-GAU	22
20	trnP-UGG	trnP-UGG	trnP-UGG	trnP-UGG	trnP-UGG	trnP-UGG	*cob* (4i)	21
37'				trnR-UCG			trnP-UGG	20
21	cob	*cob* (3i)	*cob* (6i)	*cob* (5i)	*cob* (4i)	*cob* (6i)	trnN-GUU	19

(continued)

Table 4.2 (continued)

Gene order#	*G. lucidum*	*G. meredithiae*	*G. applanatum*	*G. sinense*	*G. tsugae*	*G. calidophilum*	*G. leucocontextum*	Gene order 2#
22	trnI-GAU	trnI-GAU	trnI-GAU	trnI-GAU	trnI-GAU	trnI-GAU	trnL-UAA	18
23	nad1	nad1	nad1	nad1	*nad1* (2i)	*nad1* (1i)	trnM-CAU	17
24	trnR-UCU	trnR-UCU	trnR-UCU	trnR-UCU	trnR-UCU		trnE-UUC	16
25	trnC-GCA	trnC-GCA		trnC-GCA		trnC-GCA	trnS-GCU	2
26	atp9	atp9	atp9	atp9	atp9	atp9	trnM-CAU	4
27	trnD-GUC	trnD-GUC	trnD-GUC	trnD-GUC	trnD-GUC	trnD-GUC	trnH-GUG	5
28	trnW-CCA	trnW-CCA	trnW-CCA	trnW-CCA	trnW-CCA	trnW-CCA	trnS-UGA	6
29	trnG-UCC	trnG-UCC	trnG-UCC	trnG-UCC	trnG-UCC	trnG-UCC	trnM-CAU	7
30	nad3	nad3	nad3	nad3	nad3	nad3	trnY-GUA	8
31	nad2	nad2	nad2	nad2	nad2	nad2	trnK-UUU	9
32	trnS-GCU	trnS-GCU	trnS-GCU	trnS-GCU	trnS-GCU	nad6	trnQ-UUG	10
33	atp8	atp8	atp8	atp8	atp8	rps3	trnT-UGU	11
34	nad6	nad6	nad6	nad6	nad6	**trnS-GCU**	trnF-GAA	12
35	rps3	rps3	rps3	rps3	rps3	**atp8**	trnA-UGC	13
36	*cox2* (1i)	*cox2* (1i)	*cox2* (1i)	*cox2* (3i)	cox2	cox2	cox2	36
45		trnI-UAU						
44'		trnR-UCU						
37	trnR-UCG		trnR-UCG	trnR-UCG				
38	trnL-UAG	trnL-UAG	trnL-UAG	trnL-UAG	trnL-UAG	trnL-UAG	trnL-UAG	38
39	trnV-UAC	trnV-UAC	trnV-UAC	trnV-UAC	trnV-UAC	trnV-UAC	trnV-UAC	39
40	rns	rns	rns	*rns* (2i)	rns	rns	rns	40
37'					trnR-UCG	trnR-UCG	trnR-UCG	37'
37"				trnR-UCG$				
41	*nad5* (2i)	*nad5* (2i)	*nad5* (5i)	*nad5* (2i)	*nad5* (4i)	*nad5* (4i)	*nad5* (2i)	41
42	nad4L	nad4L	nad4L	nad4L	nad4L	nad4L	nad4L	42
43	*cox3* (1i)	cox3	*cox3* (1i)	*cox3* (2i)	*cox3* (1i)	*cox3* (1i)	cox3	43

(continued)

Table 4.2 (continued)

Gene order[#]	*G. lucidum*	*G. meredithiae*	*G. applanatum*	*G. sinense*	*G. tsugae*	*G. calidophilum*	*G. leucocontextum*	Gene order 2[#]
44			trnR-UCU	trnR-UCU		trnR-UCU		44
25”		trnC-GCA[$]						

The first column [#] Gene order is for the former six species; gene order 2 is for the last one, *G. leucocontextum*, showing the mitogenome's large rearrangements

[*]Number in the blankets denotes how many introns exist in the gene

[$]The gene is one more duplicate

Gene in bold of *G. calidophilum* denotes a rearrangement

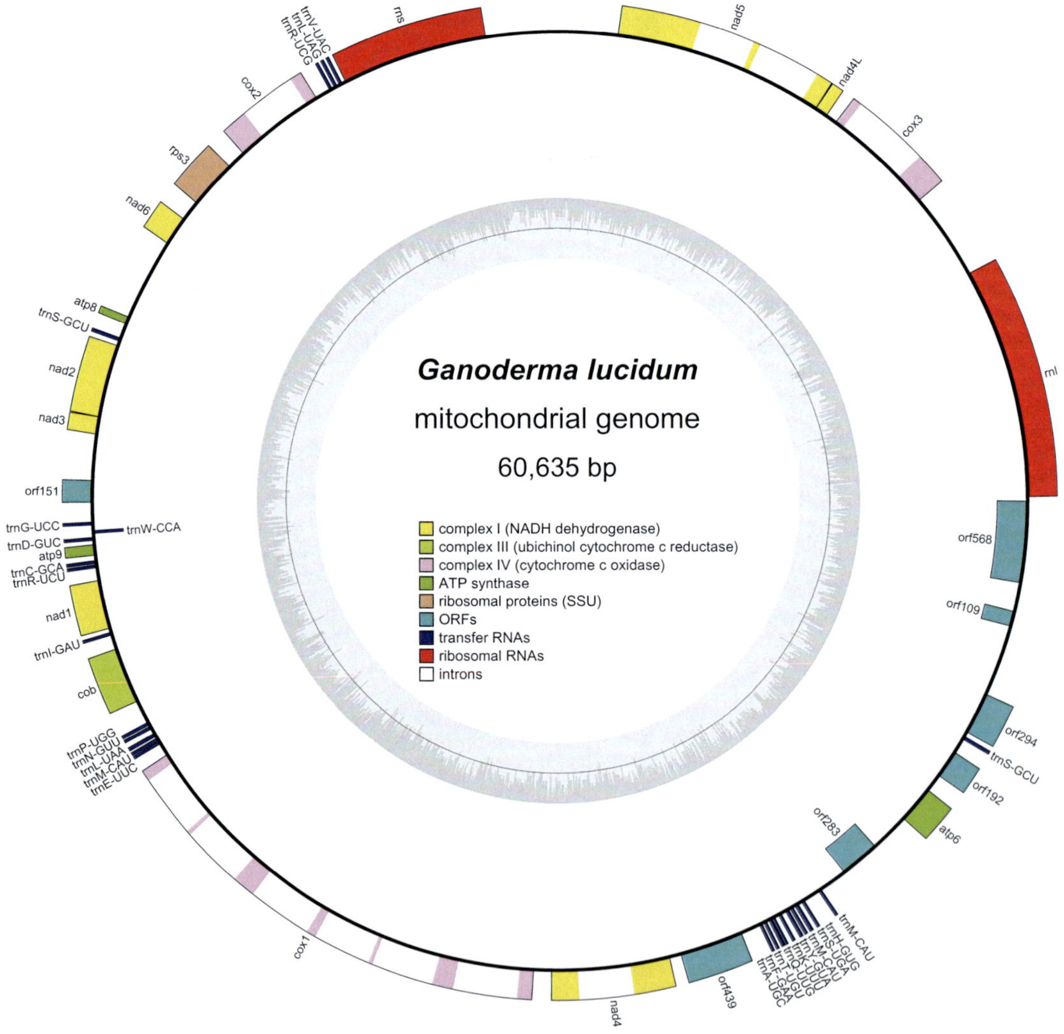

Fig. 4.2 Mitochondrial genome map of *G. lucidum* CGMCC 5.26 (NC_021750)

specimen of *G. meredithiae* Adask. & Gilb. (JEA 345). JEA 345 is deposited in US National Fungus Collections (BPI, Beltsville, Maryland). The mitogenome of this species was reported by Wang et al. (2016b).

The 100 bp paired-end reads were assembled using ABySS-pe. The mitogenome (HF570115) of *G. lucidum* CGMCC 5.26 served as a reference sequence. A total of 1.35 GB clean data of 100 bp paired-end reads were produced. After assembly, 14,054 contigs were obtained. All the contigs were aligned with the reference sequence. And two contigs were identified as a mitochondrial origin. Contig 133 was 46,593 bp long, and contig 223 was 31,145 bp long. The two contigs shared a 390 bp long repeat sequence at both ends. The two contigs were assembled according to the reference sequence. Four gaps were detected in the preliminary mitochondrial genome sequence. They were then filled by comparing the assembly results produced by AbySS-pe software with parameters of k-mer 25 and k-mer 64. The authors mapped reads onto this genome sequence to validate this mitogenome's assembly and check its coverage. A total of 546,887 reads were mapped, accounting for 4.0% of all the reads produced (546,887/13,512,066). The average coverage depth was 693.5, and the maximum coverage depth was 1,552.

This mitogenome is a circular mtDNA with a total length of 78,447 bp and a GC content of 26.14%. It encodes a set of mitochondrial protein and RNA genes, including 15 conserved proteins, 29 tRNAs, large and small ribosomal RNAs, and 18 homing endonucleases. All structural genes are located on the same strand except *trnW-CCA* (Fig. 4.3). The gene order is listed in Table 4.2. The tRNA genes code for all 20 standard amino acids. Most amino acids are coded by only one tRNA gene; however, two *trnC-GCA*, two *trnI* (*trnI-GAU* and *trnI-UAU*), two *trnL* (*trnL-UAA* and *trnL-UAG*), three *trnM-CAU*, three *trnR* (one *trnR-UCG* and two *trnR-UCU*), and three *trnS* (one *trnS-UGA* and two *trnS-GCU*) are found in this mitogenome. Eighteen introns are detected in six genes: *cob* (3), *cox1* (10), *cox2* (1), *nad4* (1), *nad5* (2), and *rnl* (1). For each intron, there is one homing endonuclease. Three of the 18 endonucleases are GIY-YIG type (ip4, ip10, and ip17), and the others are LAGLIDADG type.

Compared with the mitogenome of *G. lucidum* (NC_021750), *G. meredithiae* has larger genome content, lower GC content, three more tRNA genes (*trnC-GCA*, *trnI-UAU*, and *trnR-UCG*), and seven more introns in the protein-coding or rRNA genes.

4.2.2.2 Mitochondrial Genome of *Ganoderma applanatum*

The strain CGMCC 5.249 of *G. applanatum* (Pers.) Pat. was isolated from a basidiocarp collected from Changbai Mountain, Jilin Province, China. The mitogenome of this species was reported by Wang et al. (2016a).

The 125 bp pair-end reads were assembled using CLC Genomics Workbench (CLC Bio, version 7.5.1, Aarhus, Denmark). The authors used the mitogenome of *G. meredithiae* CGMCC 5.766 as the reference to identify the scaffolds and to determine their order. Manual comparison of the ends of the ordered scaffolds helped to assemble the scaffolds into a circular molecule.

A total of 1.54 GB clean data of 125 bp paired-end reads were produced. After assembly, 21,718 contigs were obtained. All the contigs were aligned with the reference sequence, and three contigs were identified as a mitochondrial origin. Contigs 21, 22, and 47 were 35,819 bp, 83,529 bp, and 304 bp long, respectively. The contigs were assembled in the order: 22-47-21-47. A 287 bp long repeat sequence was shared by the tail of contig 22 and the head of contig 47; a 21 bp long repeat sequence was shared by the tail of contig 47 and the heads of contigs 21 and 22; another 21 bp long repeat sequence was shared by the tail of contig 21 and the head of contig 47. No gap was detected in the assembled mitochondrial genome sequence. To validate the mitogenome assembly, the authors mapped reads onto this genome sequence. A total of 2,000,647 reads were mapped, accounting for 16.2% of all

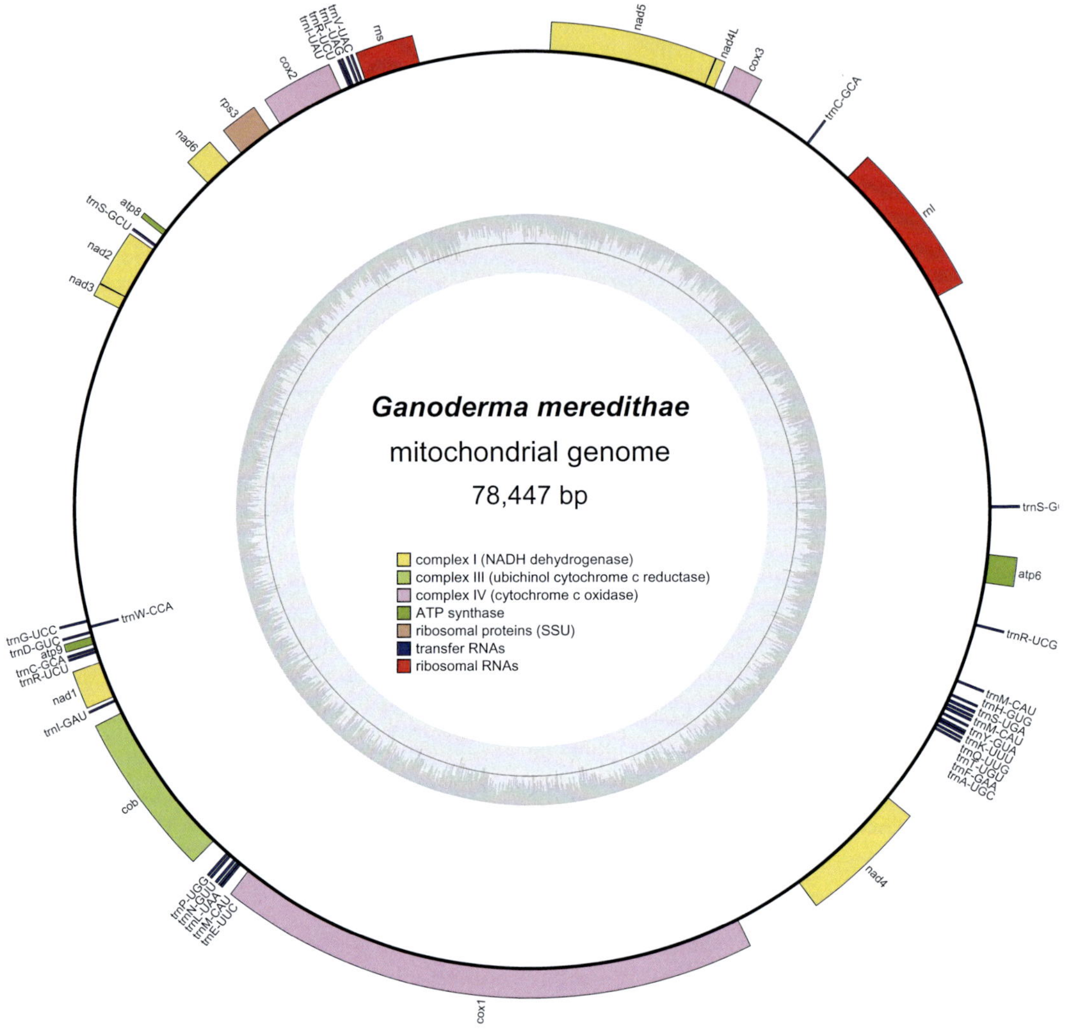

Fig. 4.3 Mitochondrial genome map of *G. meredithae*

the reads produced. The average coverage depth was 2,086.9, and the maximum coverage depth was 3,458.

The mitogenome of *G. applanatum* is 119,803 bp long, with the GC content being 26.66%. It encodes 15 conserved proteins, 25 tRNAs, and the large and small ribosomal RNAs. All structural genes are located on the same strand except *trnW-CCA* (Fig. 4.4). The gene order is listed in Table 4.2. The tRNA genes contain codons for all 20 standard amino acids except Cysteine. Most amino acids are represented by only one tRNA gene; however, two *trnL* (*trnL-UAA* and *trnL-UAG*), two *trnS* (*trnS-GCU* and *trnS-UGA*), three *trnM-CAU*, and three *trnR* (one *trnR-UCG* and two *trnR-UCU*) genes are found in this mitogenome. Thirty-five introns are detected in seven genes, i.e., *cob* (6), *cox1* (14), *cox2* (1), *cox3* (1), *nad4* (2), *nad5* (5), and *rnl* (6).

Compared with the mitogenome of *G. lucidum* (NC_021750), *G. applanatum* has larger genome content, almost the same GC content, one less tRNA gene, and 24 more introns in the protein-coding or rRNA genes.

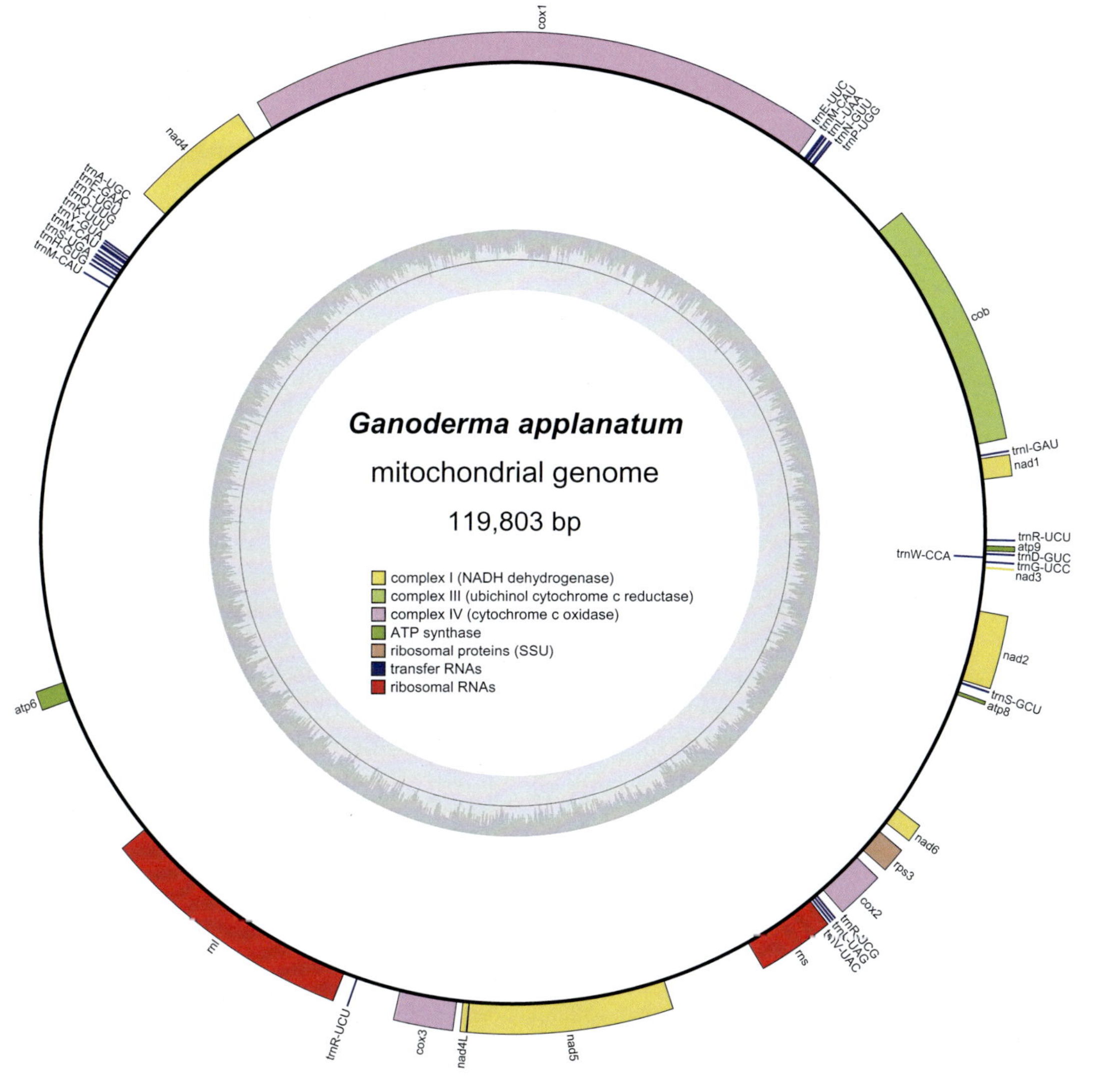

Fig. 4.4 Mitochondrial genome map of *G. applanatum*

4.2.2.3 Mitochondrial Genome of *Ganoderma sinense*

The strain CGMCC 5.69 of *G. sinense* J.D. Zhao, L.W. Hsu, and X.Q. Zhang was isolated from a basidiocarp collected from Hainan Province, China. This mitogenome was submitted by J. Qian, Institute of Medicinal Plant Development, Beijing, China.

The mitogenome of *G. sinense* is 86,451 bp long, with the GC content being 26.78%. It encodes 15 conserved proteins, 27 tRNAs, and the large and small ribosomal RNAs. All of them are located on the same strand except *trnW-CCA*. The gene order is listed in Table 4.2. A total of 30 introns exist in this mitogenome. Compared with the mitogenome of *G. lucidum* (NC_021750), *G. sinense* has larger genome content, higher GC content, one more tRNA gene, and 19 more introns in the protein-coding or rRNA genes. In detail, *trnS-GCU* is missing; two *trnR-UCG* and one *trnR-UCU* are added; three more introns in *rnl*, six more in *cox1*, five more in *cob*, two more in *cox2* and *rns,* respectively, and one more in *cox3*.

4.2.2.4 Mitochondrial Genome of *Ganoderma tsugae*

The strain s90 of *G. tsugae* Murrill was isolated from a basidiocarp collected from Wangqing County, Jilin Province, China. This mitogenome was published by Li et al. (2019).

The mitogenome of *G. tsugae* is 92,511 bp long, with the GC content being 26.67%. It encodes 15 conserved proteins, 25 tRNAs, and the large and small ribosomal RNAs. All structural genes are on the same strand except *atp6*, *trnS-GCU*, and *trnW-CCA*. The gene order is listed in Table 4.2. A total of 24 introns exist in this mitogenome. Compared with the mitogenome of *G. lucidum* (NC_021750), *G. tsugae* has larger genome content, the same GC content, one less tRNA gene, and 13 more introns in the protein-coding or rRNA genes. That is, *trnC-GCA* is missing; the locations of *trnS-GCU* and *trnR-UCG* are changed; one more intron in *rnl*, five more in *cox1*, four more in *cob*, two more in *nad1* and *nad5*, respectively.

4.2.2.5 Mitochondrial Genome of *Ganoderma calidophilum*

The strain s136 of *G. calidophilum* J.D. Zhao, L. W. Hsu, and X.Q. Zhang was isolated from a basidiocarp collected from Qiongzhong County, Hainan Province, China. This mitogenome was published by Li et al. (2019).

The mitogenome of *G. calidophilum* is the largest one in this genus up to now. It is 124,588 bp long, with the GC content of 25.43%. It encodes 15 conserved proteins, 25 tRNAs, and the large and small ribosomal RNAs. All of them are on the same strand except *trnW-CCA*. The gene order is listed in Table 4.2. A total of 31 introns exist in this mitogenome. Compared with the mitogenome of *G. lucidum* (NC_021750), *G. calidophilum* has larger genome content, lower GC content, one less tRNA gene, and 20 more introns in the protein-coding or rRNA genes. Specifically, *trnS-GCU* is missing, and the locations of *trnR-UCU* and *trnR-UCG* are changed; *trnS-GCU* and *atp8* are shifted to the region between *rps3* and *cox2*. In particular, there are four more introns in *rnl*, eight more in *cox1*, six more in *cob*, one more in *nad1*, and two more in *nad5*.

4.2.2.6 Mitochondrial Genome of *Ganoderma leucocontextum*

The strain s116 of *G. leucocontextum* TH Li et al. was isolated from a basidiocarp collected from Huidong County, Sichuan Province, China. This mitogenome was published by Li et al. (2019).

The mitogenome of *G. leucocontextum* is 88,194 bp long, with the GC content being 27.08%. It encodes 15 conserved proteins, 25 tRNAs, and the large and small ribosomal RNAs. A total of 22 genes are on the same strand, and 20 genes are on the anti-sense strand. The gene order is listed in Table 4.2. A total of 19 introns exist in this mitogenome. Compared with the mitogenome of *G. lucidum* (NC_021750), the gene order of *G. leucocontextum* is much different, as shown in Table 4.2. The tRNA genes (gene order 4 to 13) are not located close to *atp6* (gene order 3); in contrast, they are shifted to the region before *cox2* (gene order 36). And there is a big reverse region including genes from order 16 to 35. This arrangement of mitochondrial genes has not been seen in other *Ganoderma* species. It has larger genome content, higher GC content, one less tRNA gene, and eight more introns in the protein-coding or rRNA genes. In detail, *trnC-GCA* is missing. There are two more introns in *rnl*, three more in *cox1*, one more in *nad3*, and four more in *cob*.

4.2.2.7 Mitochondrial Phylogenomics of *Ganoderma*

To investigate the mitochondrial phylogeny of *Ganoderma*, the accessible mitogenomes in Polyporales were all included: *Phlebia radiate* (NC_020148), *Taiwanofungus camphoratus* (NC_042771), *Trametes cingulata* (NC_013933), *Trametes hirsute* (NC_037239), and *Wolfiporia cocos* (NC_050681). Amino acid sequences of the 15 conserved protein-coding genes were extracted from the mitogenomes. The 15 proteins included subunits of the respiratory chain complexes (*cox1*, *cox2*, *cox3*, and *cob*), ATPase subunits (*atp6*, *atp8*, and *atp9*), NADH:quinone reductase subunits (*nad1*, *nad2*, *nad3*, *nad4*, *nad4L*, *nad5*, and *nad6*), and ribosomal protein S3 (*rps3*). The alignment of the concatenated 15 genes is 5837 amino acid long.

As shown in Fig. 4.5, the genus *Ganoderma* is monophyly and contains nine sequenced members. *Ganoderma lucidum* and *G.* sp. s8 clustered in the same branch with very short branch lengths and very strong support, hinting that they are the same species. *Ganoderma meredithae* has the closest relationship with *G. lucidum*. *Trametes* is the sister group of *Ganoderma*, consistent with the previous results (Wang et al. 2016a, b; Li et al. 2019).

Ganoderma calidophilum has the largest mitogenome among the *Ganoderma* members, while *G.* sp. s8 has the smallest one. The former is more than twice in mitogenome size as the latter (Table 4.1). Except *for G. lucidum* (including *G.* sp. s8), *G. meredithae* has the smallest mitogenome. This might suggest that the evolution of the mitochondrial genome of *Ganoderma* correlates with the phylogeny in some branch to some extent. For GC content, *G. leucocontextum* has the highest value, while *G. calidophilum* has the lowest one. For the number of tRNAs, *G. meredithae* has the most genes (29), while *G. applanatum*, *G. calidophilum*, *G. leucocontextum*, and *G. tsugae* have the least genes (25). For the number of introns, *G. applanatum* has the most (35); in contrast, *G. lucidum* has the least (11–13). The mitogenomes of *Ganoderma* presented highly genetic diversity in terms of genome content, gene order and structure, and so on.

4.3 Future Perspective

4.3.1 Problems

Ganoderma species have high medicinal values (Wang et al. 2012). There are three *Ganoderma* species recorded in the Chinese pharmacopeia, i.e., *G. lucidum*, *G. sinense*, and *G. tsugae*. The mitogenomes of them had been published or recovered in the previous researches and compared here. Nevertheless, most of the mitogenomes of the species in this genus *Ganoderma*

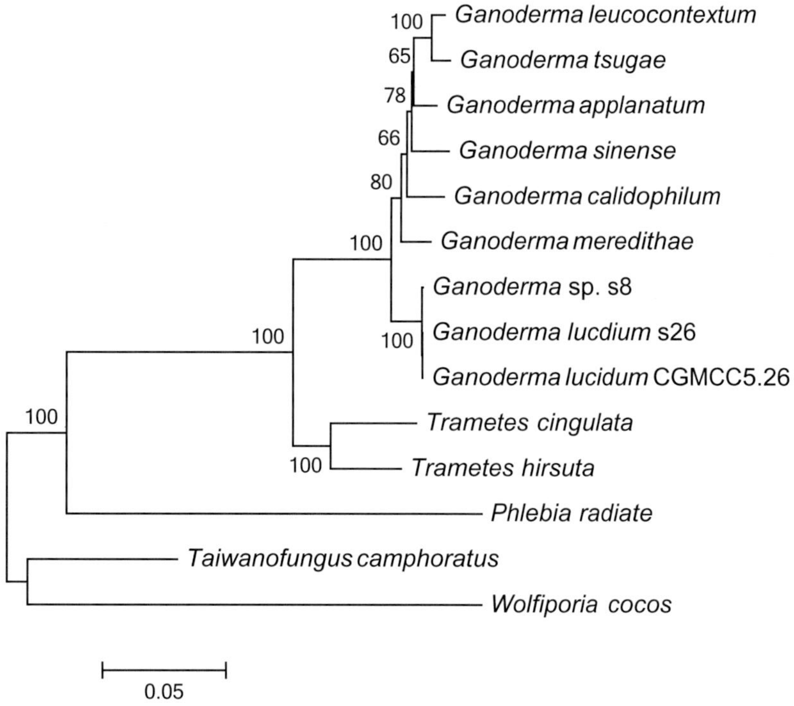

Fig. 4.5 Mitochondrial phylogenomic tree of Polyporales

have not been studied. More than 80 phylogenetic species of *Ganoderma* were recognized in a meta-analysis of internal transcribed spacer (ITS) rDNA sequences (Fryssouli et al. 2020), and many of them have been proven to be with significant medicinal or ecological values. Only seven *Ganoderma* species (Table 4.1 and Fig. 4.5) were investigated for their mitochondrial genomes. More intense sampling is in need for the future mitogenome researches of *Ganoderma* at both inter- and intra-specific levels.

4.3.2 Topics of Interests

ITS has demonstrated high efficacy in resolving relationships amongst most of the *Ganoderma* taxa; however, it was not equally useful at elucidating species boundaries across the entire genus (Fryssouli et al. 2020). Mitochondrial small subunit (*mtSSU* = *rns*) gene was used to infer this genus's phylogeny (Hong and Jung 2004). However, the incongruence between mitochondrial and nuclear phylogenies of the genus was detected based on three mitochondrial genes and seven nuclear genes (Wang 2012).

On the other hand, phylogenetic analyses based on different single genes of mitogenome recovered incongruent tree topologies. But the BI phylogeny based on the *cox1* gene was mostly consistent with that generated using combined datasets, indicating this gene could be a potential molecular marker for phylogenetic analysis (Li et al. 2019). These findings suggest that selecting suitable molecular markers is essential for studying phylogenetic relationships in the *Ganoderma* genus.

Acknowledgements This work was supported by the National Natural Science Foundation of China (31700014, 31750001), Key Research Program of Frontier Science, Chinese Academy of Sciences (QYZDY-SSW-SMC029), Beijing Natural Science Foundation (7154224), and China Postdoctoral Science Foundation (2014M550659).

References

Altschul SF, Gish W, Miller W, Myers EW, Lipman DJ (1990) Basic local alignment search tool. J Mol Biol 215(3):403–410

Anderson S, Bankier AT, Barrell BG, de Bruijn MH, Coulson AR, Drouin J, Eperon IC, Nierlich DP, Roe BA, Sanger F, Schreier PH, Smith AJ, Staden R, Young IG (1981) Sequence and organization of the human mitochondrial genome. Nature 290 (5806):457–465

Andersson SG, Zomorodipour A, Andersson JO, Sicheritz-Ponten T, Alsmark UC, Podowski RM, Naslund AK, Eriksson AS, Winkler HH, Kurland CG (1998) The genome sequence of *Rickettsia prowazekii* and the origin of mitochondria. Nature 396 (6707):133–140

Bankevich A, Nurk S, Antipov D, Gurevich AA, Dvorkin M, Kulikov AS, Lesin VM, Nikolenko SI, Pham S, Prjibelski AD, Pyshkin AV, Sirotkin AV, Vyahhi N, Tesler G, Alekseyev MA, Pevzner PA (2012) SPAdes: a new genome assembly algorithm and its applications to single-cell sequencing. J Comput Biol 19(5):455–477

Basse CW (2010) Mitochondrial inheritance in fungi. Curr Opin Microbiol 13(6):712–719

Bernt M, Donath A, Juhling F, Externbrink F, Florentz C, Fritzsch G, Putz J, Middendorf M, Stadler PF (2013) MITOS: improved de novo metazoan mitochondrial genome annotation. Mol Phylogenet Evol 69(2):313–319

Brown WM (1980) Polymorphism in mitochondrial-DNA of humans as revealed by restriction endonuclease analysis. P Natl Acad Sci-Biol 77(6):3605–3609

Brown WM, George M Jr, Wilson AC (1979) Rapid evolution of animal mitochondrial DNA. Proc Natl Acad Sci U S A 76(4):1967–1971

Brown WM, Prager EM, Wang A, Wilson AC (1982) Mitochondrial DNA sequences of primates: tempo and mode of evolution. J Mol Evol 18(4):225–239

Chinnery PF, Hudson G (2013) Mitochondrial genetics. Br Med Bull 106:135–159

Dujon B (2020) Mitochondrial genetics revisited. Yeast 37(2):191–205

Edgar RC (2004) MUSCLE: multiple sequence alignment with high accuracy and high throughput. Nucleic Acids Res 32(5):1792–1797

Foury F, Roganti T, Lecrenier N, Purnelle B (1998) The complete sequence of the mitochondrial genome of *Saccharomyces cerevisiae*. FEBS Lett 440(3):325–331

Friedman JR, Nunnari J (2014) Mitochondrial form and function. Nature 505(7483):335–343

Fryssouli V, Zervakis GI, Polemis E, Typas MA (2020) A global meta-analysis of ITS rDNA sequences from material belonging to the genus *Ganoderma* (Basidiomycota, Polyporales) including new data from selected taxa. Mycokeys 75:71–143

Gray MW, Burger G, Lang BF (1999) Mitochondrial evolution. Science 283(5407):1476–1481

Hong SG, Jung HS (2004) Phylogenetic analysis of *Ganoderma* based on nearly complete mitochondrial small-subunit ribosomal DNA sequences. Mycologia 96(4):742–755

Katoh K, Misawa K, Kuma K, Miyata T (2002) MAFFT: a novel method for rapid multiple sequence alignment based on fast Fourier transform. Nucleic Acids Res 30 (14):3059–3066

Krumsiek J, Arnold R, Rattei T (2007) Gepard: a rapid and sensitive tool for creating dotplots on genome scale. Bioinformatics 23(8):1026–1028

Lackner LL (2014) Shaping the dynamic mitochondrial network. BMC Biol 12:35

Lang BF, Jakubkova M, Hegedusova E, Daoud R, Forget L, Brejova B, Vinar T, Kosa P, Fricova D, Nebohacova M, Griac P, Tomaska L, Burger G, Nosek J (2014) Massive programmed translational jumping in mitochondria. Proc Natl Acad Sci U S A 111(16):5926–5931

Li J, Zhang J, Chen H, Chen X, Lan J, Liu C (2013) Complete mitochondrial genome of the medicinal mushroom *Ganoderma lucidum*. PLoS ONE 8(8): e72038

Li Q, Xiang D, Wan Y, Wu Q, Wu X, Ma C, Song Y, Zhao G, Huang W (2019) The complete mitochondrial genomes of five important medicinal *Ganoderma* species: features, evolution, and phylogeny. Int J Biol Macromol 139:397–408

Lohse M, Drechsel O, Bock R (2007) OrganellarGenomeDRAW (OGDRAW): a tool for the easy generation of high-quality custom graphical maps of plastid and mitochondrial genomes. Curr Genet 52(5–6):267–274

Lowe TM, Chan PP (2016) tRNAscan-SE On-line: integrating search and context for analysis of transfer RNA genes. Nucleic Acids Res 44(W1):W54–W57

Malina C, Larsson C, Nielsen J (2018) Yeast mitochondria: an overview of mitochondrial biology and the potential of mitochondrial systems biology. FEMS Yeast Res 18(5):foy040

Milne I, Stephen G, Bayer M, Cock PJA, Pritchard L, Cardle L, Shaw PD, Marshall D (2013) Using tablet for visual exploration of second-generation sequencing data. Brief Bioinform 14(2):193–202

Nass MM, Nass S (1963) Intramitochondrial fibers with DNA characteristics. I. Fixation and electron staining reactions. J Cell Biol 19:593–611

Rozas J, Ferrer-Mata A, Sanchez-DelBarrio JC, Guirao-Rico S, Librado P, Ramos-Onsins SE, Sanchez-Gracia A (2017) DnaSP 6: DNA sequence polymorphism analysis of large data sets. Mol Biol Evol 34(12):3299–3302

Simpson JT, Wong K, Jackman SD, Schein JE, Jones SJ, Birol I (2009) ABySS: a parallel assembler for short read sequence data. Genome Res 19(6):1117–1123

Tamura K, Stecher G, Peterson D, Filipski A, Kumar S (2013) MEGA6: molecular evolutionary genetics analysis version 6.0. Mol Biol Evol 30(12):2725–2729

Thawthong A, Hapuarachchi KK, Wen TC, Raspe O, Thongklang N, Kang JC, Hyde KD (2017) *Ganoderma sichuanense* (Ganodermataceae, Polyporales) new to Thailand. Mycokeys 22:27–43

van de Vossenberg B, Brankovics B, Nguyen HDT, van Gent-Pelzer MPE, Smith D, Dadej K, Przetakiewicz J, Kreuze JF, Boerma M, van Leeuwen GCM, Andre Levesque C, van der Lee TAJ (2018) The linear mitochondrial genome of the quarantine chytrid *Synchytrium endobioticum*; insights into the evolution and recent history of an obligate biotrophic plant pathogen. BMC Evol Biol 18(1):136

van der Giezen M, Tovar J (2005) Degenerate mitochondria. EMBO Rep 6(6):525–530

Wang XC (2012) Phylogenetic study on Ganodermataceae Donk. University of Chinese Academy of Sciences, Beijing

Wang XC, Xi RJ, Li Y, Wang DM, Yao YJ (2012) The species identity of the widely cultivated *Ganoderma*, '*G. lucidu*' (Ling-zhi), in China. PLoS ONE 7(7):e40857

Wang XC, Shao J, Liu C (2016a) The complete mitochondrial genome of the medicinal fungus *Ganoderma applanatum* (Polyporales, Basidiomycota). Mitochondrial DNA A DNA Mapp Seq Anal 27 (4):2813–2814

Wang XC, Wu K, Chen H, Shao J, Zhang N, Chen X, Lan J, Liu C (2016b) The complete mitochondrial genome of the white-rot fungus *Ganoderma meredithiae* (Polyporales, Basidiomycota). Mitochondrial DNA A DNA Mapp Seq Anal 27(6):4197–4198

Yao YJ, Wang XC, Wang B (2013) Epitypification of *Ganoderma sichuanense* J.D. Zhao & XQ. Zhang (Ganodermataceae). Taxon 62(5):1025–1031

Yao YJ, Li Y, Du Z, Wang K, Wang XC, Kirk PM, Spooner BM (2020) On the typification of *Ganoderma sichuanense* (Agaricomycetes)-the widely cultivated Lingzhi medicinal mushroom. Int J Med Mushrooms 22(1):45–54

Transcriptome of Lingzhi

5

Haimei Chen, Yang Ni, and Heyu Yang

Abstract

The transcriptome is the set of all RNA molecules in one cell or a population of cells. There are several different methods to obtain the transcriptome, including cDNA-amplified fragment length polymorphism (cDNA-AFLP), expressed sequence tags (EST), serial analysis of gene expression (SAGE), microarrays, and RNA-Seq. In this chapter, we summarize the recent progress in the transcriptome analysis of *Ganoderma* species. Besides, we described the entire procedure for RNA-seq analysis, including quality control of raw data, processing of raw reads, genome-guided transcriptome assembly, and quantification, identification of differential expression genes, enrichment analysis of differential expression genes, alternative splicing analysis, single nucleotide polymorphism identification, and insertion-deletion analysis.

H. Chen (✉)
Key Laboratory of Bioactive Substances and Resource Utilization of Chinese Herbal Medicine from the Ministry of Education, Institute of Medicinal Plant Development, Chinese Academy of Medical Sciences, Peking Union Medical College, Beijing 100193, China
e-mail: hmchen@implad.ac.cn

Y. Ni
College of Agriculture, Fujian Agriculture and Forestry University, Fuzhou 350002, China

H. Yang
School of Environmental Science and Engineering, Tianjin University, Tianjin 300072, China

5.1 Introduction

Ganoderma lucidum (Leyss. ex Fr.) Karst, also known as "Lingzhi", is a species of white-rot fungus that belongs to the family Polyporaceae (Zhou et al. 2018). More than 600 compounds were isolated and identified from the genus *Ganoderma*. They include polysaccharides, triterpenes, meroterpenoids, steroids, alkaloids, nucleosides, and nucleobases. Among them, triterpenes and polysaccharides are the main active components (Gong et al. 2019). *G. lucidum* has many beneficial effect, such as antitumor, regulating immunity, antioxidation, and protecting cells, and so on (Xu and Li 2019). The identification of genes responsible for the biosynthesis of active components and their regulation have been active areas of research.

The transcriptome is the set of all RNA molecules in one cell or a population of cells. Transcriptome analysis is a common method to discover the gene contents, identify differentially expressed genes (DEGs), and identify repeat elements for molecular markers' development. In recent years, the advent of next-generation sequencing (NGS) technologies, such as Roche 454, Illumina, SOLiD, Pacbio, and Nanopore,

C. Liu (ed.), *The Lingzhi Mushroom Genome*, Compendium of Plant Genomes,
https://doi.org/10.1007/978-3-030-75710-6_5

has provided powerful platforms for transcriptome analysis. This chapter will summarize the recent advances in the transcriptomic studies of *Ganoderma* species, including the discovery of genes and the understanding of the regulation of secondary metabolism.

5.2 Method for Transcriptome Analysis of *G. Lucidum*

5.2.1 cDNA-Amplified Fragment Length Polymorphism Method for Transcription Profiling

The cDNA-amplified fragment length polymorphism (cDNA-AFLP) method is an mRNA fingerprinting technique based on restriction enzymes and PCR reactions, which were performed to analyze the transcriptome of a wild species of *Ganoderma* with Cd tolerance (Chuang et al. 2009). The cDNA-AFLP was performed according to the following protocol (Fig. 5.1). The total RNA was first reverse transcribed. The resulting complementary DNA (cDNA) was digested first with the restriction enzymes *Eco*RI and *Mse*I. After digestion, the restriction fragments were ligated with *Eco*RI-adaptor and *Mse*I-adaptor. The ligated cDNA fragments were subsequently amplified by PCR using nonselective primers. This PCR reaction was carried out using one cycle of 95 °C for 2 min, followed by 20 cycles of 95 °C for 30 s, 50 °C for 30 s, and 72 °C for 1 min. The amplified cDNA was diluted 500-fold and then used for selective PCR amplification using selective *Eco*RI and *Mse*I primers. The PCR products were separated on a 4% denaturing polyacrylamide gel running at 80 W for 2.5 h and visualized by silver staining. These cDNA fragments showing different expression levels between control and treatment were used for sequencing analysis, and their putative function was predicted by BLASTX search against the Nucleotide sequence databases. This method has found widespread use as an attractive technology for gene discovery based on fragment detection and for temporal quantitative gene expression analysis. It takes 3–4 d to complete the protocol.

5.2.2 Expressed Sequence Tags from a cDNA Library

Expressed sequence tag (EST) analysis allows rapid and large-scale identification of uniquely expressed genes (Luo et al. 2010; Huang et al. 2013). Simple sequence repeats (SSRs) identified from EST can be used as new molecular markers for genetic mapping. The construction process of the EST library is as following (Fig. 5.2). The mRNA of the sample was isolated and purified. The cDNA library was constructed from purified mRNA with a Creator™ SMART™ cDNA Library Construction kit (Clontech, USA). The double-stranded cDNA was ligated into the pDNR-lib vector (Clontech, USA) and electroporated into a DH5α Escherichia coli strain. Randomly selected clones were cultured in liquid LB medium containing chloramphenicol. The plasmid DNA was prepared from these clones with an Axyprep-96 plasmid kit (Axygen, USA) and submitted for direct sequencing on the ABI 3730 DNA sequencer.

The first cDNA library for *G. lucidum* was constructed from the fruiting body (Luo et al. 2010) to discover the transcripts involved in secondary metabolite biosynthesis and the developmental regulation of *G. lucidum*. A total of 1,023 clones were randomly selected from the *G. lucidum* library and sequenced, yielding 879 high-quality ESTs. The high-quality ESTs were assembled into unique sequences, which were annotated according to sequence similarities to genes or proteins available in public databases. Among them, a total of 173 unique sequences showed similarities to known genes involved in the biosynthesis of secondary metabolites and the developmental regulation, including squalene epoxidase (SE) and farnesyl-diphosphate synthase (FPS). Several candidate genes, such as

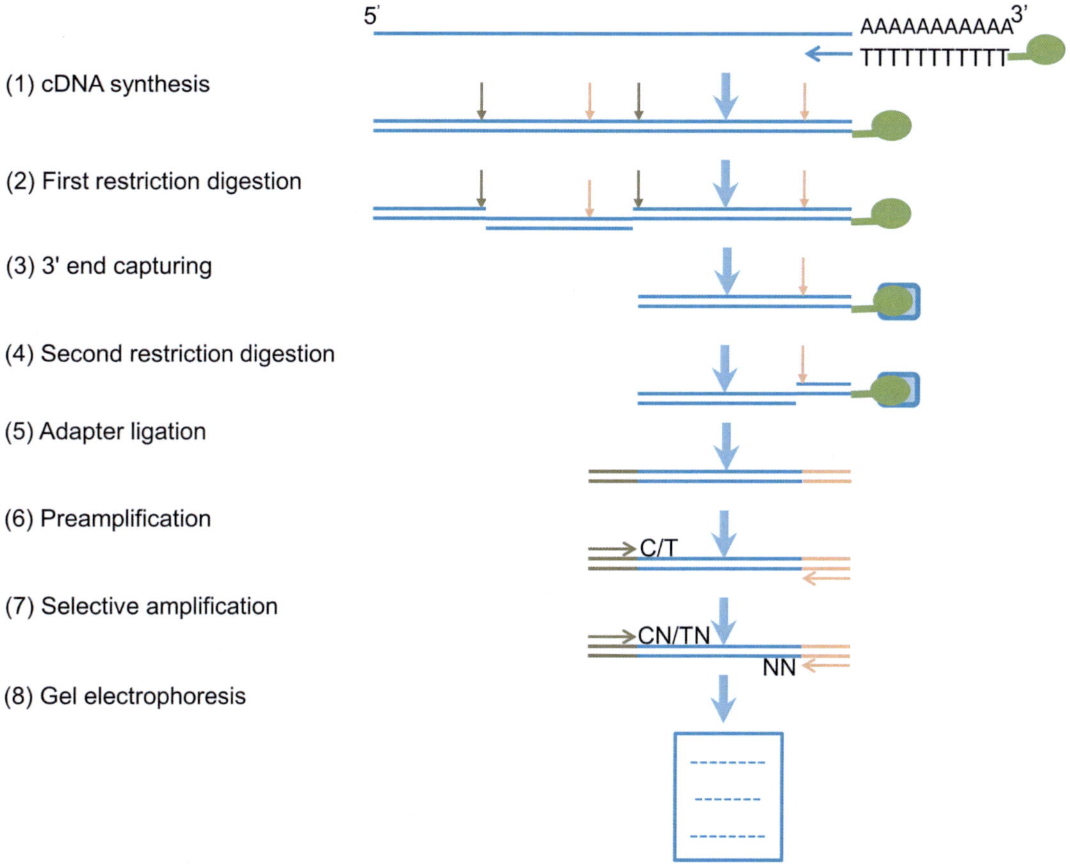

Fig. 5.1 Outline of the cDNA-AFLP procedure using the *Eco*RI/*Mse*I restriction enzyme combination. (1) mRNA is converted into double-stranded cDNA using a biotinylated (represented by a green circle) oligo-dT primer. Red and brown arrows represent *Eco*RI and *Mse*I restriction enzyme sites, respectively; (2) first digestion with *Eco*RI; (3) recovery of the 3' termini of the cDNA by binding biotin to streptavidin-coated beads, resulting in a single transcript-derived fragment (TDF) per transcript; (4) second digestion with *Mse*I; (5) ligation of the double-stranded *EcoR*I (brown)-and *Mse*I (orange)-specific adapters to the TDF ends to generate PCR templates; (6) reduction of the template mixture complexity by selective pre amplification of particular subsets of TDFs, using either the *Eco*RI+C or the *Eco*RI+ T primer in combination with an *Mse*I primer with no selective nucleotides; (7) final selective amplification of subsets of TDFs using *Eco*RI + T/C and *Mse*I primers, each with either one or two selective nucleotides (represented by N), with the *Eco*RI primer being labeled to allow subsequent detection of the TDF; and (8) electrophoretic size fractionation and display on denaturing polyacrylamide gels of the *Eco*RI/*Mse*I TDFs

hydrophobin, MOB2, profilin, and PHO84, were identified for the first time in *G. lucidum*. Also, there were thirteen potential SSR microsatellite loci which identified the ESTs of *G. lucidum*.

To support gene structure prediction, the Consortium of *Ganoderma lucidum* (CGL) initiated the expressed sequence tag (EST) project, which was a companion project of the *G. lucidum* genome project (Huang et al. 2013). In this study, a total of 47,285 ESTs were obtained from in vitro cultures of *G. lucidum* mycelium at various durations. These ESTs were merged into 7,774 non-redundant expressed loci. These expressed contigs were classified into three groups: over-representation, alternative splicing, and natural antisense transcripts. 262 ESTs contained alternative splice forms, among which 16 genes have skipped exon(s), and 191 genes contain retained intron(s). There were 101 pairs of antisense transcripts, of which 68 pairs are

Fig. 5.2 The workflow of Creator™ SMART™ cDNA Library Construction Kit

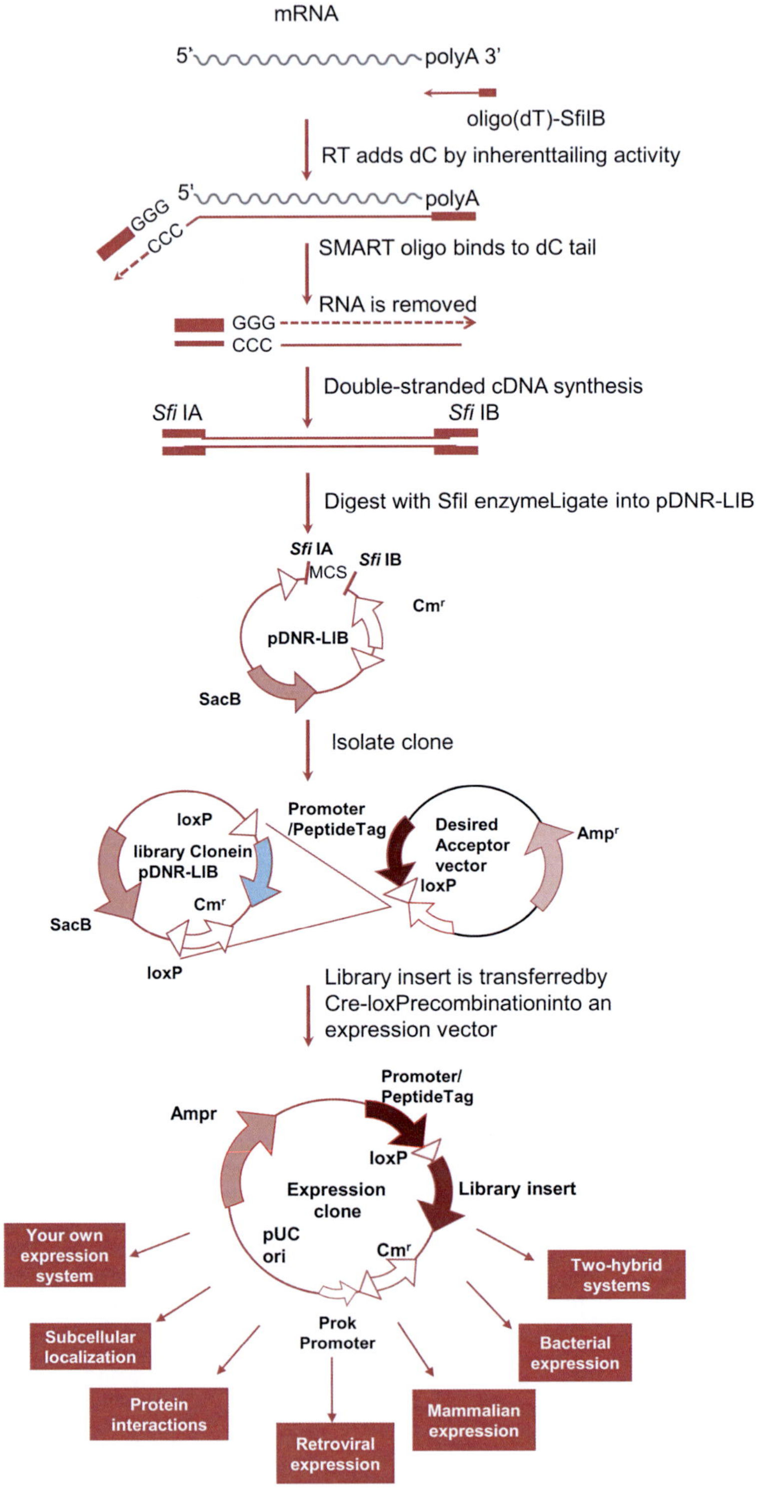

convergent, and seven pairs are divergent. For the other 28 pairs of antisense transcripts, it was found that one forward transcript fully covers the complement transcripts. Besides, 51 pairs of antisense transcripts consist of protein-coding genes and 35 pairs belong to the 3'-UTR type. Besides, 108 over-represented genes were supported by at least 50 ESTs, where significantly more age-specific ESTs supported 44 predicted genes. These results indicate that the over-represented genes in fast-growing dikaryotic mycelium are closely related to growth, such as bioactive compounds and cell wall synthesis. These results provide an invaluable resource for exploring the *G. lucidum* transcriptome and its regulation. A dedicated website has been set up at http://csb2.ym.edu.tw/est/.

5.2.3 RNA Sequencing

RNA-sequencing (RNA-seq) is a method that can examine the sequences and quantity of RNA in a sample using Next Generation Sequencing (NGS) technologies. It analyzes the transcriptome to identify gene expression patterns encoded within the target RNA, including mRNA, rRNA, and tRNA. RNA-seq is widely regarded as superior to other technologies, such as cDNA-AFLP, EST, and microarray hybridization. RNA-seq has several advantages: (1) Not limited to genomic sequences. RNA-seq can detect transcripts from organisms with previously undetermined genomic sequences. (2) Low background signal. The cDNA sequences from RNA-seq can be mapped to targeted regions on the genome, which will remove experimental noise. (3) More quantifiable. RNA-seq data is quantifiable for both low and high abundance genes. The RNA-seq process includes the following three steps: cDNA library preparation, cDNA sequencing, and RNA-Seq data analysis. RNA-seq is well established as the preferred method for transcriptome analysis nowadays.

5.3 The Workflow of RNA-Seq Data Analysis

The cDNA library construction and sequencing for Illumina RNA-Seq including the following steps. Briefly, mRNA was enriched from total RNA using Oligo(dT) magnetic beads and fragmented physically or chemically into small pieces. These fragmented sequences were used for the first-strand cDNA synthesis with reverse transcriptase and random hexamers primers, followed by the second strand cDNA synthesis with DNA polymerase I and RNase H. The double-strand cDNA then went through purification, end repair, the addition of a single "A" base and ligation of the adapters, and finally enriched by PCR amplication to be ready for sequencing. The cDNA libraries were sequenced on the Illumina sequencing platform. The original image process to sequences, base-calling, and quality value calculation were performed by the Illumina GA Pipeline, in which paired-end reads or single-end reads were obtained. In this section, we provide short descriptions of the RNA-seq data analysis pipeline in Fig. 5.3.

5.3.1 Quality Control of Reads

The analysis starts with raw sequence reads in FASTQ format. FASTQ format file stores sequences and Phred qualities in a single file. Phred quality scores Q are defined as a property that is logarithmically related to the base-calling error probabilities P:

$$Q = -10log_{10}P.$$

The Phred qualities were then converted to a character based on the ASCII coding scheme. A FASTQ file uses four lines per sequence. Line 1 begins with a '@' character followed by a sequence identifier and an optional description (like the title line in a FASTA file). Line 2 is the raw sequence letters. Line 3 begins with a '+'

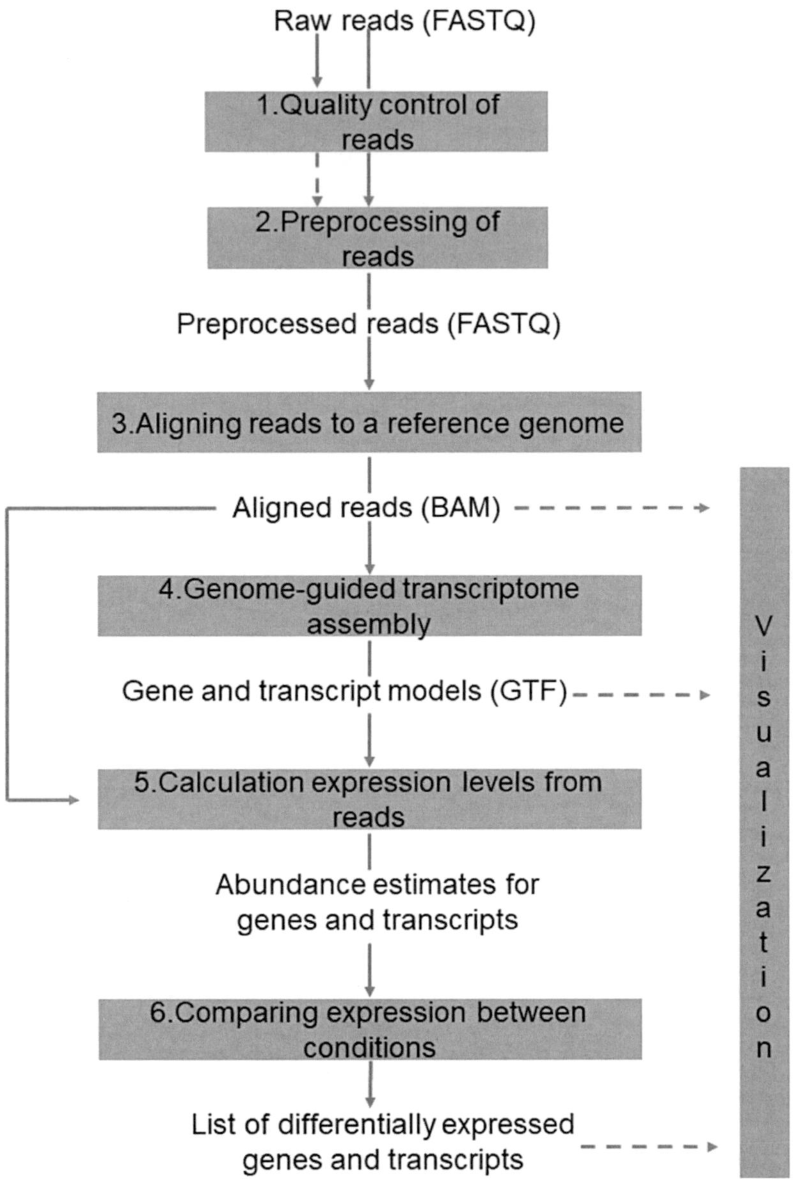

Fig. 5.3 The workflow of RNA-seq data analysis (reproduced from Korpelainen et al. 2014 with permission).The entire process contains six steps shown in gray shades. The input and output data are shown. The format for the corresponding data is shown in parenthesis

character and is optionally followed by the same sequence identifier (and any description) again. Line 4 encodes the quality values corresponding position-to-position to the nucleotides in Line 2.

FastQC (http://www.bioinformatics.babraham.ac.uk/projects/fastqc/) can analyze the overall quality of the raw data. FastQC provides a simple way to do quality control checks on raw sequence data from high throughput sequencing pipelines. It provides a modular set of analyses that users can use to give a quick impression of whether your data has any problems before further analyses. The main functions of FastQC are as follows: (1) Importing data from BAM, SAM, or FASTQ files, (2) providing a quick overview to show in which areas there may be problems, (3) providing summary graphs and tables to assess the data quickly, (4) exporting results to an

HTML based report, and (5) offline operation to allow automated generation of reports without running the interactive application.

The following command produces a quality report:

```
> fastqc reads.fastq
```

The file fastqc_report.html contains the analysis results. In addition to reporting several quality metrics, FastQC also gives a judgment on them as pass, warning, and fail, shown as traffic lights in the HTML report (Fig. 5.4). The FastQC report is composed of the following 12 sections: (1) Basic Statistics (Fig. 5.4a), (2) Per Base Sequence Quality (Fig. 5.4b), (3) Per Tile Sequence Quality (Fig. 5.4c), (4) Per Sequence Quality Scores (Fig. 5.4d), (5) Per Base Sequence Content (Fig. 5.4e), (6) Per Sequence GC Content (Fig. 5.4f), (7) Per Base N Content (Fig. 5.4g), (8) Sequence Length Distribution (Fig. 5.4h), (9) Duplicate Sequences (Fig. 5.4i), (10) Overrepresented Sequences (Fig. 5.4g), (11) Adapter Content (Fig. 5.4k), and (12) Kmer Content (Fig. 5.4l).

5.3.2 Processing of Reads

The preprocessing is to remove low-quality bases and artifacts such as adapter. Trimmomatic is a versatile Java-based tool for preprocessing reads (Bolger et al. 2014).

The following Trimmomatic command for paired-end reads (PE) trims bases from the 3′ end when the base quality is below 20 (TRAILING:20) and filters out reads which are shorter than 50 bases after trimming (MINLEN:50).

```
> java -jar trimmomatic-0.32.jar PE -phred64
    reads1.fastq.gz reads2.fastq.gz
    paired1.fq.gz unpaired1.fq.gz paired2.fq.gz
    unpaired2.fq.gzTRAILING:20MINLEN:50
```

The detail parameters are as follows:

ILLUMINACLIP: Cut adapter and other Illumina-specific sequences from the read.
SLIDINGWINDOW: Perform a sliding window trimming, cutting once the average quality within the window falls below a threshold.
LEADING: Cut bases off the start of a read, if below a threshold quality.
TRAILING: Cut bases off the end of a read, if below a threshold quality.
CROP: Cut the read to a specified length.
HEADCROP: Cut the specified number of bases from the start of the read.
MINLEN: Drop the read if it is below a specified length.
TOPHRED33: Convert quality scores to Phred-33..
TOPHRED64: Convert quality scores to Phred-64.

5.3.3 Mapping Reads to the Reference Genome

Due to the limitation of the sequencer's read length, the DNA molecular must be fragmented first in constructing the library, and the sequence obtained by the sequencer is only a partial sequence on the genome. To determine the positions of the reads on the genome, one needs to align the reads back to the reference genome. This step is called reads mapping to reference genome.

The following factors should be considered when performing read mapping: (1) Consumption of computer resources. In general, the larger the genome, the more memory it takes. For large genomes, such as the human genome, optimizing memory consumption is a key point; (2) Running speed. With the decline in sequencing prices and the demand for in-depth data mining, the throughput of sequencing increases, and the read mapping requires a fast enough speed; (3) Accuracy. For SNP/indel, sequencing error rate and other factors make the read mapping to the genome have several bp errors. The mapping algorithm must support base mismatches or the existence of gaps. Besides, a read might be

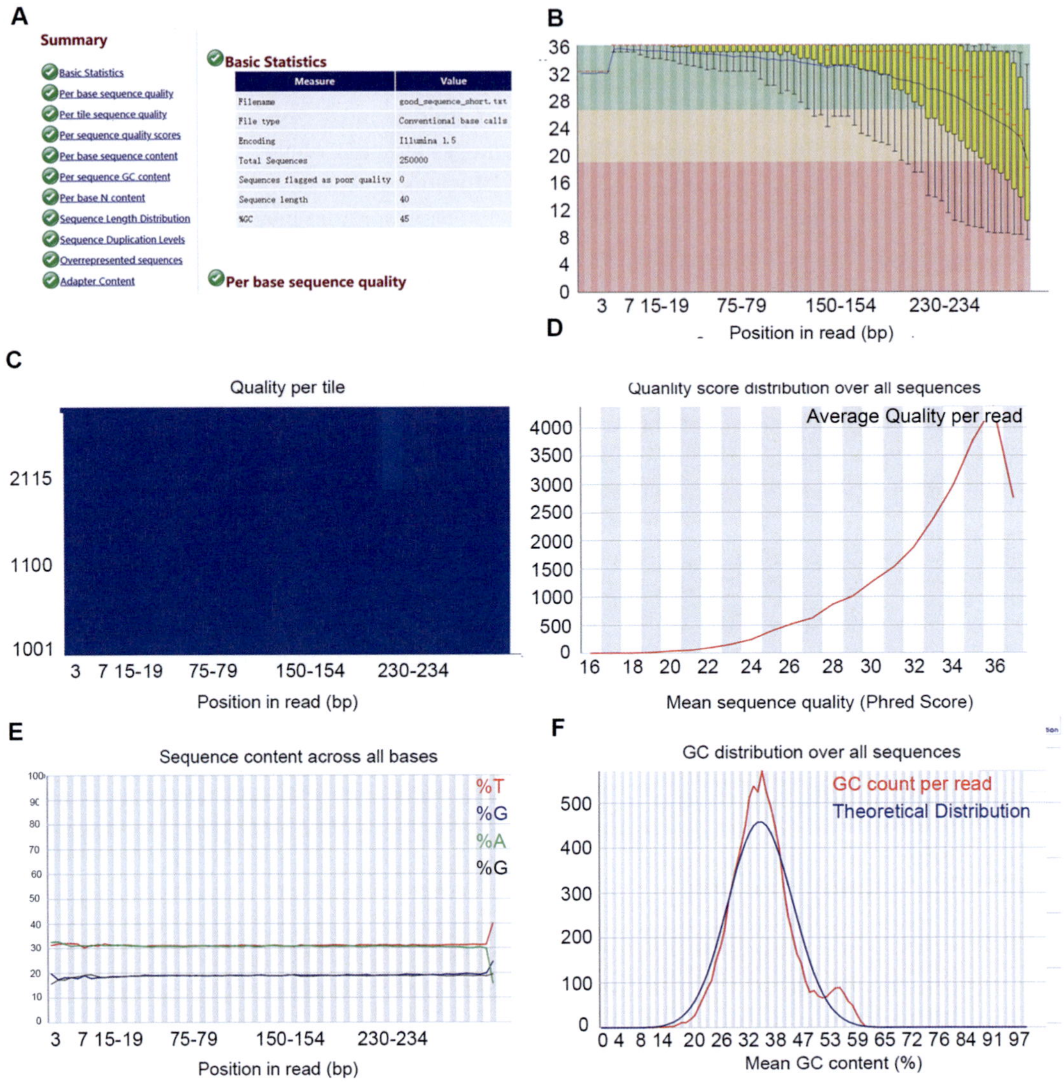

Fig. 5.4 The FastQC report. **a** The FastQC quality report's beginning offers basic statistics (right) and a judgment on the different quality aspects measured (left). **b** The overview of the range of quality values across all bases at each position. **c** The deviation from the average quality for each tile, **d** The overall sequence's quality score distribution. **e** Per base sequence content plot. **f** The GC content across the whole sequence length. **g** The percentage of N base calls at each position. **h** The distribution of fragment sizes overall sequences. **i** The relative number of sequences with different degrees of duplication. **j** The list of overrepresented sequences makes up more than 0.1% of the whole dataset. **k** The percentage of adapters. **l** The assumption that any small fragment of the sequence should not have a positional bias

aligned to multiple positions in the genome. As a result, different algorithms need to be compared to produce the mapping result's reliability; (4) The accuracy of exon–intron boundaries. Introns' existence in eukaryotes results in the mapping of cDNA fragments to the genome is not continuous. Placing spliced reads across introns and determining exon–intron boundaries is difficult because sequence signals at splice sites are limited, and introns can be thousands of bases long. At present, there are many mapping tools, such as BWA, HISAT2 (Kim et al. 2019), and

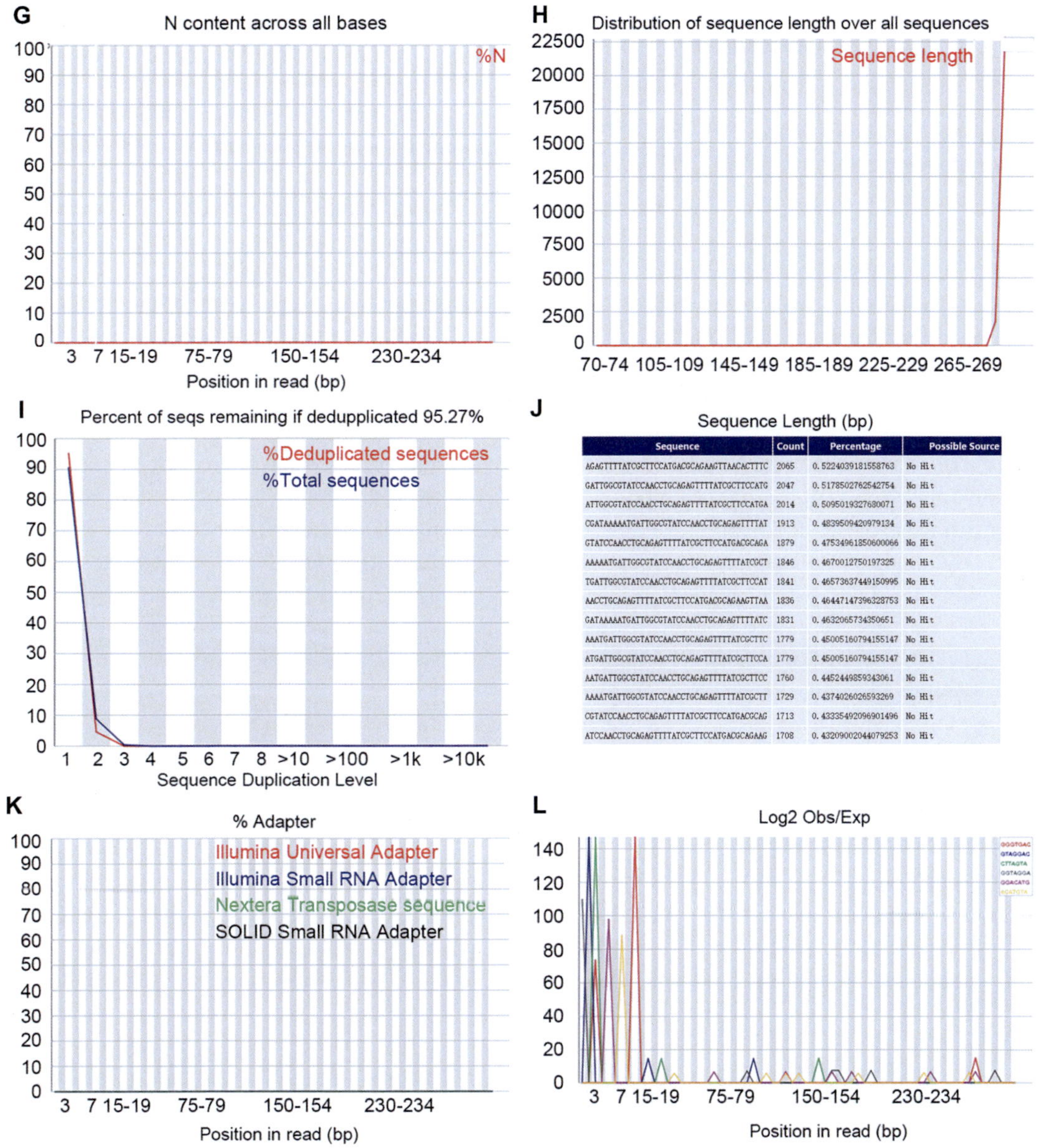

Fig. 5.4 (continued)

STAR (Dobin et al. 2013). HISAT2 is the fastest among them. It is an upgraded version of TopHat2 software using the improved FM index algorithm.

The following commands map RNA-seq reads to the reference genome using HISAT2. The alignments can be used for discovering new genes and splice variants using Cufflink for reference-based analysis.

1. Build index for the reference genome

   > hisat2-build -p 4 genome.fa genome

2. Align the RNA-Seq reads to the genome

   > hisat2 -p 4 -x genome - 1 R1.fastq
   -2 R2.fastq -S align.sam

5.3.4 New Transcript Prediction and Quantification

Reads are assembled into transcripts based on the location information of the reads on the reference genome using StringTie (Pertea et al. 2015), which applies a network flow algorithm and optional de novo assembly to assemble and quantitate transcripts representing alternative splice variants for each gene locus. Compared with software such as Cufflinks, StringTie uses a genome-guided transcriptome assembly approach along with concepts from de novo genome assembly to improve transcript assembly, which is more accurate and faster for isoform transcripts. Transcript assembly pipelines for StringTie are as following:

(1) The reads are assembled into "Super-reads." If there is a k-mer overlap between two reads, they are extended until no more bits can be extended in both directions and finally assembled into longer sequences called "Super-reads (SR)."
(2) Matching reads to the reference genome. The StringTie + SR method uses mixed reads, i.e., super-reads and unassembled-reads, while StringTie uses all reads obtained from sequencing.
(3) The mapped reads are clustered, and each cluster of reads corresponds to a variable shear map. Each variable shear map represents all possible transcript isoforms of a gene.
(4) Identify the path with the highest number of reads covered from the variable shear graph and construct a stream network for that path.
(5) Use the network flow algorithm to assign reads to transcripts to cover the maximum number of reads.
(6) Remove the reads used by the clipping graph in step (5) and iterate (4)–(5) until there are no paths left to follow.

The spliced transcripts obtained are then compared with the annotation information using GffCompare. And new transcripts or new genes are found from them after excluding transcripts containing only a single exon.

5.3.5 Identification of Differential Expression Genes (DEGs)

The gene expression level obtained from RNA-seq data should be normalized to enable comparisons between samples and genes. RPKM (FPKM) stands for Reads (Fragments) Per Kilobase per Million (mapped) reads, and it corrects the raw counts both to the gene or transcript length and sequencing depth, which was designed to enable comparisons of the same gene's expression levels across samples or of different genes' expression levels in the same sample (Mortazavi et al. 2008).

Cuffdiff (Trapnell et al. 2012) and DESeq2 (Love et al. 2014) are two popular programs for differential expression analysis, representing two typical workflows in differential expression analysis. The Cuffdiff program is part of the popular Cufflinks package for assembly, quantification, and differential expression analysis. It can assess differential expression on the gene and isoform transcript levels simultaneously. Cuffdiff does not support more complex experimental designs than DESeq2, a differential expression analysis package in R's BioConductor project. The GLMs used in DESeq2 assume that the read counts are distributed according to the negative binomial distribution. By contrast, DESeq2 did not consider isoforms, which will lead to bias in differential expression analysis.

There were many ways to visualize the results of differential expression analysis, such as MA plot, volcano plot, and Cluster analysis.

The MA plot visualizes the overall associations of gene expression levels and the degree of differential expression (Fig. 5.5a). The vertical coordinate indicates the M value, representing the logarithm of the difference in the fold change unit ($\log_2$Fold Change). The horizontal coordinate indicates the A value, representing the mean of normalized counts of gene expression in the two samples; the red points represent genes with FDR less than 0.05.

The volcano plot displays the overall distribution of DEGs in terms of fold changes and p-value measuring the fold change's significance between the two sets of samples (Fig. 5.5b). The horizontal coordinate indicates the gene

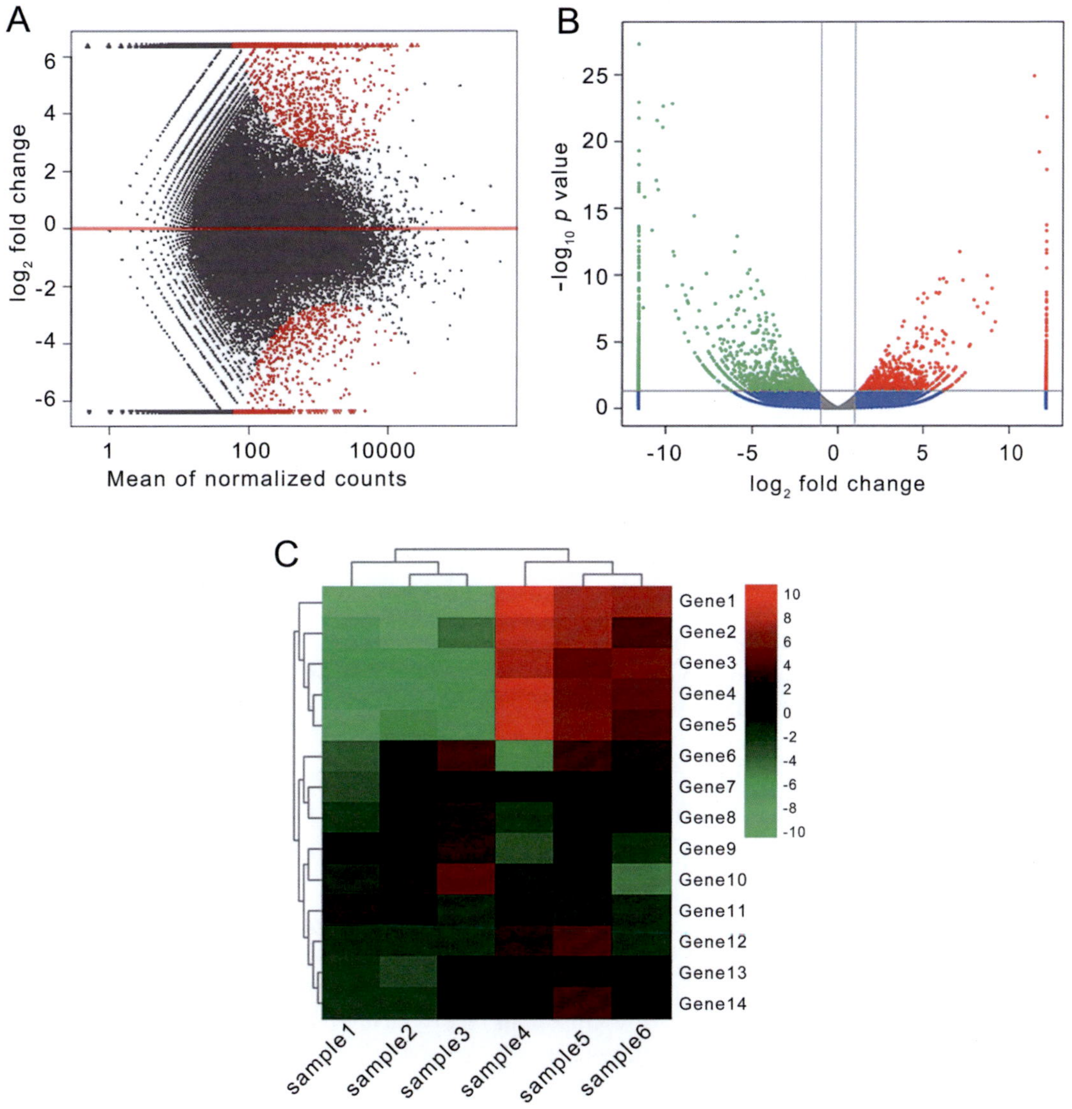

Fig. 5.5 The visualization of differential expression analysis results. **a** MA plot. The Y-axis is the M value, representing the difference between the expression levels of the same gene in the experiment and control group, respectively. The X-axis is the A value, representing the mean of gene expression levels in the experimental and control groups. The red points represent genes with an FDR value less than 0.05. **b** Volcano map. The X-axis is the difference in gene expression levels represented by fold change. The Y-axis represents the significance level of DEGs. Red dots represent upregulated DEGs, green dots represent down-regulated DEGs, and blue dots represent non-differentially expressed genes. **c** Heat map for DEGs. Horizontal coordinates indicate samples and the hierarchical clustering results based on the expression profile of all DEGs in each sample, and vertical coordinates indicate DEGs and the hierarchical clustering results based on the expression profile of each DEG across all samples. Red indicates higher expression levels compared with the mean expression level. Green indicates lower expression

expression fold change ($\log_2$ Fold Change), and the vertical coordinate indicates the significance level of the differential expression ($-\log_{10}$ *p*-value). Red indicates significant up-regulation of gene expression compared to the mean gene expression levels, while green indicates significant down-regulation. And black indicates no significant difference in gene expression.

Cluster analysis is used to determine the expression patterns of DEGs under different experimental conditions. Genes with similar expression patterns may have similar functions or be involved in the same biological process. Therefore, clustering analysis can be used to infer the functions of unknown genes or new functions of known genes. DEGs' FPKM values were extracted for hierarchical clustering analysis. The clustering of the samples are based on the expression profiles of all DEGs in each sample. And the clustering of the DEGs are based on the expression profiles of each DEG across all the samples. The color of each cell indicates whether or not the expression level is higher or lower compared with the mean expression levels of all DEGs across all the samples (Fig. 5.5c).

5.3.6 Enrichment Analysis of DEGs

5.3.6.1 KEGG Pathway Annotation of DEGs

DEG products in an organism perform biological functions through interactions. And pathway annotation analysis of DEGs helps to study the gene functions further. The Kyoto Encyclopedia of Genes and Genomes (KEGG, https://www.genome.jp/kegg) is a comprehensive database that integrates genomic, biological pathway, disease, drug, and chemical information (Kanehisa et al. 2007). KEGG organically combines genomic information with high-level functional details to provide a systematic analysis of data generated by genome sequencing and other high-throughput experimental techniques. The sequence of cDNA or protein can be matched to the KEGG database using sequencing comparison tools like BLAST (Altschul et al. 1990; Camacho et al. 2009). The pathway id (KO number) is assigned to each gene. This step is also called pathway annotation. After annotating genes to the KEGG database, the number of DEGs contained in each KEGG pathway was counted and plotted in a bar chart (Fig. 5.6). The "X" axis is the percentage of total genes mapped to each KEGG pathway; the "Y" axis is the corresponding KEGG pathway. The numbers to the right of each bar represent the number of genes mapped to the pathway.

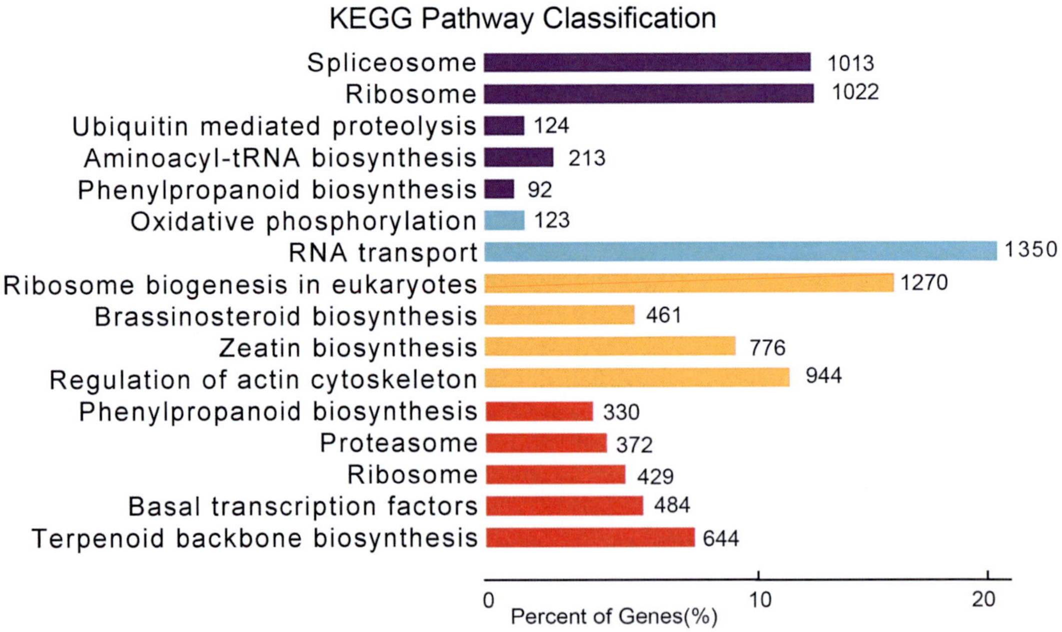

Fig. 5.6 KEGG annotation of all unique sequences. Purple is Cellular Processes; Blue is Environmental Information Processing; Orange is Genetic Information Processing; D: Red is Metabolism. The "X" axis is the percentage of DEGs to total genes mapped to the corresponding KEGG pathway; the "Y" axis is the KEGG pathway. The numbers to the right of each bar represent the numbers of DEGs mapped to the pathway

5.3.6.2 KEGG Enrichment Analysis of DEGs

The KEGG enrichment analysis was based on hypergeometric distribution and corrected for multiple hypothesis testing probabilities (p-value) using the Benjamini–Hochberg method. The screening condition for enrichment results was that the corrected p-value was less than 0.05. The formula for the hypergeometric distribution is shown below.

$$P = 1 - \sum_{i-0}^{m-1} \frac{\binom{M}{i}\binom{N-M}{n-i}}{\binom{N}{n}},$$

where N represents the number of all genes in a genome, n represents the number of all DEGs, M represents the number of all genes that is annotated to a KEGG pathway, and m represents the number of DEGs that is annotated to a KEGG pathway.

The scatter plot of KEGG pathway enrichment results can be used to visualized the degree of KEGG pathway enrichment (Fig. 5.7). The vertical coordinate represents the KEGG pathway. The horizontal coordinate represents the enrichment score, which is the ratio of the number of DEGs annotated to a KEGG pathway to the total number of genes annotated to the same KEGG pathway. The size of the dot represents the number of genes enriched in the pathway. The larger the number of dots, the greater the enrichment. The dot's color represents the significance of the pathway's enrichment, as assessed by the corrected p-value. The closer the p-value to 0, the redder the color, the more significant the enrichment is.

After annotating DEGs to the KEGG pathway, the up- and down-regulation genes are marked in the pathway map for visualization (Fig. 5.8). The nodes containing upregulated DEGs are marked in red, the nodes containing only down-regulated DEGs are marked in green, and the nodes containing both up- and down-regulated genes are marked in yellow.

5.3.6.3 GO Term Annotation of DEGs

Gene Ontology (GO, https://geneontology.org) is an ontology system established by the Gene Ontology Consortium. The system is used to describe gene, gene product, and their functions in a three-class structure, including molecular function, biological process, and cellular component (Botstein et al. 2000; Consortium GO 2017). The query gene or protein is compared with entries in the Swiss-Prot database using BLAST (Wu and Chen 2011; Consortium U 2018), and then the GO terms of the best hit sequences were transferred to the query sequences. This process of finding the GO terms for query sequence is called GO term annotation. After GO annotation, the number of DEGs contained in the GO terms was counted.

GO term annotation can be performed at different tiers. Usually, GO term annotation is performed to at least the second level. For example, the number of DEGs in GO terms' secondary levels was counted and then plotted in a bar graph (Fig. 5.9). The "X" axis shows the GO terms of genes. The "Y" axis shows the number of DEGs and its percentage of all genes annotated to the same GO terms.

5.3.6.4 GO Term Enrichment Analysis of DEGs

Enrichment analysis identifies GO terms that are significantly enriched in DEGs compared to the genomic background, thus identifying DEGs that are significantly associated with biological functions. The enrichment of GO terms can be performed in the same way as that for KEGG pathways. The GO enrichment analysis has been implemented in the R language package clusterProfiler (Yu et al. 2012a).

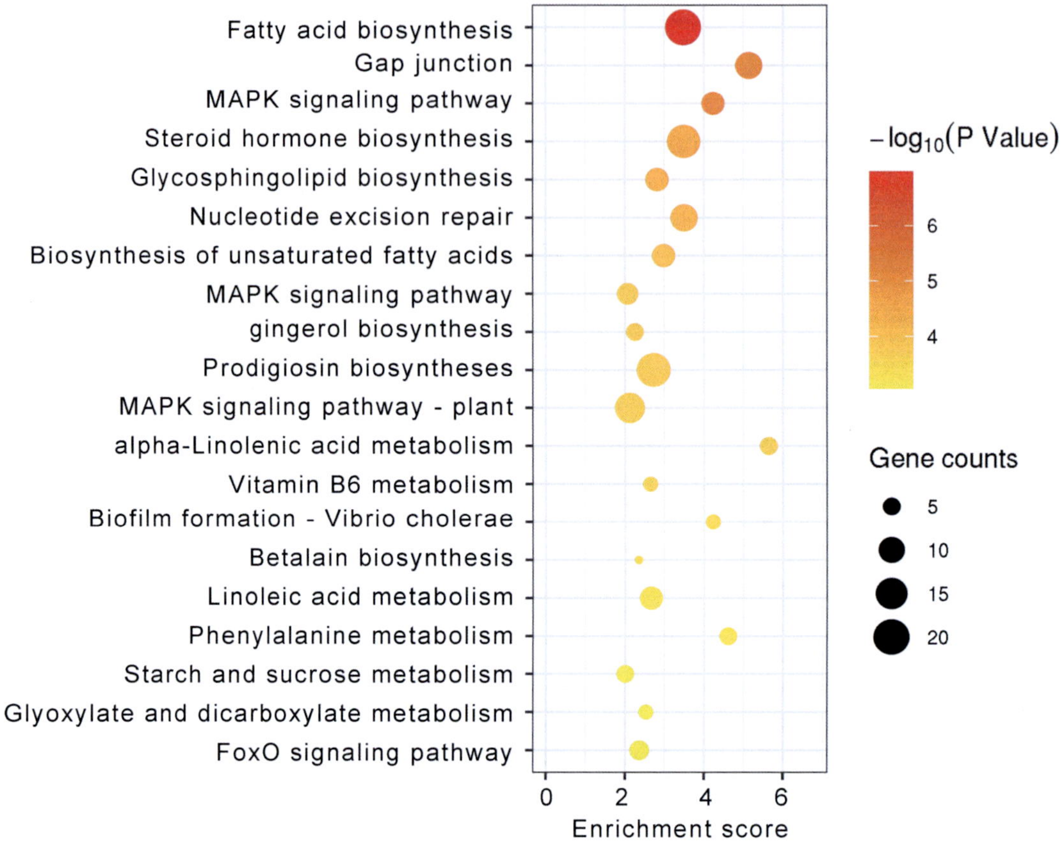

Fig. 5.7 KEGG enrichment of DEGs. The "X" axis is the number of DEGs mapped to each KEGG pathway. The "Y" axis shows the KEGG pathways. The color of the circle reflects the P-value for the enrichment. The size of the circle reflects the count of DEGs

The scatter plot can be used to visualize the enrichment level of GO entries (Fig. 5.10). The vertical coordinate represents the GO terms. The horizontal coordinate represents the Gene Radio, which is the ratio of the number of DEGs annotated to a GO term to the total number of genes annotated to the same GO term. The larger the Gene Radio, the greater the enrichment. The dot's size represents the number of genes enriched for the GO term. The larger the dots, the more significant the enrichment. The dot's color represents the significance of the GO term enrichment, as assessed by the *p*-value. The closer the *p*-value to 0, the redder the color, the more significant the enrichment.

5.3.7 Alternative Splicing Analysis

Alternative Splicing (AS) is when precursor mRNAs excise introns and join different exons to produce multiple mature mRNA molecules by selecting various splice sites. Variable splicing is widespread in eukaryotes and is an essential mechanism for regulating gene expression and generating proteomic diversity.

AS analysis of transcriptomic data was implemented using rMATS (Shen et al. 2014). rMATS quantifies the expression of variable splicing events in different biological replicate samples. It calculates the *p*-value using the likelihood-ratio test to indicate the difference between various variable splicing events in two

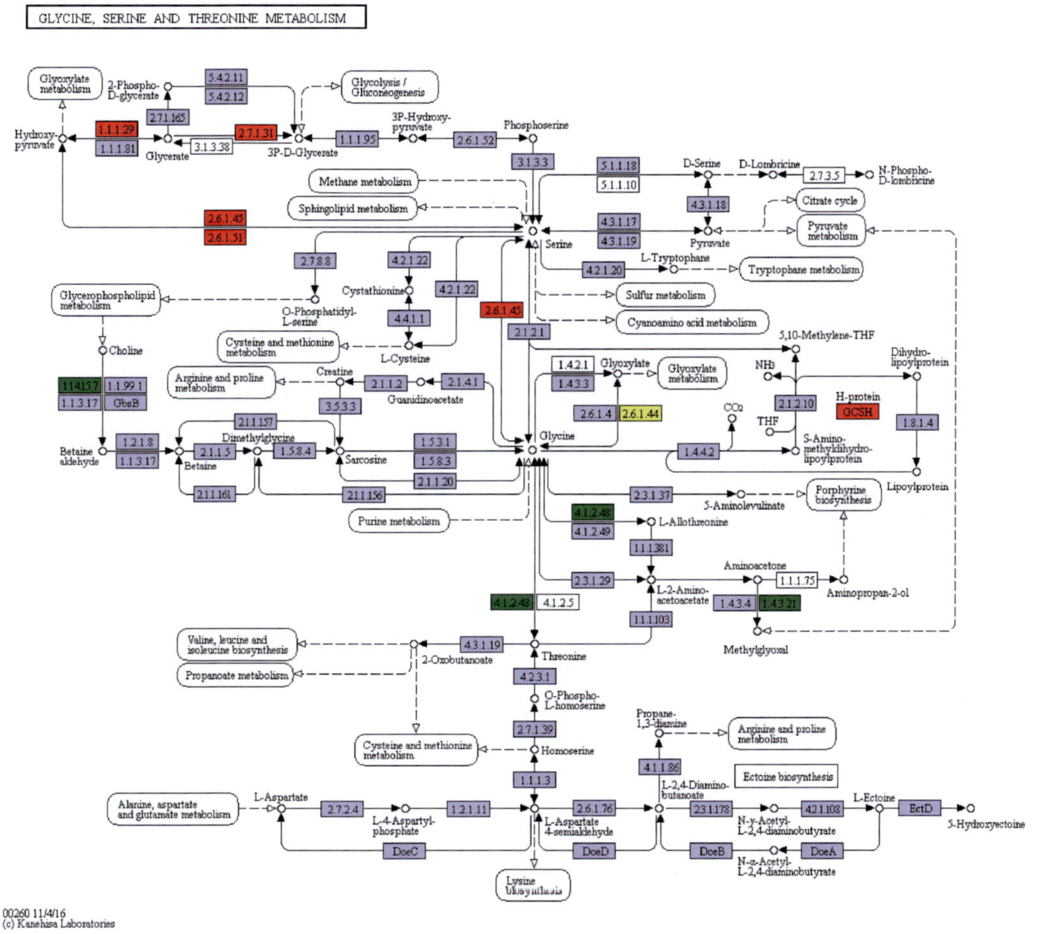

Fig. 5.8 KEGG pathways for Glycine, Serine and Threonine metablisms are shown here as an example. Individual KEGG pathway is labeled in the curved square. The enzymes that catalyze particular reactions are shown in squares. Red indicates that the enzyme's expression level was upregulated in the study. Green indicates that the enzyme's expression level was down-regulated in the study. Yellow means that the enzyme was either upregulated or downregulated in the study

groups of samples. It then uses the variable splicing events with FDR less than 0.05 were considered as differential variable splicing.

The rMATS can identify the following five variable splicing events (Fig. 5.11):

(1) Exon skipping (Skipped Exon, SE)
(2) 5'-end variable splice (Alternative 5' Splice Site, A5SS)
(3) 3'-end variable splicing (Alternative 3' Splice Site, A3SS)
(4) Mutually Exclusive Exons (MXE)
(5) Intron retention (retained intron, RI).

5.3.8 Single Nucleotide Polymorphism and Insertion-Deletion Analysis

Single nucleotide polymorphism (SNP) refers to a genetic marker formed by a single nucleotide variation on the genome. The SNPs are important markers to study the relationship between biological phenotypes and genotypes. Insertion-deletion (InDel) refers to the insertion or deletion of nucleotide fragments of different sizes at the same

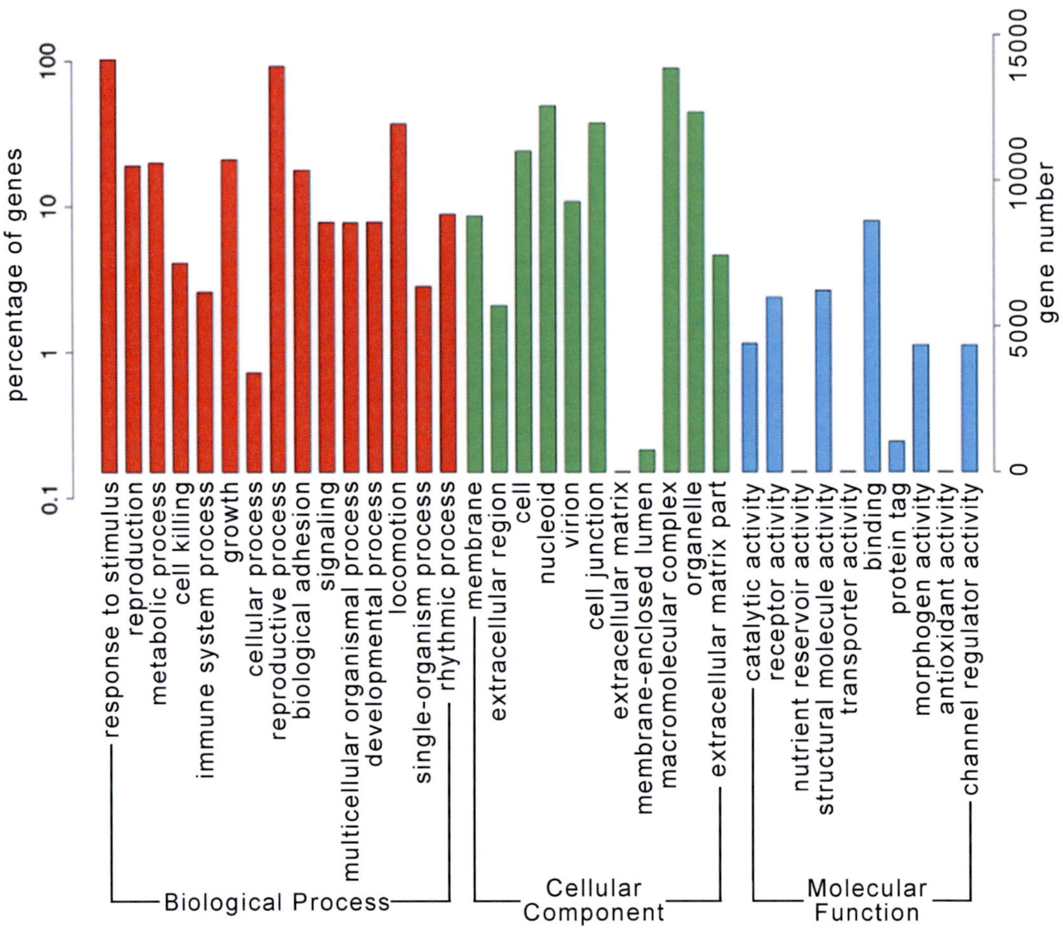

Fig. 5.9 GO annotation of the unigene sequences. Annotated unique sequences were classified into "Biological process," "Cellular component," and "Molecular function," respectively. The "X" axis shows the GO terms. The "Y" axis shows the percentage and number of DEGs mapped to particular GO terms

locus of the genome between individuals of closely related species or the same species. InDel markers have been widely used in genetic analysis of plant and animal populations and molecular marker-assisted breeding because of their stability, high polymorphism, and simple typing system. In general, SNPs are single nucleotide variants with variation frequency greater than 1%, and the length of InDel is mostly within 50 bp.

Since some mRNAs undergo RNA editing (RNA editing), i.e., the transcribed RNA undergoes base insertions, deletions, or substitutions in the coding region to produce polymorphic gene expression products. It is difficult to determine whether a polymorphic site results from RNA editing or transcription of a polymorphic site. Therefore, SNPs and InDel detected from transcriptome sequencing data inevitably contain the products of RNA editing. Detection of SNPs and InDel was implemented in GATK (McKenna et al. 2010), and annotation was implemented in ANNOVAR (Wang et al. 2010; DePristo et al. 2011). The workflow of SNP and InDel detection is shown in Fig. 5.12.

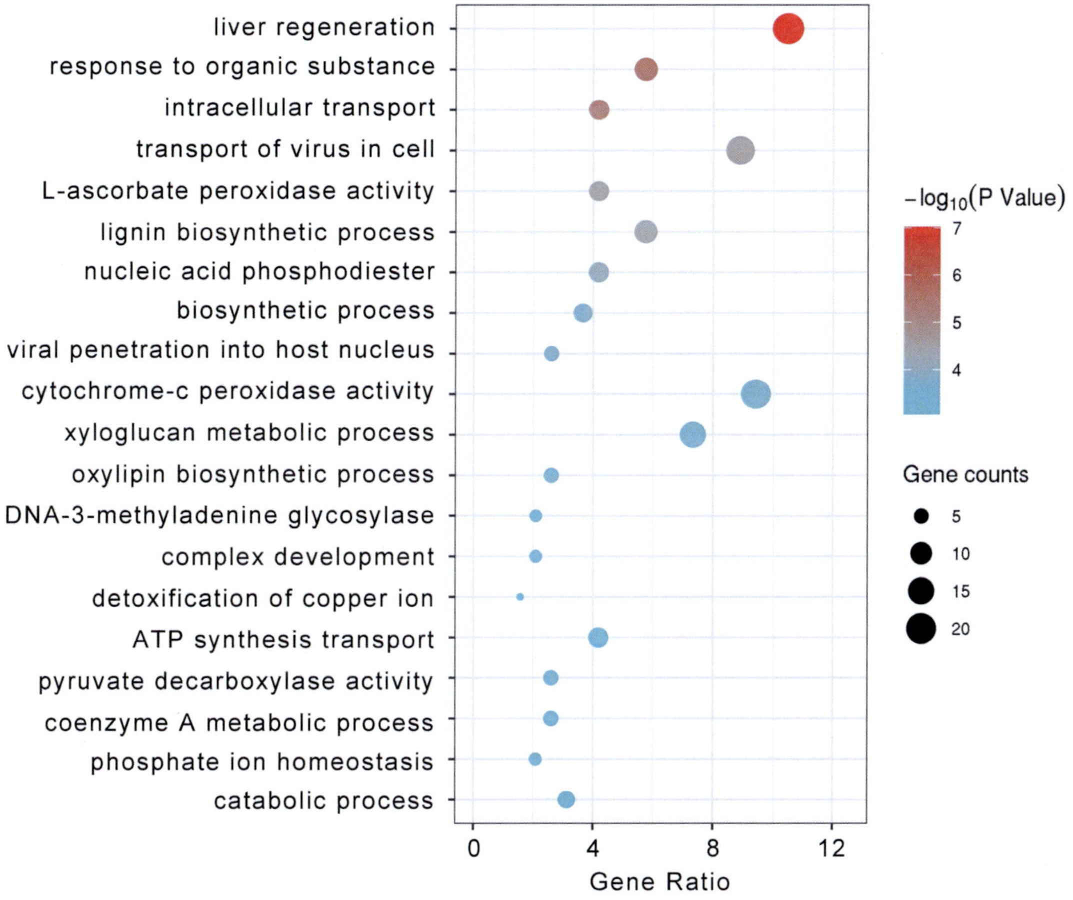

Fig. 5.10 Scatter diagram of GO enrichment results. The x-axis is gene counts, and the y-axis is GO Term. The circle's size indicates the number of genes, and the circle's color represents the p-value for the significance of the enrichment

5.4 Transcriptome Analysis of Genus *Ganoderma*

Several studies reported the discovery of key genes for active compounds' biosynthesis using cDNA-AFLP, EST, and RNA-seq technologies in *Ganoderma*. Also, environmental adaptability is another important research direction of *G. lucidum* using transcriptome analysis. Table 5.1 summarizes the advance in transcriptomic studies of *G. lucidum*. Below, we describe these studies in detail.

5.4.1 Responses to Environmental Factors

Increasing land has been polluted by toxic metals such as copper (Cu), mercury (Hg), and cadmium (Cd). Exposure to heavy metals results in an increased production of reactive oxygen species (ROS), which leads to unspecific oxidation of proteins and membrane lipids or causes DNA injury. The following studies intend to determine the effects of environmental factors on the gene expression of *Ganoderma* species.

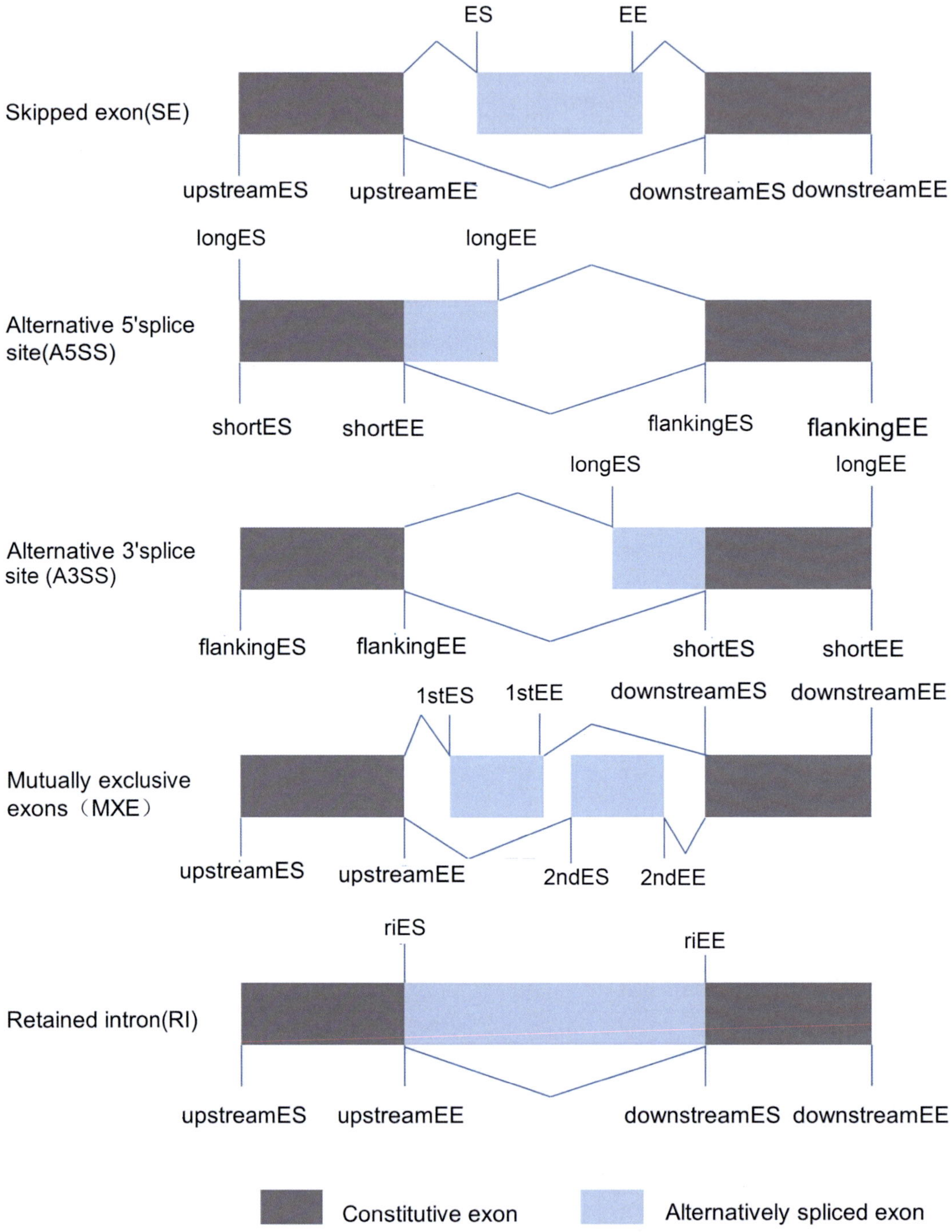

Fig. 5.11 Types of alternative splice events recognized by rMATS.**ES: ex**on start site, EE: exon end site.

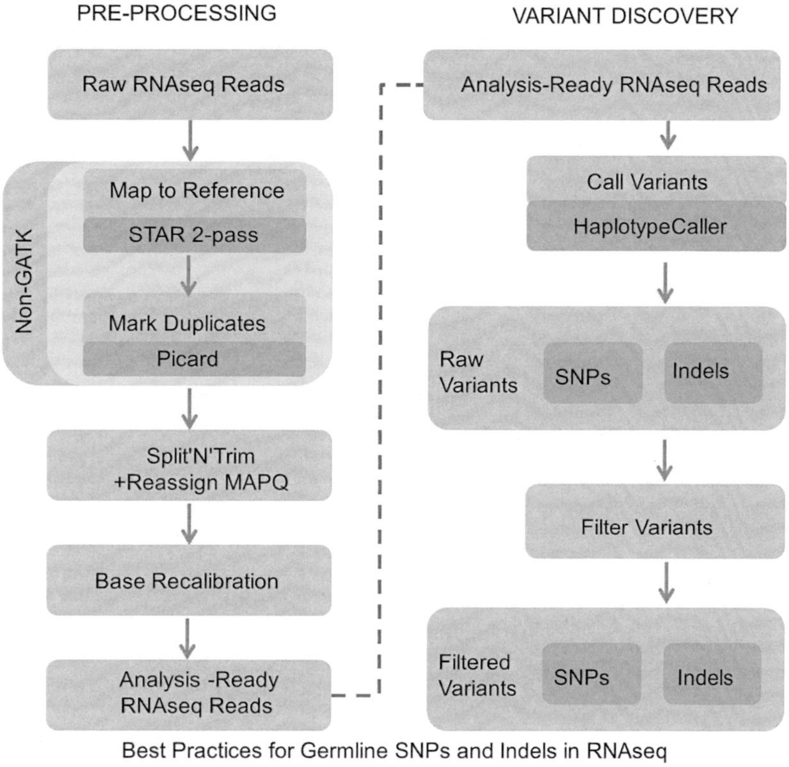

Fig. 5.12 Schematic diagram of the SNP and InDel detection process. The part on the left is preprocessing the data. First, the sequence reads were mapped to the reference genome, and the multiple mapped reads were marked. Afterward, the "N" is removed, and the mapping is recalibrated for further analysis. The right part is the discovery of variants. SNPs and InDel are detected using GATK. The raw variant data are filtered to obtain the final variant information

Cd

Cd is one of the most accumulated heavy metals in the mycelia of *Ganoderma* species (Gabriel et al. 1994). Cd enters the cells via membrane translocators that transfer essential ion elements for cellular necessity. A well-known Cd detoxification mechanism is the binding of Cd to a cysteine-rich protein such as glutathione, phytochelatins, and metallothioneins (Mehra and Winge 1991). The cDNA-AFLP method was performed to analyze the transcription profiling of *Ganoderma* species in response to Cd treatment. A total of 12,925 transcript-derived fragments (TDFs) were amplified using 256 primer combinations. Among them, forty-nine differentially expressed TDFs were validated by DNA dot-blot analysis. Then the expression levels of 34 Cd-inducible TDFs were verified using Northern blot analysis. Sequence analysis showed that upregulated genes are involved in several processes, including reactive oxygen species generation, synthesis of sulfur-containing metabolites, DNA repair, translation machinery transporting system, proteolysis pathway, and cell wall biosynthesis. These results provide an overall picture of gene expression profile in response to Cd in *Ganoderma* species.

Copper

Copper is another essential abiotic factor during the growth of fungi. Some extracellular enzymes such as laccases, lignin peroxidase, and manganese peroxidase are copper-containing. They are induced in the presence of copper in *G. lucidum* (Altschul et al. 1990; Camacho et al. 2009). This extracellular enzyme uses molecular oxygen as an electron acceptor to catalyze the oxidation of phenolic compounds and aromatic amines, contributing to the utilization of biomass, the bleaching of fibers, and the degradation of organic compounds (Altschul et al. 1990; Camacho et al. 2009). Cu^{2+} also affects the induction of wood-degrading enzymes, which stabilize laccases and are also responsible for the induction of species-specific laccase isoforms in *Ganoderma* (Altschul et al. 1990).

Table 5.1 Advance in transcriptomic studies of *G. lucidum*

BioProject	Tissue	Sequencing platform	Read length (bp)	Raw data (bp)	Assembly software	Assembly result	Reference
NA	*G. lucidum* fruiting body	Applied Biosystems BigDye Terminator v3.1	150–1500	NA	Phrap	879 high-quality ESTs	Luo et al. (2010)
NA	*G. lucidum* mycelium and fruiting body	Illumina HiSeq™ 2000	90	1,157,072,400	SOAP de novo and TGICL	18,892 and 27,408 unigenes	Yu et al. (2012a)
DDBJ: HO710205–HO757489	*G. lucidum* mycelium after cultured at 30 °C for 5, 14, 18, and 30 days	Applied Biosystems BigDye Terminator v3.1, Applied Biosystems 3730xl DNA analyzer and GE Healthcare MegaBACE 1000	50–949	NA	ClusterMerge	47,285 ESTs	Huang et al. (2013)
PRJNA376106	*G. oregonense* mycelia treated at 28 °C, 32 °C, and at 28 °C or 32 °C with an NO donor	Illumina HiSeq 2000	150	39G	Trinity	186,159 contigs, 110,631 transcripts, and 58,550 unigenes	Chen et al. (2017)
PRJNA419168	Control and heat-shock-treated samples	HiSeq™ 2000	150	50G	Trinity	99,899 transcripts and 59,136 unigenes	Tan et al. (2018)
PRJNA406843	Mycelium has fully grown, primordium, young fruiting body, mature fruiting body, massive spore production	Illumina HiSeq 3000	150	50G	NA	NA	Zhou et al. (2018)
PRJNA514399	*G. boninense* at the axenic condition and planta condition	Illumina HiSeq1000	100	69	NA	NA	Wong et al. (2019)
SRP127479	Fungal mycelia (three flasks were pooled from both control and copper-induced)	Illumina Hi Seq 2500	100	6.0G	Trinity	48,693 transcripts	Jain et al. (2020)

(continued)

Table 5.1 (continued)

BioProject	Tissue	Sequencing platform	Read length (bp)	Raw data (bp)	Assembly software	Assembly result	Reference
PRJNA643653 SUB7707164	*G. lucidum* in liquid superficial-static culture and submerged culture	Illumina HiSeq 2000	150	203G	NA	NA	Wang et al. (2020)

NA not available

To reveal the influence of copper ions on the gene expression pattern in *G. lucidum*, researchers compared RNA-Seq analysis results between Cu^{2+} treatment and control samples. The transcriptome was analyzed using the Illumina sequencing platform, followed by *de novo* transcriptome assembly (Jain et al. 2020). A total of 26,083,372 and 35,713,076 high-quality reads were obtained from induced and uninduced cultures, respectively. The authors predicted 194 transcripts coding for oxidoreductases and 402 transcripts coding for CAZymes. Further, analyses of the secretome revealed the enrichment of GO terms related to plant cell wall degrading. These enzymes were presumed to help *Ganoderma* in survival and colonization. The increased Cu^{2+} concentrations are also associated with higher secretion of lignocelluloses and increased phenolics and antioxidants production. The differences in the transcriptomic and proteomic signatures for lignocellulolytic enzymes provide vital clues about Cu^{2+} mediated gene regulation. The results suggest that the presence of copper may have triggered apoptosis. This, in turn, leads to increased laccase production, which might be a self-defense mechanism of the fungus to survive from oxidative burst caused by copper (Jain et al. 2020) (Fig. 5.13).

Heat Stress

Zhang et al. (2016a) reported that 42 °C heat stress (HS) inhibited mycelium growth, reduced hyphal branching, induced the accumulation of heat shock proteins (HSPs), and increased ganoderic acid (GA) biosynthesis in *G. lucidum*. Further evidence showed that Ca^{2+} might be a factor in the HS-mediated regulation of hyphal branching, GA biosynthesis, and the accumulation of HSPs. The results suggest that the cytosolic Ca^{2+} participates in HS signal transduction and regulates downstream events in *G. lucidum* (Fig. 5.14). HS enhanced the polysaccharide production after treatment of 42 °C for 2 h, resulting in the highest polysaccharide yield of 10.50%, 45.63% higher than the control. In contrast, 37 and 45 °C HS had no significant effect on the production (Zhang et al. 2016a). Moreover, the in-vitro antioxidant activity of heat-treated polysaccharides was higher than untreated ones. Taken together, HS is supposed to influence the growth of *G. lucidum* (Zhang et al. 2016a) significantly.

To further study the genes involved in the HS response and to explore the mechanism, Tan et al. (2018) used the RNA-Seq method to analyze the differences between control and heat-treated mycelium. Six cDNA libraries were obtained, of which three were from the control group (mycelium cultured at 28 °C), and three were from the heat treatment group (mycelium cultured at 42 °C for 2 h). A total of 99,899 transcripts were obtained using the Trinity de novo assembly method. Among them, 59,136 single genes were annotated in 7 public databases. Two thousand seven hundred ninety genes were identified as DEGs. In particular, 1991 were upregulated, and 799 were down-regulated.

According to their putative functions and possible metabolic pathways, 176 DEGs were divided into five main categories. These categories include stress resistance-related factors; protein assembly, transportation, and degradation; signal transduction; carbohydrate

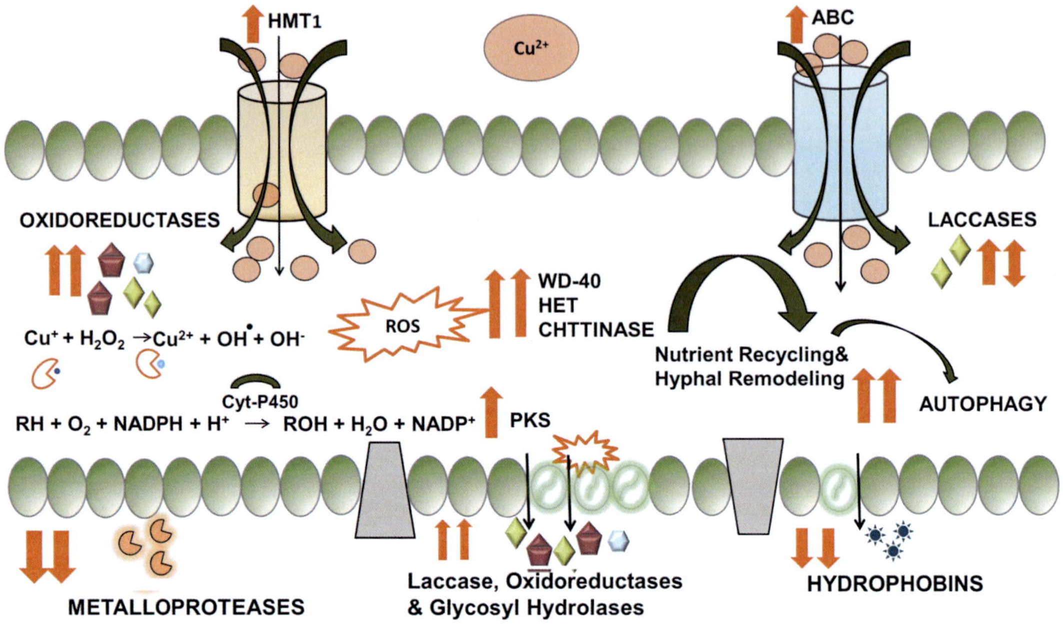

Fig. 5.13 A model representing the cellular, molecular responses to copper. It shows that NADPH-oxidase plays a role in secondary metabolite biosynthesis by producing ROS, which eventually affects the Ca^{2+} signaling pathway (reproduced from Jain et al. 2020 with permission)

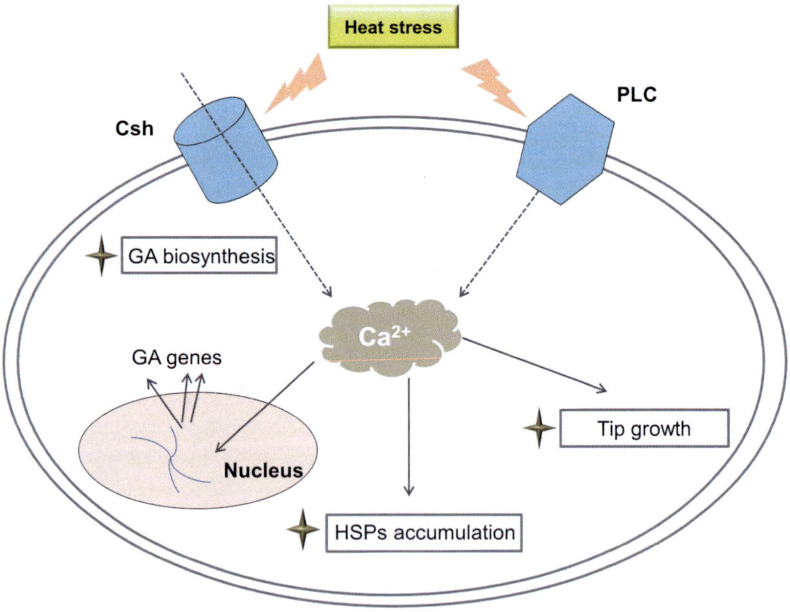

Fig. 5.14 Schematic representation showing that HS regulates HSP expression, hyphal branching, and GA biosynthesis via cytosolic Ca^{2+} in *G. lucidum*. HS-induced cytosolic Ca^{2+} regulates GA biosynthesis, HSP accumulation, and hyphal branching. The solid black arrows indicate data supported by experiments, and the dotted arrows indicate data obtained from other fungal systems (reproduced from Zhang et al. 2016 with permission)

metabolism and energy provision-related processes; and other related functions. The results suggest that HS activates a series of metabolic pathways in *G. lucidum*, and the reaction mechanism involves a complex molecular network that needs further study. It is worth noting that 48 DEGs have been found to regulate carbohydrate metabolism, both in carbohydrate hydrolysis to provide energy and polysaccharide synthesis. In summary, this comprehensive transcriptome analysis provides a wealth of information on the molecular responses to HS conditions.

Nitric Oxide

Nitric oxide (NO) is a simple, small gas molecule that can easily diffuse through cell membranes, playing an important role in alleviating fungal HS (Chen et al. 2017). NO also plays a crucial role in preventing the mycelia of *G. lucidum* from Cd toxicity (Guo et al. 2016). Chen et al. (2017) used RNA-Seq technology to generate large-scale transcriptome data from *G. oregonense* mycelia subjected to HS (32 °C) and exposed to exogenous NO. After HS treatment, heat shock proteins (HSPs), probable stress-induced proteins, and unigenes involved in D-amino-acid oxidase activity and oxidoreductase activity were all significantly upregulated in *G. oregonense*. These results provide insights into the transcriptional response of *G. oregonense* to HS and the mechanism by which NO enhances the HS tolerance at the gene expression level.

5.4.2 Transcriptome Analysis of Particular Tissues and Developmental Stages

The growth cycle of *G. lucidum* has divided into five typical stages according to the morphology of each stage: the first stage, full growth of mycelium (3 months after inoculation); the second stage, primordium (2 weeks after the end of the first stage); the third stage, young fruiting bodies (2 weeks after the end of the second stage); the fourth stage, mature fruiting bodies (1 week) one week after the end of the third stage); and the fifth stage, massive spore production (one month after the end of the fourth stage).

To identify the differential expression of genes involved in metabolic pathways and lignocelluloses degradation, researchers studied the transcriptomes of two developmental stages (mycelium and fruiting body) from *G. lucidum* (Yu et al. 2012b). Over 12 million high-quality DNA fragments with 90 bp were obtained. After filtering, a total of 6,439,690 and 6,416,670 high-quality reads from the mycelium and fruiting body of *G. lucidum* were obtained. These high-quality reads were then assembled into 18,892 and 27,408 unigenes, respectively. A similarity search was conducted against the NCBI non-redundant nucleotide database and a customized database composed of 5 fungal genomes. Using NCBI non-redundant nucleotide database and customized database, 11,098 and 8775 single genes were identified. All unigenes are further annotated with GO, Eukaryotic Homologous Group Terms (EOG), and KEGG. The DEGs from the *G. lucidum* mycelium and fruiting body stage were analyzed. 13, 22, and 120 unigenes were identified based on sequence similarity to be involved in the terpenoid backbone biosynthesis pathway, putative fungal oxidative lignin enzymes (FOLymes), and carbohydrate-active enzymes (CAZymes), respectively.

G. lucidum is a white-rot fungus that degrades plant biomass by secreting enzymes. Similar to other economically valuable mushrooms, *G. lucidum* is cultivated artificially in logs or compost. Generally, the compost used to grow mushroom chambers contains about 60–70% dry weight of lignocellulose (Cosgrove 1997), which is degraded and converted into fungal biomass. Therefore, the ability of *G. lucidum* to degrade the substrate may be a factor affecting its growth. Lignocellulolytic enzymes are carbohydrate-active enzymes (CAZy), which play an essential role in organisms' carbohydrate metabolism. Lignocellulolytic enzymes are classified into cellulose, hemicellulose, and lignin-modifying enzymes. Researchers studied the transcriptome in the mycelial growth cycle to identify the enzymes degrading lignocellulose in *G. lucidum* (Zhou et al. 2018).

The cellulase is expressed higher than hemicellulase and lignin-modifying enzyme in fruit body. The abundance and activity of cellulase and hemicellulase continue to increase until the end of the growth cycle. Besides, the abundance of lignin-modifying enzymes and the expression of their corresponding genes, including laccase and lignin-degrading heme peroxidase, are highest when the mycelium is fully diffused in the compost bag. Type I cellobiohydrolase is the most abundant extracellular lignocellulose hydrolase produced by *G. lucidum*. Heme peroxidase of the AA2 family is the main lignin-modifying enzyme expressed during mycelial growth. There may be several laccases functioning in the process of primordium formation. This study revealed the changes in the degradation of lignocellulose during *G. lucidum* growth, which helps explore new approaches to promote the development of *G. lucidum*.

5.4.3 Transcriptome Analysis of Oil Palm Pathogen *Ganoderma Boninense*

Basal stem rot disease causes severe economic losses to oil palm production in Southeast Asia, and the pathogenic fungi are the basidiomycete *G. boninense*. To identify both housekeeping and pathogenicity genes of *G. boninense,* the RNA-seq data of *G. boninense* obtained from two different conditions (axenic and in planta) (Wong et al. 2019). The first set of data was collected from a 7-day-old *G. boninense* sample under axenic conditions, which will provide insights into the genes responsible for the maintenance, growth, and development of *G. boninense*. Another set of *G. boninense* collected from oil palm-*G. boninense* pathosystem (in planta condition) at one month post-inoculation. This information will provide a comprehensive understanding of *G. boninense* pathogenesis and infection, especially molecular mechanisms and pathways.

5.4.4 Transcriptome Analysis of Ganoderic Acid Biosynthesis

GA, an important secondary metabolite of *G. lucidum*, is synthesized via the MVA pathway in *G. lucidum*. Lanosterol was the common cyclic intermediate of triterpenoids and ergosterol in *G. lucidum*. The GAs biosynthesis was investigated by comparing metabolites and transcriptome dynamics during liquid superficial-static culture (LSSC) and submerged culture (SC) (Wang et al. 2020) (Fig. 5.15). LSSC has been shown to be an efficient strategy to produce *G. lucidum* GAs and total triterpenoid (Wang et al. 2020). In SC, ergosterol was the major triterpenoid, and its content was higher than that in LSSC. Transcriptome analysis showed that CYP5150L8 was the key gene regulating lanosterol culture time in LSSC and could be potential candidate genes related to the biosynthesis of different GAs. Six of the ten expressed genes involved in the ergosterol biosynthetic pathway were shown upregulated at certain time points in SC.

5.4.5 RNA-Editing Events in the *G. Lucidum*

RNA editing is a widespread, post-transcriptional molecular phenomenon that changes RNA molecules' nucleotide sequence diversity relative to their genomic templates. Computational analysis of RNA sequencing (RNA-Seq) data can be used to discover RNA editing events. To date, a total of 8906 possible RNA-editing sites were identified within the *G. lucidum* genome using a customized pipeline in Fig. 5.16 (Zhu et al. 2014). RNA-editing sites were found on the exon and intron sequences and the 5'-/3'-untranslated regions of 2991 genes and the intergenic regions. The major editing types included C-to-U, A-to-G, G-to-A, and U-to-C transformations. A total of 97 putative editing events were randomly selected and validated using PCR and Sanger sequencing. The genes containing RNA-editing events were functionally classified by the KEGG and GO annotation. The enrichment analysis showed that laccase genes involved in lignin degradation, key enzymes involved in triterpenoid biosynthesis, and transcription factors were enriched. Four putative RNA-editing enzymes were also identified, including three adenosine deaminases and a deoxycytidylate deaminase. This study shed light on the mechanism responsible for RNA-editing events in the

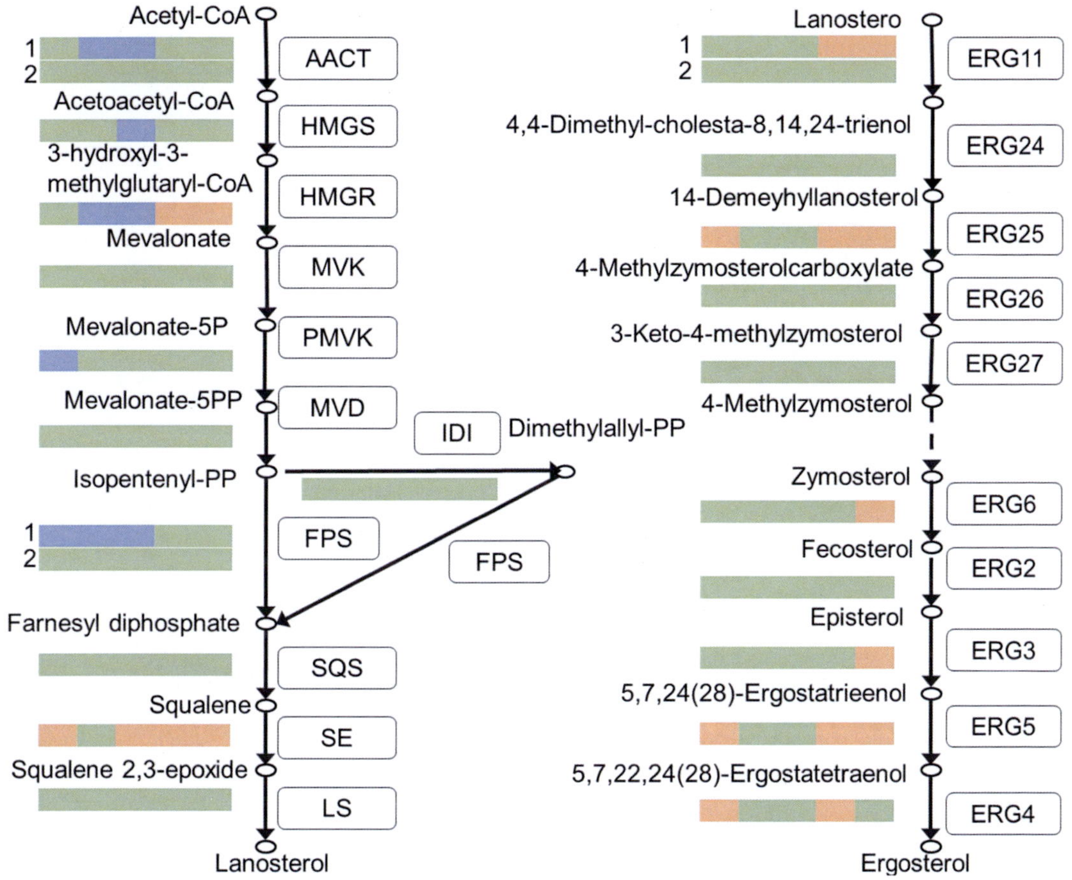

Fig. 5.15 DEGs involved in the ergosterol biosynthetic pathway in *G. lucidum*. Red: LSSC upregulated; yellow: not DEGs; and blue: LSSC downregulated. The genes involved in the ergosterol biosynthetic pathway include acetyl-CoA C-acetyltransferase (AACT), hydroxymethylglutaryl-CoA (HMGS), 3-hydroxy-3-methylglutaryl coenzyme A reductase (HMGR), mevalonate kinase (MVK), phosphomevalonate kinase (PMVK), mevalonate pyrophosphate decarboxylase (MVD), isopentenyl diphosphate isomerase (IDI), farnesyl diphosphate synthases (FPS), squalene synthase (SQS), lanosterol synthase (LS), and genes involved in ergosterol biosynthetic pathway (ERG) (reproduced from Zhang et al. 2016 with permission)

growth and development of *G. lucidum* and its adaptation to the environment and the regulation of valuable secondary metabolite pathways.

5.5 Concluding Remarks

Transcriptome analysis has been used widely for the discovery of genes from *Ganoderma* species. Based on the transcriptome data, key genes in biosynthesis pathways for active components have been cloned and subjected to functional validation. Furthermore, transcriptome analysis has been used to understand the molecular mechanisms in response to biological stress and abiotic stress. While transcriptome analysis can be used to address diverse biological questions, it is more powerful when combined with other methods such as proteomics, metabolomics, and molecular biology. Experimental validation is essential to get definite answers to a biological question.

Acknowledgements This work was supported by the Chinese Academy of Medical Sciences, Innovation Funds for Medical Sciences (CIFMS) [2016-I2M-3-016, 2017-I2M-1-013], National Science Foundation Funds

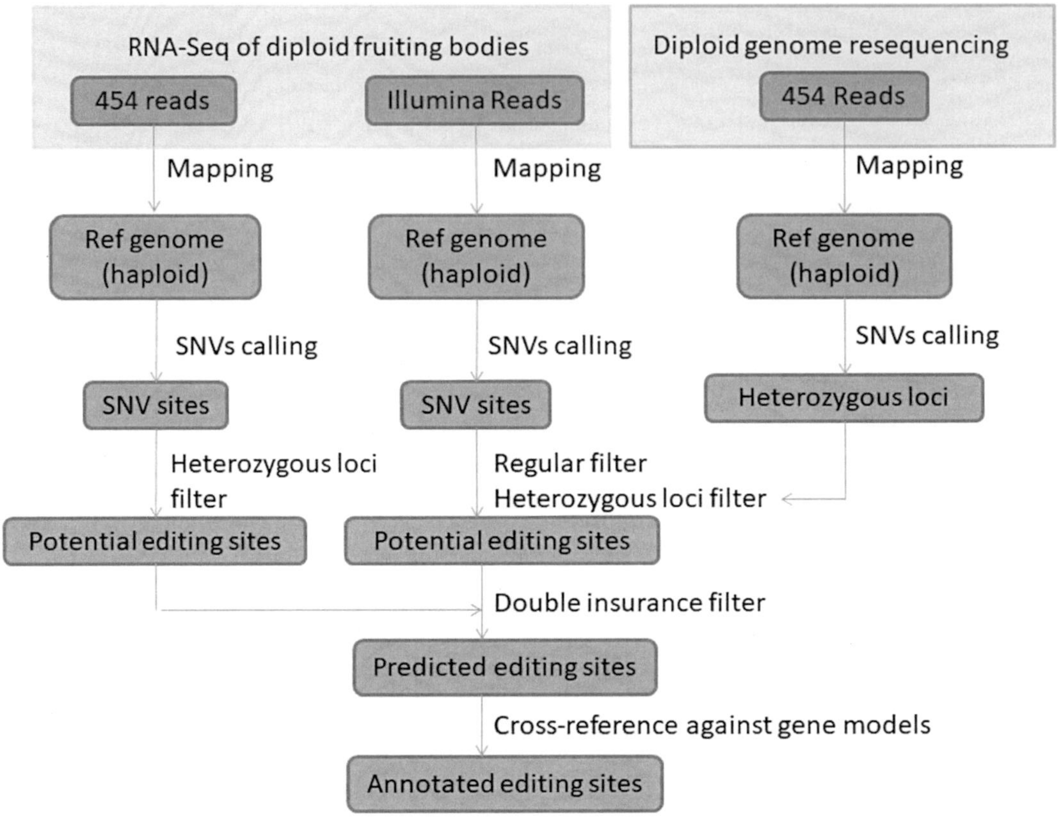

Fig. 5.16 Flowchart for calling RNA-editing sites. Raw inputs data include RNA-Seq reads from 454 and Illumina and genomic 454 reads produced by diploid genome resequencing. Then read mapping and single nucleotide variant detection were conducted. The pipeline of identification of RNA-editing sites includes three filtering steps: a "regular filter," a "heterozygous loci filter," and a "double-insurance filter" (reproduced from Zhu et al. (2014) with permission)

[81872966], National Science and Technology Fundamental Resources Investigation Program of China [2018FY100705], National Mega-Project for Innovative Drugs of China [2019ZX09735-002]. The funders were not involved in the study design, data collection, and analysis, decision to publish, or manuscript preparation.

References

Altschul SF, Gish W, Miller W, Myers EW, Lipman DJ (1990) Basic local alignment search tool. J Mol Biol 215(3):403–410

Bolger AM, Lohse M, Usadel B (2014) Trimmomatic: a flexible trimmer for Illumina sequence data. Bioinformatics 30(15):2114–2120. https://doi.org/10.1093/bioinformatics/btu170

Botstein D, Cherry JM, Ashburner M, Ball C, Blake J, Butler H, Davis A, Dolinski K, Dwight S, Eppig J (2000) Gene ontology: tool for the unification of biology. Nat Genet 25(1):25–29

Camacho C, Coulouris G, Avagyan V, Ma N, Papadopoulos J, Bealer K, Madden TL (2009) BLAST+: architecture and applications. BMC Bioinform 10(1):421

Chen C, Li Q, Wang Q, Lu D, Zhang H, Wang J, Fu R (2017) Transcriptional profiling provides new insights into the role of nitric oxide in enhancing *Ganoderma oregonense* resistance to heat stress. Sci Rep 7 (1):15694. https://doi.org/10.1038/s41598-017-15340-6

Chuang HW, Wang IW, Lin SY, Chang YL (2009) Transcriptome analysis of cadmium response in *Ganoderma lucidum*. FEMS Microbiol Lett 293 (2):205–213. https://doi.org/10.1111/j.1574-6968.2009.01533.x

Consortium GO (2017) Expansion of the gene ontology knowledgebase and resources. Nucleic Acids Res 45 (D1):D331–D338

Consortium U (2018) UniProt: the universal protein knowledgebase. Nucl Acids Res 46(5):2699

Cosgrove DJ (1997) Assembly and enlargement of the primary cell wall in plants. Annu Rev Cell Dev Biol 13:171–201. https://doi.org/10.1146/annurev.cellbio.13.1.171

DePristo MA, Banks E, Poplin R, Garimella KV, Maguire JR, Hartl C, Philippakis AA, del Angel G, Rivas MA, Hanna M, McKenna A, Fennell TJ, Kernytsky AM, Sivachenko AY, Cibulskis K, Gabriel SB, Altshuler D, Daly MJ (2011) A framework for variation discovery and genotyping using next-generation DNA sequencing data. Nat Genet 43(5):491–498. https://doi.org/10.1038/ng.806

Dobin A, Davis CA, Schlesinger F, Drenkow J, Zaleski C, Jha S, Batut P, Chaisson M, Gingeras TR (2013) STAR: ultrafast universal RNA-seq aligner. Bioinformatics 29(1):15–21. https://doi.org/10.1093/bioinformatics/bts635

Gabriel MM, J. Bílý , P. Rychlovský (1994) Accumulation of heavy metals by some wood-rotting fungi. Folia Microbiol 39:115–118

Gong T, Yan R, Kang J, Chen R (2019) Chemical components of Ganoderma. Adv Exp Med Biol 1181:59–106. https://doi.org/10.1007/978-981-13-9867-4_3

Guo S, Yao Y, Zuo L, Shi W, Gao N, Xu H (2016) Enhancement of tolerance of *Ganoderma lucidum* to cadmium by nitric oxide. J Basic Microbiol 56(1):36–43. https://doi.org/10.1002/jobm.201500451

Huang YH, Wu HY, Wu KM, Liu TT, Liou RF, Tsai SF, Shiao MS, Ho LT, Tzean SS, Yang UC (2013) Generation and analysis of the expressed sequence tags from the mycelium of *Ganoderma lucidum*. PloS ONE 8(5):e61127. https://doi.org/10.1371/journal.pone.0061127

Jain KK, Kumar A, Shankar A, Pandey D, Chaudhary B, Sharma KK (2020) De novo transcriptome assembly and protein profiling of copper-induced lignocellulolytic fungus *Ganoderma lucidum* MDU-7 reveals genes involved in lignocellulose degradation and terpenoid biosynthetic pathways. Genomics 112(1):184–198. https://doi.org/10.1016/j.ygeno.2019.01.012

Kanehisa M, Araki M, Goto S, Hattori M, Hirakawa M, Itoh M, Katayama T, Kawashima S, Okuda S, Tokimatsu T (2007) KEGG for linking genomes to life and the environment. Nucl Acids Res 36(1):D480–D484

Kim D, Paggi JM, Park C, Bennett C, Salzberg SL (2019) Graph-based genome alignment and genotyping with HISAT2 and HISAT-genotype. Nat Biotechnol 37(8):907–915. https://doi.org/10.1038/s41587-019-0201-4

Korpelainen E, Tuimala J, Somervuo P, Huss M, Wong G (2014) A practical approach. RNA-seq data analysis. CRC Press, Abingdon, Oxfordshire

Love MI, Huber W, Anders S (2014) Moderated estimation of fold change and dispersion for RNA-seq data with DESeq2. Genome Biol 15(12):550. https://doi.org/10.1186/s13059-014-0550-8

Luo H, Sun C, Song J, Lan J, Li Y, Li X, Chen S (2010) Generation and analysis of expressed sequence tags from a cDNA library of the fruiting body of *Ganoderma lucidum*. Chin Med 5:9. https://doi.org/10.1186/1749-8546-5-9

McKenna A, Hanna M, Banks E, Sivachenko A, Cibulskis K, Kernytsky A, Garimella K, Altshuler D, Gabriel S, Daly M, DePristo MA (2010) The genome analysis toolkit: a MapReduce framework for analyzing next-generation DNA sequencing data. Genome Res 20(9):1297–1303. https://doi.org/10.1101/gr.107524.110

Mehra RK, Winge DR (1991) Metal ion resistance in fungi: molecular mechanisms and their regulated expression. J Cell Biochem 45(1):30–40. https://doi.org/10.1002/jcb.240450109

Mortazavi A, Williams BA, McCue K, Schaeffer L, Wold B (2008) Mapping and quantifying mammalian transcriptomes by RNA-Seq. Nat Meth 5(7):621–628. https://doi.org/10.1038/nmeth.1226

Pertea M, Pertea GM, Antonescu CM, Chang TC, Mendell JT, Salzberg SL (2015) StringTie enables improved reconstruction of a transcriptome from RNA-seq reads. Nat Biotechnol 33(3):290–295. https://doi.org/10.1038/nbt.3122

Shen S, Park JW, Lu Z, Lin L, Henry MD, Wu YN, Zhou Q, Xing Y (2014) rMATS: robust and flexible detection of differential alternative splicing from replicate RNA-Seq data. Proc Natl Acad Sci U S A 111(51):5593–5601

Tan X, Sun J, Ning H, Qin Z, Miao Y, Sun T, Zhang X (2018) De novo transcriptome sequencing and comprehensive analysis of the heat stress response genes in the basidiomycetes fungus *Ganoderma lucidum*. Gene 661:139–151. https://doi.org/10.1016/j.gene.2018.03.093

Trapnell C, Roberts A, Goff L, Pertea G, Kim D, Kelley DR, Pimentel H, Salzberg SL, Rinn JL, Pachter L (2012) Differential gene and transcript expression analysis of RNA-seq experiments with TopHat and Cufflinks. Nat Protoc 7(3):562–578. https://doi.org/10.1038/nprot.2012.016

Wang K, Li M, Hakonarson H (2010) ANNOVAR: functional annotation of genetic variants from high-throughput sequencing data. Nucl Acids Res 38(16):e164. https://doi.org/10.1093/nar/gkq603

Wang Q, Xu M, Zhao L, Wang F, Li Y, Shi G, Ding Z (2020) Transcriptome dynamics and metabolite analysis revealed the candidate genes and regulatory mechanism of ganoderic acid biosynthesis during liquid superficial-static culture of *Ganoderma lucidum*. Microb Biotechnol. https://doi.org/10.1111/1751-7915.13670

Wong MY, Govender NT, Ong CS (2019) RNA-seq data of *Ganoderma boninense* at axenic culture condition and under in planta pathogen-oil palm (Elaeis guineensis Jacq.) interaction. BMC Res Notes 12(1):631. https://doi.org/10.1186/s13104-019-4652-y

Wu CH, Chen C (2011) Protein bioinformatics databases and resources. Meth Mol Biol 694(1):3

Xu J, Li P (2019) Researches and application of Ganoderma spores powder. Adv Exp Med Biol 1181:157–186. https://doi.org/10.1007/978-981-13-9867-4_6

Yu G, Wang LG, Han Y, He QY (2012a) clusterProfiler: an R package for comparing biological themes among gene clusters. Omics J Integr Biol 16(5):284–287

Yu GJ, Wang M, Huang J, Yin YL, Chen YJ, Jiang S, Jin YX, Lan XQ, Wong BH, Liang Y, Sun H (2012) Deep insight into the *Ganoderma lucidum* by comprehensive analysis of its transcriptome. PloS ONE 7 (8):e44031. https://doi.org/10.1371/journal.pone.0044031

Zhang X, Ren A, Li MJ, Cao PF, Chen TX, Zhang G, Shi L, Jiang AL, Zhao MW (2016a) Heat stress modulates mycelium growth, heat shock protein expression, ganoderic acid biosynthesis, and hyphal branching of ganoderma lucidum via cytosolic Ca2. Appl Environ Microbiol 82(14):4112–4125. https://doi.org/10.1128/AEM.01036-16

Zhang X, Ren A, Li MJ, Cao PF, Chen TX, Zhang G, Shi L, Jiang AL, Zhao MW (2016b) Heat stress modulates mycelium growth, heat shock protein expression, ganoderic acid biosynthesis, and hyphal branching of *Ganoderma lucidum* via cytosolic Ca^{2+}. Appl Environ Microbiol 82(14):4112–4125. https://doi.org/10.1128/AEM.01036-16

Zhou S, Zhang J, Ma F, Tang C, Tang Q, Zhang X (2018) Investigation of lignocellulolytic enzymes during different growth phases of *Ganoderma lucidum* strain G0119 using genomic, transcriptomic, and secretomic analyses. PloS ONE 13(5):e0198404. https://doi.org/10.1371/journal.pone.0198404

Zhu Y, Luo H, Zhang X, Song J, Sun C, Ji A, Xu J, Chen S (2014) Abundant and selective RNA-editing events in the medicinal mushroom *Ganoderma lucidum*. Genetics 196(4):1047–1057. https://doi.org/10.1534/genetics.114.161414

Proteomic Characterization of Lingzhi

6

Ang Ren, Liang Shi, Jing Zhu, Rui Liu, Ailiang Jiang, and Mingwen Zhao

Abstract

In the postgenomic era, proteomics is an important part of systems biology research. Two-dimensional electrophoresis, iTRAQ, and mass spectrometry were used to analyze thousands of proteins in cells and tissues comprehensively. Based on the examples of the application, this chapter reviews the researches on the complex system of *G. lucidum* using proteomics technology. The main techniques and applications of proteomics in the *G. lucidum* complex system research were reviewed and discussed here. The conventional proteomics analysis process usually includes sample preparation, chromatographic separation, mass spectrometry detection, and data analysis. However, with the increasing demands for large-scale proteomics analysis, high-coverage, short-time, and fast 'high-throughput' proteomics analysis is becoming one of the critical technical challenges that need to be resolved in *G. lucidum*. In this part, the research strategy and future research direction of proteomics of *G. lucidum* are proposed and discussed, which will provide some ideas for researchers engaged in proteomics of *G. lucidum.*

A. Ren · L. Shi · J. Zhu · R. Liu · A. Jiang · M. Zhao (✉)
College of Life Sciences, Nanjing Agricultural University, Nanjing 210095, Jiangsu, People's Republic of China
e-mail: mwzhao@njau.edu.cn

6.1 Overview of the *Ganoderma lucidum* Proteomics

The proteome is defined as all the proteins in a cell (Wilkins et al. 1998). Proteomics is a branch of science that takes proteome as the research object and studies the protein composition and changing patterns (Wilkins et al. 1998; Jensen et al. 1999). Proteomics essentially refers to the study of protein characteristics on a large scale, including protein expression level, post-translational modification, protein–protein interaction, and so on. Consequently, a comprehensive understanding of biological processes on the protein level was obtained. Current proteomics techniques mainly include two-dimensional gel electrophoresis, isoelectric focusing, and biological mass spectrometry (Dominguez et al. 2007).

Various proteins in the physiological regulation network of *G. lucidum* were systematically analyzed, and the essential regulatory proteins that regulated Ganoderic acids (GA) biosynthesis were investigated. This analysis may help establish a foundation for further research attempting to improve GA biosynthesis through the genetic engineering method. The extracellular secretory proteome, fruitbody primordium proteome, and proteome under MeJA induction have been previously investigated in *G. lucidum* (detailed results

C. Liu (ed.), *The Lingzhi Mushroom Genome*, Compendium of Plant Genomes,
https://doi.org/10.1007/978-3-030-75710-6_6

are presented below). However, proteomics analyses of *G. lucidum* are still in their infancy. This gap in the research is primarily attributed to two factors. One factor is that *G. lucidum*, as a medicinal fungus commonly consumed in Asian countries, mainly in East Asia, China, Japan, and South Korea, is not an ideal model for proteomic research. The other is that the total protein extraction methods employed for *G. lucidum* are different from those for most plants and microorganisms. The low extraction rate of total proteins also directly affects the results of proteomics analysis (Chandramouli and Qian 2009).

6.2 Differential Proteomics Research

Widely utilized proteomics techniques

An important objective of life sciences research is to study life phenomena and laws for protein types and expressions. This research requires high-throughput identifications and quantifications of proteins, which accelerates the development of Proteomics. At present, mainstream proteomic techniques are developed based on biological mass spectrometry technology. The three most widely used proteomics techniques are described below.

iTRAQ

Isobaric tags for relative and absolute quantification (iTRAQ) are employed most commonly in quantitative proteomics. The principle of this technology is peptide labeling and quantification. In particular, the content of polypeptides was labeled with four isotope markers, 114, 115, 116, and 117, or eight isotope markers, 113, 114, 115, 116, 117, 118, 119, and 121. The labeled proteins are quantified easily. The procedure is described in detail in Fig. 6.1. Firstly, proteins from different samples (S1, S2, S3, S4, S5, S6, S7, and S8) were hydrolyzed by protease. The enzymatic hydrolysis fragments were labeled with different markers and then mixed. Liquid chromatography coupled with mass spectrometry was used for primary mass spectrometry analysis. The labeled peptides of the same protein from eight different samples showed a peak on the first-order mass spectrum. Then, the labeled peptide was identified by secondary mass spectrometry analysis. The equilibrium group was removed from the report group. After the secondary mass spectrometry analysis, the reporter group generated eight report ion signals (113, 114, 115, 116, 117, 118, 119, and 121) in the low mass region of the secondary mass spectrometry. The intensities of these ion signals represent the same peptide of the eight labeled samples, respectively. The report ion's peak area ratio is the same peptide ratio among different products (Gluckmann et al. 2007; Tan et al. 2012; Wang et al. 2016; Zieske 2006; Ross et al. 2004).

This technology can detect low-abundance, cytoplasmic, membrane, nuclear, and extracellular proteins. It can be employed to analyze eight samples simultaneously and obtain the protein identification and quantitative results simultaneously. iTRAQ technology is especially suitable for the differential protein analysis of samples obtained from multiple time points.

SILAC

The basic principle of SILAC quantification is to add light, medium, or heavy stable isotope-labeled essential amino acids (lysine and arginine) to replace the corresponding amino acids in the cell culture medium. After 5–6 doubling cycles at normal conditions, the stable isotope-labeled amino acids are completely incorporated into the newly synthesized proteins to replace the original amino acids. The newly synthesized protein is labeled with a stable isotope. The lysates of various labeled cells are mixed according to the ratios of cell numbers or protein amounts. After the separation and purification, the lysates are analyzed by mass spectrometry (Fig. 6.2). The relative quantification is performed according to the area comparison of the two isotope peptides in the first-order mass spectrometry, a version of the in vivo metabolic labeling method.

SILAC is an in vivo labeling technology that reveals results closer to the sample's real state. The labeling efficiency is as high as 100%, and the labeling effect is stable. SILAC is suitable for whole-cell protein analysis and suitable for the identification and quantification of membrane proteins. Each sample only requires tens of

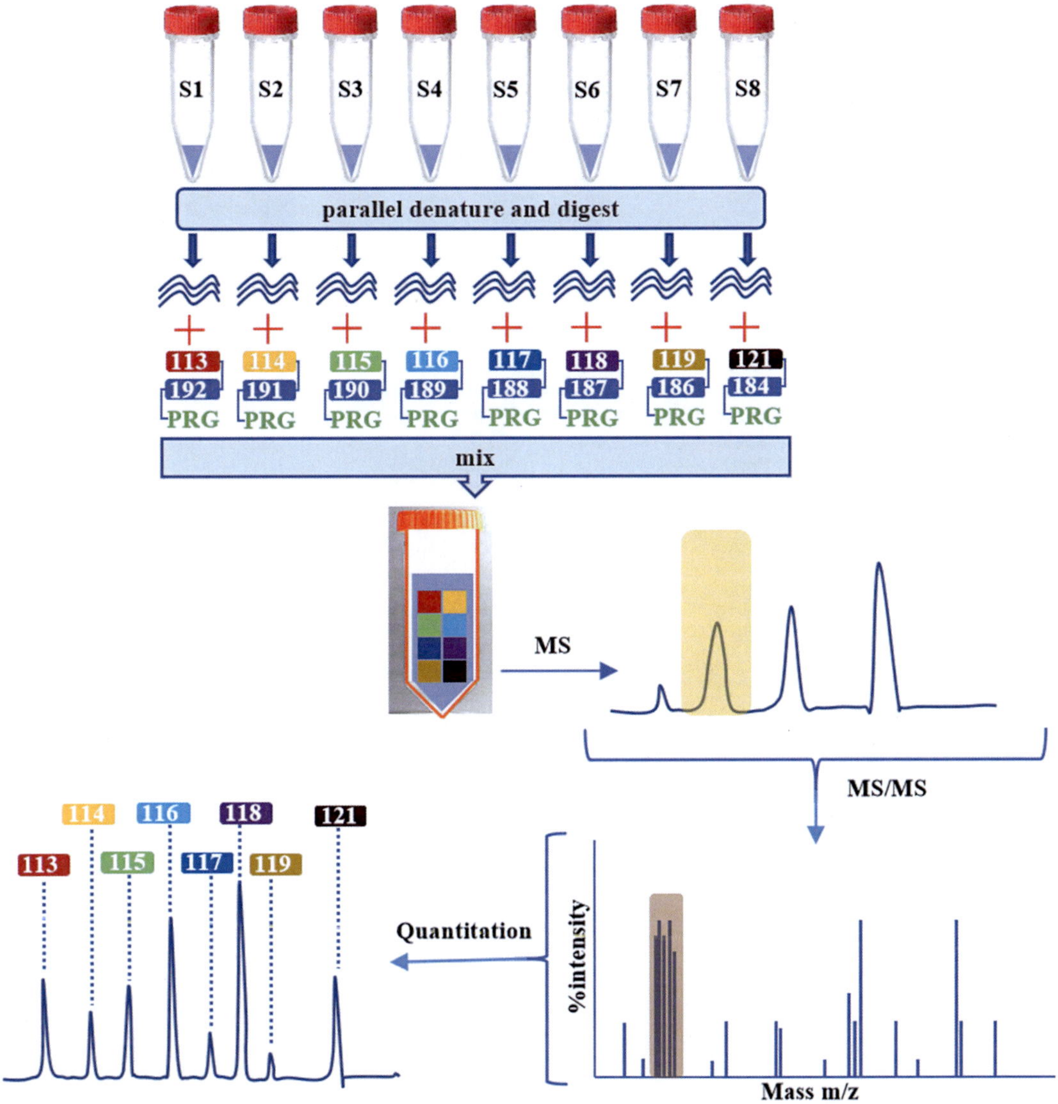

Fig. 6.1 The principle of iTraq quantitative technology in proteomics, modified from Ross et al. (2004). Proteins from different samples (S1, S2, S3, S4, S5, S6, S7, and S8) were hydrolyzed by protease. Liquid chromatography coupled with mass spectrometry was used for primary mass spectrometry. The labeled peptides of the same protein from eight different samples showed a peak on the first-order mass spectrum. Then, the labeled peptide was identified by secondary mass spectrometry analysis. The equilibrium group was removed from the report group. After the secondary mass spectrometry analysis, the reporter group generated eight report ion signals (113, 114, 115, 116, 117, 118, 119, and 121) in the low mass region of the secondary mass spectrometry. The intensities of these ion signals represent the same peptide from the eight labeled samples, respectively. The report ion's peak area ratio is the ratio of the same protein among different products

micrograms of protein. SILAC quantification is ideal for the study of cells cultured in vivo. The differences in whole-cell protein or subcellular protein from multiple samples or in the same sample under different conditions are compared (Macek et al. 2017; Polacek et al. 2010).

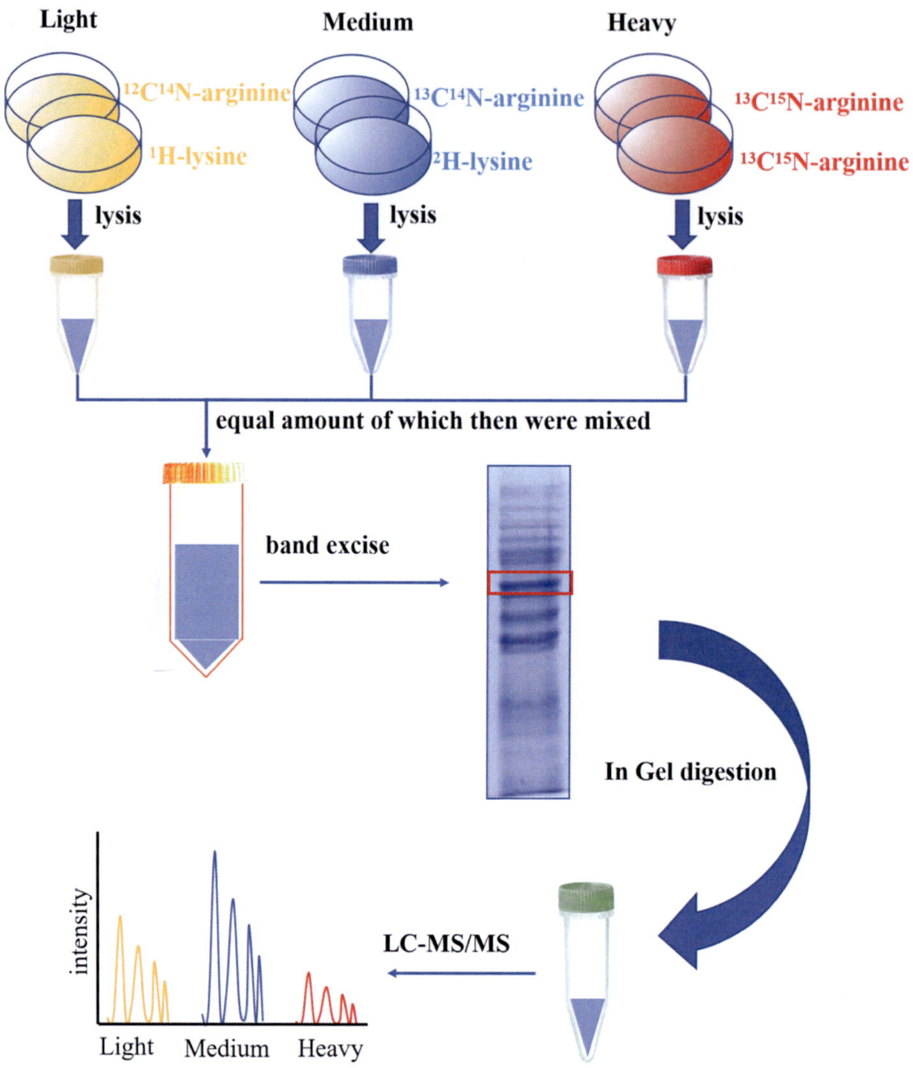

Fig. 6.2 The principle of SILAC quantitative technology in proteomics, modified from Macek et al. (2017). After adding light, medium, or heavy stable isotope-labeled essential amino acids (lysine and arginine) into the cell culture medium, the newly synthesized proteins are labeled with stable isotope. The proteins of all kinds of samples were mixed equally and analyzed by mass spectrometry after separation and purification. The relative quantifica-tion was carried out by comparing the area size of isotope peak type in the first-order mass spectrum. And the peptide was sequenced in the second-order mass spectrum for protein identification

Label-free quantification

Label-free quantification also called the unlabeled quantitative proteomics does not need specific labeling on the comparative samples (Fig. 6.3). This technique only needs to compare the GC–MS response signals of specific peptides/proteins among different samples to obtain the protein expression change among samples. Label-free quantification is suitable for large-scale protein identification and quantification. This method does not require labeling, but its operation is simple. Label-free quantification can be utilized to quantify the total protein difference of samples. The experimental stability and repeatability are higher. However, the accuracy is lower than that of labeled quantification methods.

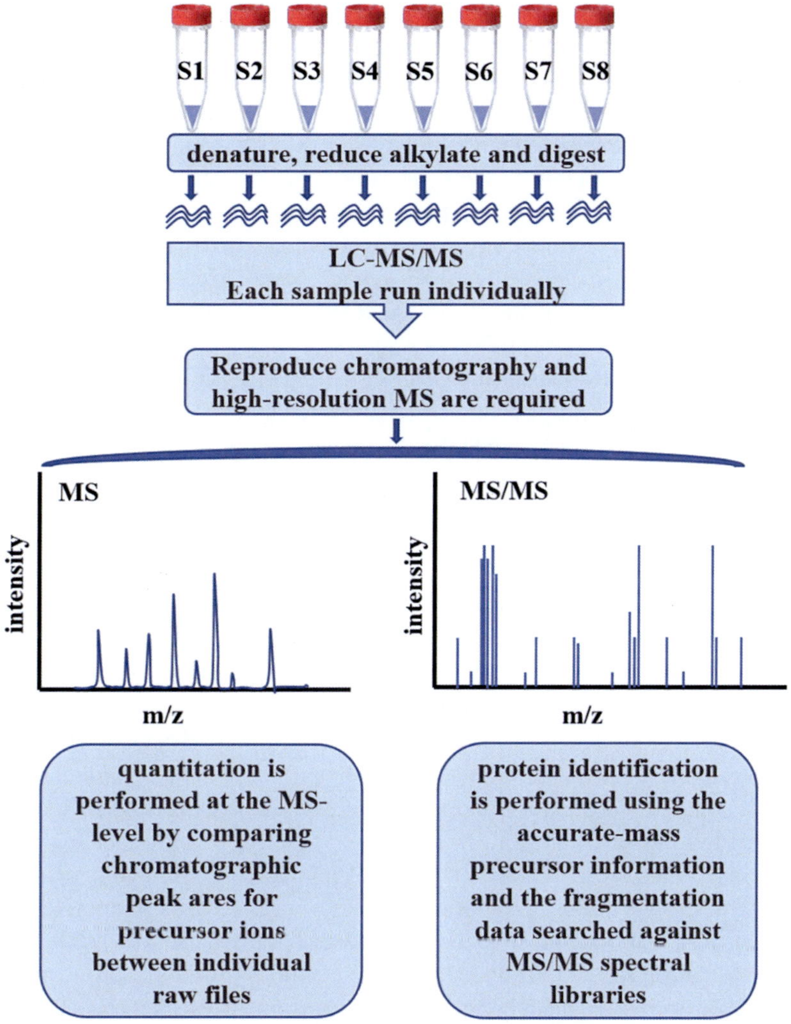

Fig. 6.3 The principle of Label-free quantification technology in proteomics, modified from Ono et al. (2006). Each sample was enzymolysis independently and then analyzed by high-resolution LC-MS/MS. Based on the first-order mass spectrometry, the area of each peak on the chromatogram was calculated as the basis of peptide signal intensity. The proteins were identified by secondary mass spectrometry information, including the total number of peptides, the total number of unique peptides, protein coverage, and peptide matching parameters

Therefore, label-free technology is suitable for performing quantitative comparisons in experiments that cannot be performed using labeled quantification experiments or have too many samples (Ono et al. 2006; Zhang et al. 2011).

Protein extraction methods

Al-Obaidi et al. conducted a comparative study on three total protein extraction methods using trichloroacetic acid, sucrose, and phenol methanol/ammonium acetate, respectively. These researchers observed that methanol/ammonium acetate had the highest extraction efficiency for total protein extraction (Al-Obaidi et al. 2016). However, there are still many parameters that could be improved. For example, in the protein extraction efficiency comparison, using 2D-PAGE as a low-precision evaluation technology means that the comparison's persuasiveness is limited. Secondly, the protein extraction methods are compared in the specific experimental design, which has little reference value. Besides, Al-Obaidi et al. detected differential protein expressions between two strains of *G. boninense* with different levels of pathogenicity in the process of infecting oil palm roots. The authors found significant differences in protein expression levels between these two strains (Al-Obaidi et al. 2017).

Case introduction in *G. lucidum*

The GAs of *G. lucidum* have attracted increasing attention in cancer research. The omics technology has been used to identify the target molecules for *Ganoderma* polysaccharides and triterpenes in tumor cells. They have also been used to elucidate the mechanism underlying the triterpene cytotoxicity. The proteomics research on *G. lucidum* is notably scarce. Manavalan studied the extracellular secretory proteome during the growth of *G. lucidum* using bagasse as a substrate (Manavalan et al. 2012). Seventy-one secretory proteins were identified by liquid chromatography-tandem mass spectrometry (LC–MS/MS) and included cellulose hydrolase, hemicellulose hydrolase, lignin-degrading enzyme, glycoside hydrolase, protease, and phosphatase. Several of these proteins were newly reported enzymes involved in lignocellulosic biomass hydrolysis.

Liu et al. optimized the extraction conditions of total proteins (Liu et al. 2009). They compared the two-dimensional electrophoresis patterns of total proteins extracted from the fruit body and primordium. Two extraction methods were used: the tris-saturated phenol method and the trichloroacetic acid (TCA)/acetone precipitation method. 565 and 273 protein spots were obtained, respectively. Results showed that the tris-saturated phenol method could generate more protein spots, obtain better isoelectric focusing, and reduce salt concentration. The tris-saturated phenol method had a better result for the extraction of total proteins from primordium of *G. lucidum*, which is also an effective protein extraction method in other medicinal mushrooms.

Zhao et al. studied the effect of light on the physiological metabolism using differential proteomics to identify the best light quality conditions. These experiments provide a theoretical basis for optimizating *G. lucidum* cultivation. Under the blue light condition, the expression level of 54 proteins was significantly changed. Remarkably, 32 proteins were up-regulated, and 22 proteins were downregulated. Among the six proteins with significantly differential expression, 14–3-3 protein and 14-putative aryl-alcohol dehydrogenase protein were identified (Zhao, 2013). The result indicated that these two proteins might be involved in the process of blue light-regulated energy metabolism. Zhao et al. further analyzed the differential expression of proteomes in the dark and four different light conditions (red, yellow, green, and blue) (Zhao, 2013). Compared with the dark condition, 58, 70, 74, and 86 proteins showed differential expressions in the red, yellow, green, and blue light. These results built the different proteome maps from mycelium under different light qualities in *G. lucidum*. Meanwhile, the differentially expressed proteins and those associated with light qualities response were found. However, because the genome of *G. lucidum* had not been published at that time and protein database information was lacking, it is difficult to identify and annotate these differentially expressed proteins.

With the continuous development of proteomics technology, especially those relative and absolute quantitative techniques based on isotope labeling, proteomics research has become common. The iTRAQ is a high-throughput peptide in vitro labeling technology. ABI developed it in 2004 (Ross et al. 2004). Compared with gel-based proteomics methods, iTRAQ has a higher detection rate for low-abundance proteins, strongly basic proteins, and proteins measuring less than 10 kD or more than 200 kD (Liu et al. 2013; Lu et al. 2012). This technology is a notably advanced, reliable, and commonly used technology for proteomics research.

It has been reported that MeJA can significantly induce GA biosynthesis (Ren et al. 2010). To further study the mechanism in this process, Jiang et al. analyzed the proteomes after MeJA treatment. In preliminary experiments, Jiang found many proteins differentially expressed between the groups with and without MeJA treatment using 2-DE PAGE (Fig. 6.4). Because of the high sensitivity and strong separation ability, iTRAQ was employed in their further study to globally identify differences in protein expression after MeJA treatment (Jiang et al. 2019). In the iTRAQ results, 307,493 secondary spectrums were obtained. After BLAST search, the number of matched

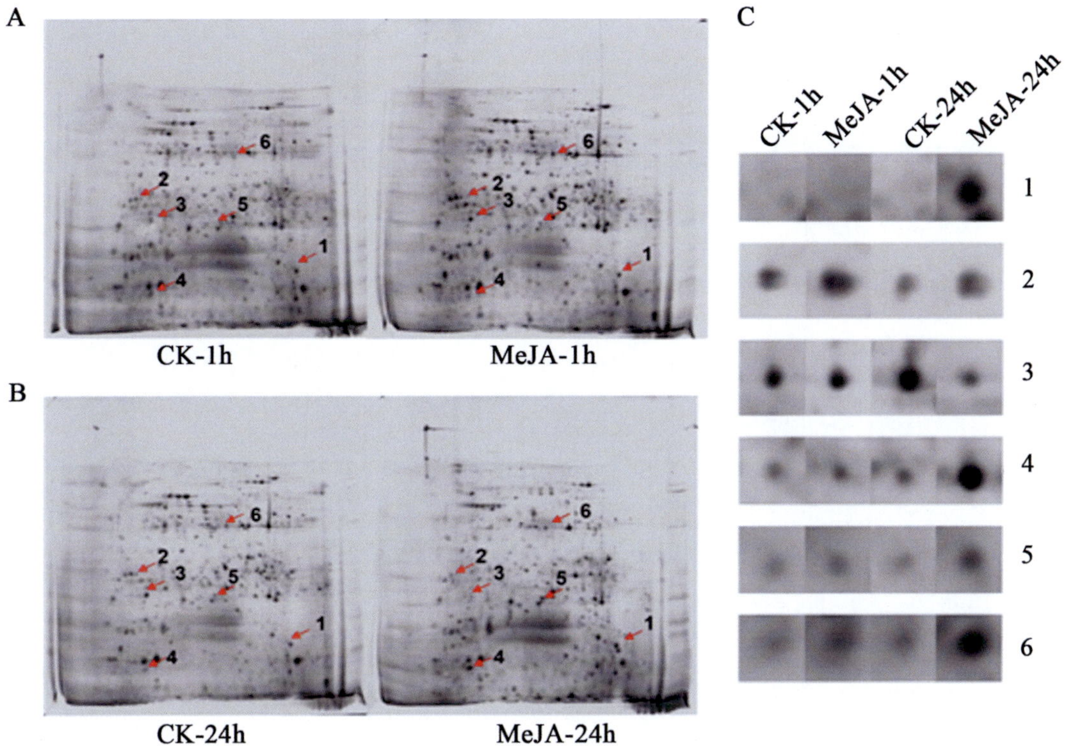

Fig. 6.4 Two-dimensional electrophoresis analysis of the whole proteins with different treatments. **a** Differential protein expression after1 h treatment with MeJA, **b** Differential protein expression after 24 h treatment with MeJA, **c** six peptides have undergone significant abundance changes after different treatments. CK-1 h, control check, wildtype cells treated with 2 μL/mL ethanol for 1 hour; MeJA-1 h, wildtype cells treated with 50 μM MeJA for 1 hour; CK-24 h, wildtype cells treated with 2 μL/mL ethanol for 24 hours; MeJA-24 h, wildtype cells treated with 50 μM MeJA for 24 hours. Targets of red arrows from A and B were the six proteins with significant expression changes identified in C (Reproduced from Jiang et al. 2019 with permission)

spectrums was 113,219, and the number of specific peptides was 106,017. Finally, 26,812 peptide segments and 5059 proteins were identified (see Table 6.1). 209 differentially abundant proteins (DAPs) were enriched in 16 pathways by KEGG analysis, including metabolic pathways, biosynthesis of secondary metabolites, carbon metabolism, fructose and mannose metabolism, and pyruvate metabolism under MeJA treatment for 15 min (M15, Fig. 6.5).

Meanwhile, they found that 202 DEPs in response to MeJA treatment for 24 h (M24). KEGG analysis (Fig. 6.6) revealed that these 202 DEPs were enriched in 21 pathways, such as biosynthesis of antibiotics, glycolysis/gluconeogenesis, fatty acid degradation, butanoate metabolism, pyruvate metabolism, fructose, mannose metabolism, tryptophan metabolism, and biosynthesis of amino acids. These DEPs are mainly involved in energy metabolism, amino acid metabolism, and genetic information processing. Comparing these results, they found a total of 352 DEPs in M15 and M24 and 59 DEPs of M15 and M24 are in common (Table 6.2). This result showed that 59 proteins were expressed simultaneously, mainly secondary metabolism synthesis, energy metabolism, and mitotic process-related proteins. It seems that MeJA induction has a

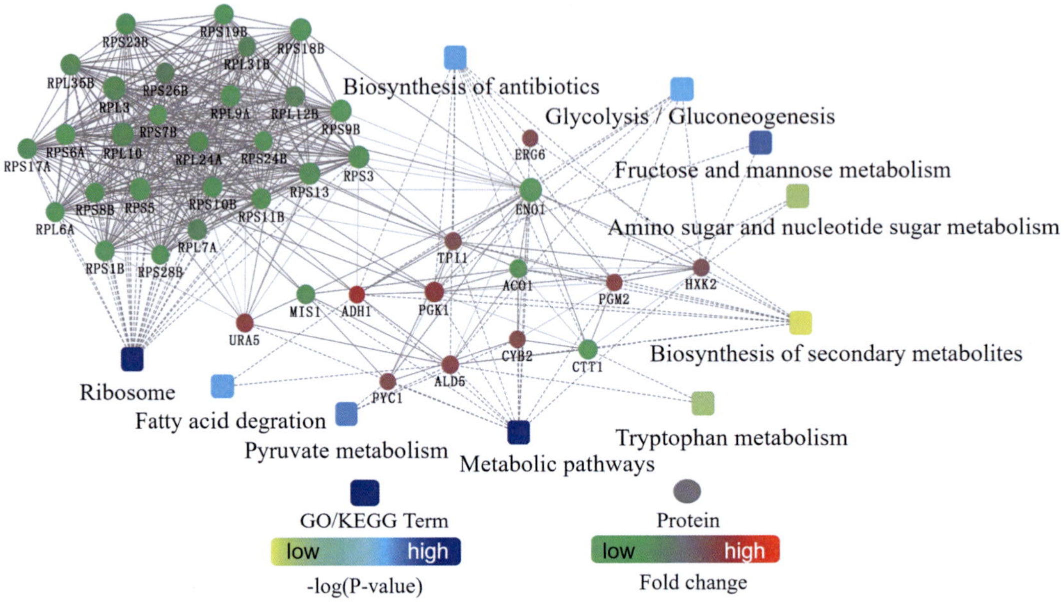

Fig. 6.5 Protein–Protein interaction analysis of 209 identified Differentially Expressed Proteins (DEPs) in Lingzhi cells treated with MeJA for 15 min (Jiang et al. 2019). Circle nodes represent proteins and rectangle represent KEGG pathway or biological process. Pathways were colored with gradient color from yellow to blue, yellow for smaller P-value and blue for bigger P-value. Biological processes were colored in red. In fold change analysis, genes/proteins were colored in red (up-regulation) and green (down-regulation). Default confidence cutoff of 400 was used: interactions with bigger confident scores shown as solid lines between proteins, otherwise as dashed lines (Repro-duced from Jiang et al. 2019 with permission)

Table 6.1 Overview of ***Ganoderma lucidum*** proteomes analyzed within iTRAQ (Jiang et al. 2019)

Unused (Conf) cutoff	Proteins detected	Proteins before grouping	Distinct peptides	Spectra identified	% Total spectra
>2.0 (99)	3874	4618	42,656	158,546	51.5
>1.3 (95)	4464	5099	43,590	160,319	52.1
>0.47 (66)	4715	5449	44,125	161,200	52.4
Cutoff Applied: >0.05 (10%)	5059	6736	44,750	162,238	52.7

continuous effect on the energy metabolism and secondary metabolite biosynthesis of *G. lucidum*.

In particular, Jiang's results demonstrated that MeJA treatment led to metabolic rearrangement that inhibited the normal glucose metabolism, energy supply, and protein synthesis of cells but promoted secondary metabolism, including GAs biosynthesis. This research may help to elucidate the underlying molecular mechanism of MeJA-induced GA biosynthesis and expand our knowledge of the metabolic processes in fungal response to MeJA.

6.3 Proteomics and Multi-omics Integration Analysis

While some progress has been made in omics research on *G. lucidum*, the overall research is still in the initial stages. Although the completion of whole-genome sequences provides a useful

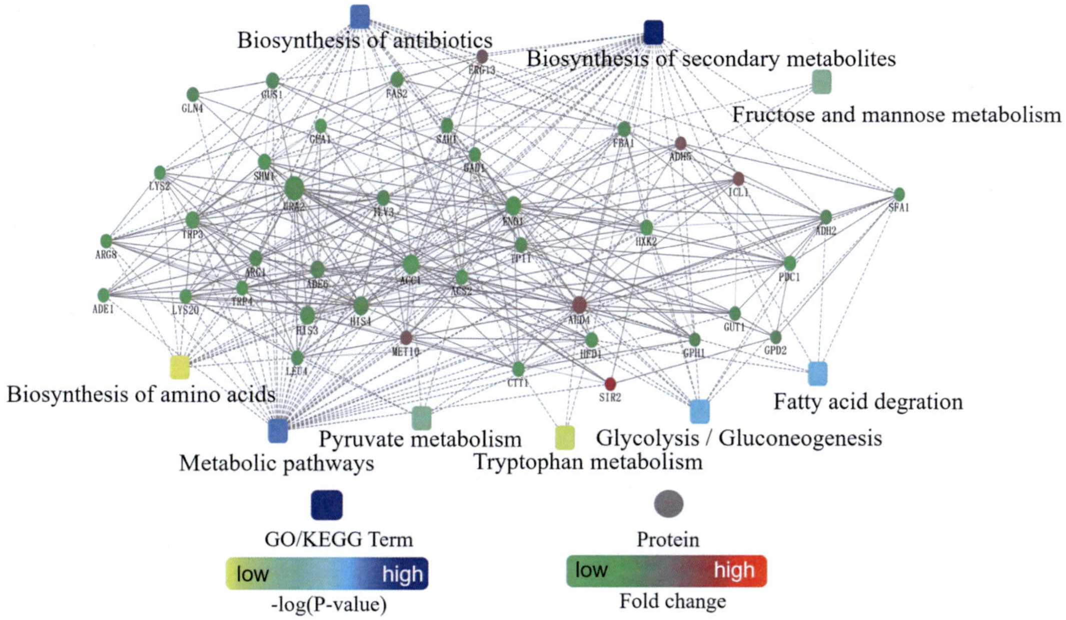

Fig. 6.6 Protein–Protein interaction analysis of 209 identified DEPs in MeJA treatment after 24 h (Jiang et al. 2019). Circle nodes represent proteins and rectangle nodes represent KEGG pathway or biological process. Pathways were colored with gradient color from yellow to blue, yellow for smaller P-value and blue for bigger P-value. Biological processes were colored in red. In fold change analysis, genes/proteins were colored in red (up-regulation) and green (down-regulation). Default confidence cutoff of 400 was used: interactions with bigger confident scores are shown as solid lines between proteins, otherwise as dashed lines. Reproduced from Jiang et al. 2019 with permission

Table 6.2 *G. lucidum* proteins with differential abundance in response to MeJA treatment and analyzed with iTRAQ (Jiang et al. 2019)

DEPs from different Treatment	Proteins detected
DEPs identified after MeJA treatment for 15 min	209
DEPs identified after MeJA treatment for 24 h	202
DEPs identified in both treatments	59

platform for future omics research, research on proteomics in *G. lucidum* is relatively lagging compared with other model fungi. On the one hand, this shortcoming is related to the difficulty in efficiently extracting the *G. lucidum* proteins. On the other hand, this lack of research is partly due to past proteomic research limitations. Due to technical limitations, traditional gel-based protein separation systems have low efficiency and can only detect proteins having high concentrations with a specific molecular weight and pH range. Thus, a large amount of adequate protein information is therefore lost.

We should actively conduct multi-omic joint analyses, including genomics, transcriptomics, proteomics, and metabolomics. Mining the genetic information of *G. lucidum* was performed to elucidate further the regulatory mechanism of *G. lucidum* growth, development, and metabolism.

It has been reported that MeJA can significantly induce GA biosynthesis (Ren et al. 2010). Ren (2013) analyzed the transcriptomic differences in cells treated with exogenous MeJA by the cDNA-AFLP method (Ren et al. 2013). MeJA treatment induced the expression of many genes, such as β-1,3-glucanase, fumarase, glucosidase, acetolactate synthase, and ceramidase. Also, many genes related to secondary metabolism are differentially expressed. These include many genes involved in GA biosyntheses, such as acetyl-CoA acetyltransferase (ACAT), cytochrome b2 (CYB2), and cytochrome P450 (ERG5). In the transcriptome analysis, these researchers also observed the transcription changes of various kinases that regulate the *G. lucidum* signal transduction pathway, such as CMGC/MAPK, Akt protein kinase, and small G protein (rho), and the ERK signaling pathway. Many serine-threonine protein kinases and MAPK kinases were further identified to have differentially expressed. For example, following 15 min of MeJA treatment, the cAMP-dependent protein kinase regulatory subunit BCY1, serine-threonine protein kinase SKY1, and IRE1 were upregulated. In contrast, MAPK kinase KSS1 and MKK2 were downregulated. Besides, the transcriptome and proteome analyses were employed to detect the differential expression of a large number of ribosome-related genes and proteins, respectively. Significantly changes at the transcription levels for ribosomal protein L13 (50S ribosomal protein L13, RPL13), ribosomal protein L10 (60S ribosomal protein L10, RPL10), and mitochondrial 50S ribosomal protein L5 (RPL5) were observed in the transcriptomic analysis. In the proteomic analysis results, a large number of ribosomal subunit proteins were significantly downregulated.

Integrating the transcriptome and proteome analysis results revealed that MeJA treatment affected many biological processes in *G. lucidum* significantly. These include signal transduction, secondary metabolism synthesis, protein transcription, stress resistance, cell cycle regulation, reactive oxygen scavenging, and other physiological processes (Jiang et al. 2019). The comparison of iTRAQ and qRT-PCR results demonstrates that most pathways undergoing significant changes are concerning the genetic information processing pathways, such as DNA replication, transcription, RNA splicing, and translation. The ribosomal subunit protein is also downregulated. In contrast, the genes related to triterpene synthesis were all upregulated. These genes include farnesyl pyrophosphate synthase (FPS) and cytochrome P450 family ERG5. These results also show that energy supply-related pathways, such as glycolysis and the TCA cycle, are continuously regulated, but the regulation direction is inconsistent. The short-term treatment group's energy supply is increased, and by 24 h, the energy supply is exhibiting a downward trend. Therefore, integrated transcriptome and proteome analysis results may help explain how MeJA regulates GA biosynthesis in *G. lucidum*.

Metabolomics is a new method that simultaneously conducts qualitative and quantitative analyses of all low-molecular-weight metabolites of cells (Fernandes et al. 2019; Rodrigues et al. 2019; Troisi et al. 2020; Williams et al. 2019; Wishart 2019). The metabolites contain information from endogenous substances (genes) and exogenous substances (nutrition, drugs, and the environment). The combination of metabolomic and proteomic analysis results can identify the key enzymes involved in regulating metabolic pathways. Also, the metabolomic data can further functionally translate proteomic data. Proteomics and metabolomics have been employed to study the genesis of many important diseases, including cancer, mental illness, and cardiovascular disease.

However, only a few studies have investigated *G. lucidum* metabolomes (Jiang et al. 2019; Wen et al. 2010), which primarily perform quantitative detection and qualitative analysis of specific metabolic components. For example, Wen used nuclear magnetic resonance spectroscopy to study *G. lucidum* produced in Korea and China (Wen et al. 2010). The detection of specific components indicated that *G. lucidum* from different origins has clear different composition and content of secondary metabolites. Some other quantitative detection methods for *G. lucidum* components have also been developed, including

NMR, proton NMR, liquid chromatography-UV, and ultra-performance convergence chromatography-SQD-MS. These methods help further to explore other secondary metabolite components in *G. lucidum*. Shi et al. discovered that MeJA partly regulates GA biosynthesis through NOX-mediated ROS signals (Shi et al. 2015). Based on the information provided by proteome and metabolome data, Jiang et al. found that MeJA can inhibit mitochondrial respiration, leading to mitochondrial ROS burst and subsequently regulating the biosynthesis of GAs (Jiang et al. 2019).

Meanwhile, they could assign structural identities to 224 metabolites, 154 by GC–MS and 70 by LC–MS/MS. These metabolites' abundance displayed significant changes ($p < 0.05$) between the samples treated with MeJA for 24 h (M24) and the control (C24). KEGG analysis revealed that the 224 metabolites were enriched in various cellular metabolism pathways, such as the biosynthesis of secondary metabolites, the biosynthesis of amino acids, protein digestion and absorption, purine metabolism, fatty acid biosynthesis, and cutin/suberine/wax biosynthesis. By integrating proteomics and metabolomics results, we can dissect the mechanism of global responses to MeJA in *G. lucidum*. Their results indicated that MeJA induction inhibited normal glucose metabolism and suggested that the intracellular energy metabolism pathways were switched from primary metabolism to secondary metabolism in response to MeJA induction (Fig. 6.7) (Jiang et al. 2019). This study has been invaluable for further analysis of the mechanisms by which environmental signals regulate GA biosynthesis.

6.4 Advantages and Disadvantages of *Ganoderma lucidum* Proteomics

Proteomics research on *G. lucidum* is still in the initial stages. The completion of whole-genome sequencing project provides a useful platform for future omics research. The shortage of proteomic analysis studies is related to the difficulty in efficiently extracting proteins in *G. lucidum* (Wang and Hanash 2003). Because of high polysaccharide and polyphenol contents, the purification of proteins is relatively cumbersome. Low-abundance proteins are easy to lose in the purification process. Furthermore, the extraction of membrane protein is another difficulty in proteomic research on *G. lucidum*. Due to membrane proteins' characteristics, such as hydrophobicity, more fat-soluble proteins than water-soluble proteins were found in the extracts. The amount and characteristics of lipids interfere with the dissolution of membrane proteins in the extraction buffer. These studies are also limited by out-of-date proteomics techniques that need to be further optimized, sensitivity, reproducibility, quantitative accuracy, and high detection costs. The traditional gel-based protein separation system has low efficiency. It can only detect high-abundance proteins within specific molecular weight and pH ranges, and a large amount of adequate protein information is therefore lost. Given that the secondary metabolites of *G. lucidum* have much medicinal value, we should actively attempt to employ new technological methods, including omics, bioinformatics, and functional genomics, to obtain information on the genome of *G. lucidum*. The results of these studies suggest that further exploration of the biosynthesis and metabolic regulation mechanism of important secondary metabolites of *G. lucidum* is needed.

6.5 Concluding Remarks

Although proteomics studies in *Ganoderma lucidum* have made preliminary progress, comprehensive qualitative and quantitative analyses of many *Ganoderma* proteins remain to be carried out. Several issues need consideration. For example, how to achieve proteomics analysis more efficiently by increasing the throughput and shorten the time? How to maintain an ideal coverage depth to discover more low-abundance functional proteins? The sensitivity of protein detection and quantification accuracy needs to be further improved to significantly distinguish between noise and low-abundance proteins.

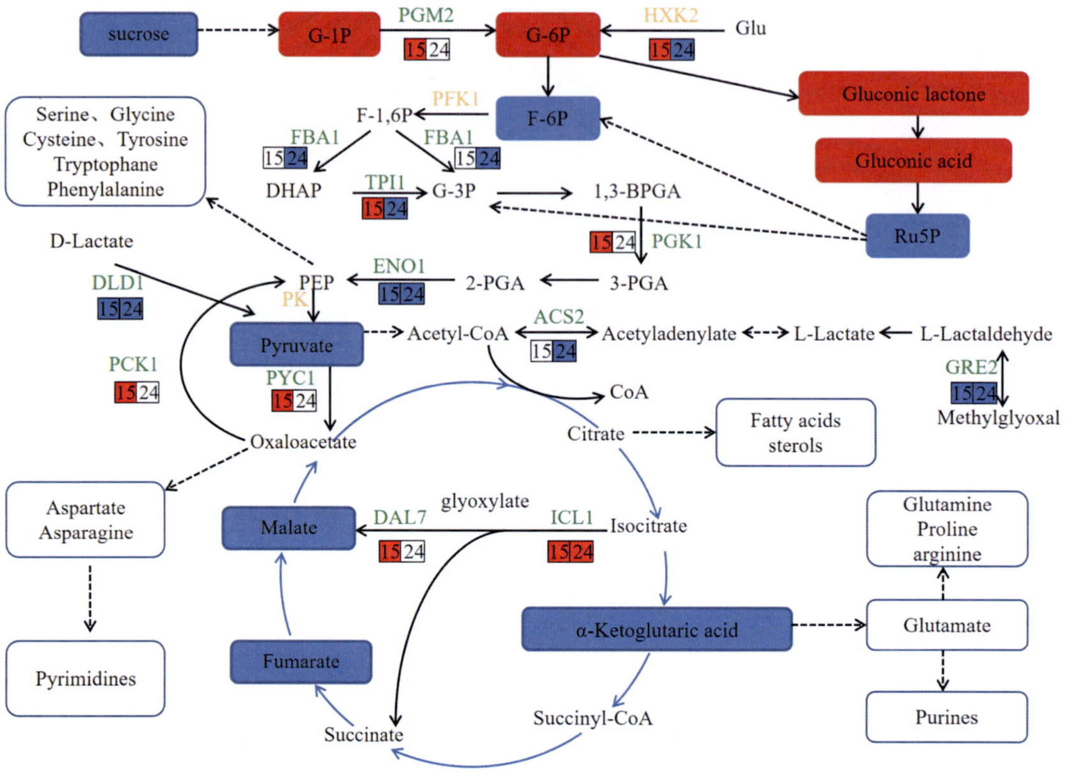

Fig. 6.7 Integrated analysis based on proteomic and metabolomic results in *G. lucidum* cells with MeJA treatment for 15 min and 24 h (Jiang et al. 2019). Metabolites are shown in curved squares. Those shaded in red are upregulated, and those shaded in blue are downregulated. The numbers "15" and "24" indicate MeJA treatment for 15 min and 24 h, respectively. Here, the red background indicates the up-regulation of the corresponding enzymes. The blue background indicates the down-regulation of the related enzymes. The white background indicates no change. The green font indicates detected functional proteins, and the golden font indicates rate-limiting enzymes. The solid line represents the one-step reaction, and the dashed line represents the multi-step reaction. They were reproduced from Jiang et al. 2019 with permission

Besides, as the main undertaker of various biological activities, proteins can provide more direct and valuable phenotypic information. The single-cell proteomics represents another frontier in *Ganoderma* research.

References

Al-Obaidi JR, Saidi NB, Usuldin SR, Hussin SN, Yusoff NM, Idris AS (2016) Comparison of Different Protein Extraction Methods for Gel-Based Proteomic Analysis of *Ganoderma* spp. Protein J 35(2):100–106

Al-Obaidi JR, Hussin SNIS, Saidi NB, Rahmad N, Idris AS (2017) Comparative proteomic analysis of *Ganoderma* species during in vitro interaction with oil palm root. Physiol Mol Plant P99:16–24.

Dominguez DC, Lopes R, Torres ML (2007) Proteomics technology. Clin Lab Sci 20(4):239–244

Gluckmann M, Fella K, Waidelich D, Merkel D, Kruft V, Kramer PJ, Walter Y, Hellmann J, Karas M, Kroger M (2007) Prevalidation of potential protein biomarkers in toxicology using iTRAQ reagent technology. Proteomics 7(10):1564–1574

Jensen PK, Pasa-Tolic L, Anderson GA, Horner JA, Lipton MS, Bruce JE, Smith RD (1999) Probing proteomes using capillary isoelectric focusing-electrospray ionization Fourier transform ion cyclotron resonance mass spectrometry. Anal Chem 71 (11):2076–2084

Jiang AL, Liu YN, Liu R, Ren A, Ma HY, Shu LB, Shi L, Zhu J, Zhao MW (2019) Integrated proteomics and metabolomics analysis provides insights into ganoderic acid biosynthesis in response to methyl jasmonate in *Ganoderma lucidum*. Int J Mol Sci 20 (24)

Liu XY, Su MS, Yang JZ, Xie XM (2009) Two-dimensional electrophoresis and extraction methods of total protein of *Ganoderma lucidum* primordium. (in Chinese). Mycosystema 28(6):802–805

Liu J, Bai J, Lu Q, Zhang L, Jiang Z, Michal JJ, He Q, Jiang P (2013) Two-dimensional liquid chromatography-tandem mass spectrometry coupled with isobaric tags for relative and absolute quantification (iTRAQ) labeling approach revealed first proteome profiles of pulmonary alveolar macrophages infected with porcine circovirus type 2. J Proteomics 79:72–86

Lu Q, Bai J, Zhang L, Liu J, Jiang Z, Michal JJ, He Q, Jiang P (2012) Two-dimensional liquid chromatography-tandem mass spectrometry coupled with isobaric tags for relative and absolute quantification (iTRAQ) labeling approach revealed first proteome profiles of pulmonary alveolar macrophages infected with porcine reproductive and respiratory syndrome virus. J Proteome Res 11(5):2890–2903

Macek B, Carpy A, Koch A, Bicho CC, Borek WE, Hauf S, Sawin KE (2017) Stable isotope labeling by amino acids in cell culture (SILAC) technology in fission yeast. Cold Spring Harb Protoc 2017(6):pdb top079814

Manavalan T, Manavalan A, Thangavelu KP, Heese K (2012) Secretome analysis of *Ganoderma lucidum* cultivated in sugarcane bagasse. J Proteomics 77:298–309

Ono M, Shitashige M, Honda K, Isobe T, Kuwabara H, Matsuzuki H, Hirohashi S, Yamada T (2006) Label-free quantitative proteomics using large peptide data sets generated by nanoflow liquid chromatography and mass spectrometry. Mol Cell Proteomics 5(7):1338–1347

Polacek M, Bruun JA, Johansen O, Martinez I (2010) Differences in the secretome of cartilage explants and cultured chondrocytes unveiled by SILAC technology. J Orthop Res 28(8):1040–1049

Ren A, Qin L, Shi L, Dong X, da Mu S, Li YX, Zhao MW (2010) Methyl jasmonate induces ganoderic acid biosynthesis in the basidiomycetous fungus *Ganoderma lucidum*. Bioresour Technol 101(17):6785–6790

Ren A, Li MJ, Shi L, Mu DS, Jiang AL, Han Q, Zhao MW (2013) Profiling and quantifying differential gene transcription provide insights into ganoderic acid biosynthesis in *Ganoderma lucidum* in response to methyl jasmonate. PLoS ONE 8(6):e65027

Rodrigues AM, Miguel C, Chaves I, Antonio C (2019) Mass spectrometry-based forest tree metabolomics. Mass Spectrom Rev 40(2):126-157

Ross PL, Huang YN, Marchese JN, Williamson B, Parker K, Hattan S, Khainovski N, Pillai S, Dey S, Daniels S et al (2004) Multiplexed protein quantitation in *Saccharomyces cerevisiae* using amine-reactive isobaric tagging reagents. Mol Cell Proteomics 3(12):1154–1169

Shi L, Gong L, Zhang X, Ren A, Gao T, Zhao M (2015) The regulation of methyl jasmonate on hyphal branching and GA biosynthesis in *Ganoderma lucidum* partly via ROS generated by NADPH oxidase. Fungal Genet Biol 81:201–211

Tan F, Jin Y, Liu W, Quan X, Chen J, Liang Z (2012) Global liver proteome analysis using iTRAQ labeling quantitative proteomic technology to reveal biomarkers in mice exposed to perfluorooctane sulfonate (PFOS). Environ Sci Technol 46(21):12170–12177

Troisi J, Cavallo P, Colucci A, Pierri L, Scala G, Symes S, Jones C, Richards S (2020) Metabolomics in genetic testing. Adv Clin Chem 94:85–153

Wang H, Hanash S (2003) Multi-dimensional liquid phase based separations in proteomics. J Chromatogr B Analyt Technol Biomed Life Sci 787(1):11–18

Wang Q, Su X, Jiang X, Dong X, Fan Y, Zhang J, Yu C, Gao W, Shi S, Jiang J et al (2016) iTRAQ technology-based identification of human peripheral serum proteins associated with depression. Neuroscience 330:291–325

Wen H, Kang S, Song Y, Song Y, Sung SH, Park S (2010) Differentiation of cultivation sources of *Ganoderma lucidum* by NMR-based metabolomics approach. Phytochem Anal 21(1):73–79

Wilkins MR, Gasteiger E, Sanchez JC, Bairoch A, Hochstrasser DF (1998) Two-dimensional gel electrophoresis for proteome projects: the effects of protein hydrophobicity and copy number. Electrophoresis 19(8–9):1501–1505

Williams C, Palviainen M, Reichardt NC, Siljander PR, Falcon-Perez JM (2019) Metabolomics applied to the study of extracellular vesicles. Metabolites 9(11):276

Wishart DS (2019) NMR metabolomics: A look ahead. J Magn Reson 306:155–161

Zhang H, Burnum KE, Luna ML, Petritis BO, Kim JS, Qian WJ, Moore RJ, Heredia-Langner A, Webb-Robertson BJ, Thrall BD et al (2011) Quantitative proteomics analysis of adsorbed plasma proteins classifies nanoparticles with different surface properties and size. Proteomics 11(23):4569–4577

Zhao Z (2013) Effect of light quality on the physiological metabolism of *Ganoderma lucidum* and related proteomic study. Peking Union Medical College Hospital

Zieske LR (2006) A perspective on the use of iTRAQ reagent technology for protein complex and profiling studies. J Exp Bot 57(7):1501–1508

7 Noncoding RNAs in Lingzhi Mushroom

Mei Jiang, Liqiang Wang, Bin Wu, and Shanfa Lu

Abstract

In this chapter, the discovery of noncoding RNAs, including lincRNA, natural antisense noncoding RNAs, circular RNAs, and microRNAs in *Ganoderma lucidum* is described. In particular, we describe the methods for validation and discovery of these noncoding RNAs, their characteristics, and their potential functions.

7.1 Introduction

Unlike protein-coding RNAs that can be translated into proteins, noncoding RNAs (ncRNAs) are a class of gene transcripts known as non-protein-coding RNAs (npcRNAs) (Storz 2002; Griffiths-Jones 2007; Wirth and Crespi 2009). They work at the RNA level and participate in various cellular processes in organisms (Li et al. 2020). Based on the expression profiles and their functions, ncRNAs are classified into two groups, including housekeeping ncRNAs and regulatory ncRNAs (Li et al. 2020; Wu et al. 2012). Housekeeping ncRNA, such as transfer RNAs (tRNAs), ribosomal RNAs (rRNAs), small nuclear RNAs (snRNAs), and small nucleolar RNAs (snoRNAs), are usually constitutively expressed (Wirth and Crespi 2009). In contrast, regulatory ncRNAs are often differentially expressed and developmentally regulated (Li et al. 2020; Wu et al. 2012; Wang et al. 2015). They include microRNAs (miRNAs)/miRNA-like RNAs (milRNAs), small interfering RNAs (siRNAs), long noncoding RNAs (lncRNAs), and circular RNAs (circRNAs). Among them, lncRNAs are the most prevalent and functionally diverse and can be linear or circular (Ma et al. 2013; Belousova et al. 2018). Based on genomic location and context, lncRNAs are further divided into intergenic lncRNAs (lincRNAs), intronic lncRNAs, sense lncRNAs, and antisense lncRNAs (natural antisense RNAs, NATs) (Ma et al. 2013). So far, lincRNAs, NATs, circRNAs, and milRNAs have been identified in lingzhi mushroom (*Ganoderma lucidum* Karst) (Li et al. 2014; Shao et al. 2017, 2019, 2020). In the following text, we will briefly describe these novel noncoding RNA molecules.

S. Lu (✉)
Institute of Medicinal Plant Development, Chinese Academy of Medical Sciences & Peking Union Medical College, Beijing 100193, China
e-mail: sflu@implad.ac.cn

M. Jiang · B. Wu
Key Laboratory of Bioactive Substances and Resource Utilization of Chinese Herbal Medicine From the Ministry of Education, Institute of Medicinal Plant Development, Chinese Academy of Medical Sciences & Peking Union Medical College, Beijing 100193, China

L. Wang
College of Pharmacy, Heze University, Heze 274015, Shandong Province, China

C. Liu (ed.), *The Lingzhi Mushroom Genome*, Compendium of Plant Genomes,
https://doi.org/10.1007/978-3-030-75710-6_7

7.2 LincRNAs in Lingzhi Mushroom

LincRNAs are long intergenic noncoding RNAs with a length usually greater than 200nt. This subset of lncRNAs plays significant regulatory roles in many cellular processes (Ulitsky and Bartel 2013; Ransohoff et al. 2018). Based on the G. *lucidum* genome's assembly and the transcriptome data obtained by high-throughput RNA-seq, Li et al. (2014) performed genome-wide identification of lincRNAs from three developmental stages (mycelia, primordia, and fruiting bodies) of *G. lucidum* using a bioinformatic pipeline. It resulted in identifying 402 lincRNA candidates, of which 46 were adjacent to cytochrome P450 genes, mating type B genes, and genes encoding carbohydrate-active enzymes or protein-coding genes involved in triterpenoid biosynthesis and lignin degradation. Further analysis of the 46 lincRNA showed that 37 were transcribed unidirectionally and nine were transcribed bidirectionally. Among the 37 unidirectionally transcribed lincRNA candidates, 16 showed positive and 5 showed negative expression correlations with the adjacent protein-coding genes across the three developmental stages (Li et al. 2014). It indicates the regulatory roles of lincRNAs in *G. lucidum* growth and development.

7.2.1 The Bioinformatic Pipeline for LincRNA Identification

Li et al. (Li et al. 2014) developed a bioinformatics pipeline (Fig. 7.1). Using the whole genome sequence as a reference (Chen et al. 2012), they first assembled short reads from high-throughput RNA-seq of transcriptome into long transcripts by Cufflinks with the default parameters (Trapnell et al. 2012). Then, a procedure consisting of 8 steps was performed. The assembled long transcripts and the predicted genes were mapped to the genome assembly of *G. lucidum* (step 1) and merged into transcript units (TUs) (step 2) (Chen et al. 2012). TUs overlapping with the predicted genes were discarded (step 3). The remaining TUs were subsequently searched for homologs by comparing with the nucleotide (Nt), non-redundant protein (Nr), and Swiss-Prot (SP) databases using the basic local alignment search tool (BLAST) with a cutoff *E*-value of $<10^{-5}$ (Altschul et al. 1997). The TUs with homologs were discarded (step 4), whereas those without homologs were subjected to open-reading frame (ORF) prediction using ESTScan (Iseli et al. 1999). TUs containing predicted ORFs that encode proteins longer than 100 amino acids were discarded (step 5) (Rymarquis et al. 2008). The remaining TUs were calculated for length. TUs with a size shorter than 200 bp were discarded (step 6) (Kapranov et al. 2007). Further analysis of the TUs for coding potential was carried out using the Coding Potential Calculator (CPC) software (step 7) (Kong et al. 2007). TUs with the coding potential score of both strands less than zero were finally searched against miRBase using BLASTN with an *E*-value cutoff 10^{-5} (step 8) (Altschul et al. 1997; Kozomara and Griffiths-Jones 2011). The remaining TUs were considered as lincRNA candidates (Li et al. 2014).

It should be noted that the pipeline developed requires complete and well-defined protein-coding gene models predicted from the genome assembly. If the predicted protein-coding genes are not complete or the predicted gene models do not contain 5'- and 3'-UTRs, some natural antisense lncRNAs and intronic lncRNAs could be included in the final dataset of lincRNA candidates.

7.2.2 The Modified 3' RACE Method for Determination of Transcriptional Orientation of LincRNAs

The RNA-seq data used for lincRNA identification was usually generated by the non-strand-specific method. In that case, determination of the transcriptional orientation of lincRNA candidates is very important for further functional

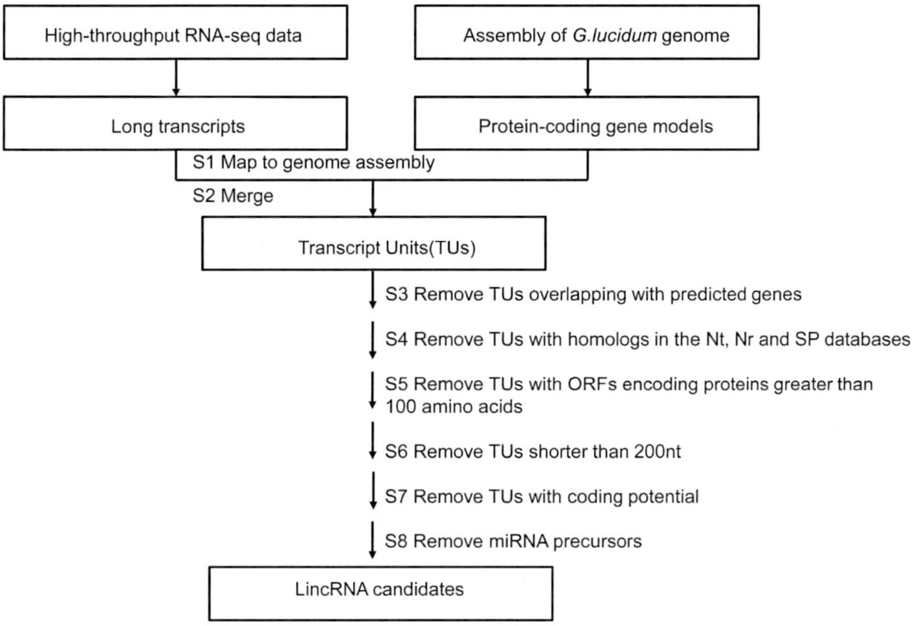

Fig. 7.1 Schematic representation of the bioinformatic pipeline for lincRNA identification

analysis. Li et al. (2014) developed a technique named the modified 3' RACE method (MRA) for this purpose. The method is outlined in Fig. 7.2. Briefly, three gene-specific primers F1, F2, and R were designed and synthesized. The relative locations of primers on a genomic locus are shown in Fig. 7.2a. Total RNA was extracted from lingzhi tissues and treated with RNase-free DNase I to remove genomic DNA contaminant. Genomic DNA-free RNA was reversely transcribed into cDNA using the GeneRacer kit (Invitrogen) and the GeneRacer oligo dT primer (UP1) in the kit (Fig. 7.2b). Then, two rounds of PCR amplification were carried out. The first-round PCR used cDNA as the amplification template, gene-specific primers F1 or R as the forward primer, and GeneRacer 3' primer (UP2) from the GeneRacer kit as the reverse primer (Fig. 7.2c). The second round of PCR used the first round of PCR products as the template, gene-specific primer F2 as the forward primer, and gene-specific primer R as the reverse primer (Fig. 7.2d). PCR products were separated by electrophoresis. The transcriptional orientation of lincRNAs was determined by the result pattern of electrophoresis (Fig. 7.2e).

7.2.3 Possible Action Mechanisms of LincRNAs in Lingzhi Mushroom

It has been shown that lincRNAs play widespread functions in organisms, such as remodeling chromatin and genome architecture, affecting RNA stabilization, and regulating gene transcription (Ulitsky and Bartel 2013; Ransohoff et al. 2018). Analysis of transcriptional direction of 46 lingzhi lincRNA candidates and the expression of these lincRNA candidates and their adjacent protein-coding genes showed that the expression of 16 unidirectionally transcribed lincRNAs was positively correlated with their adjacent protein-coding genes. It indicates that these lincRNAs can probably activate the adjacent protein-coding genes through a *cis*-transcriptional regulatory mechanism in lingzhi mushrooms (Li et al. 2014). Moreover, five unidirectionally transcribed lingzhi lincRNA candidates showed negative expression correlations with the adjacent protein-coding genes (Li et al. 2014). The action mechanism of these lincRNAs remains to be elucidated. Besides, there were nine bidirectionally transcribed

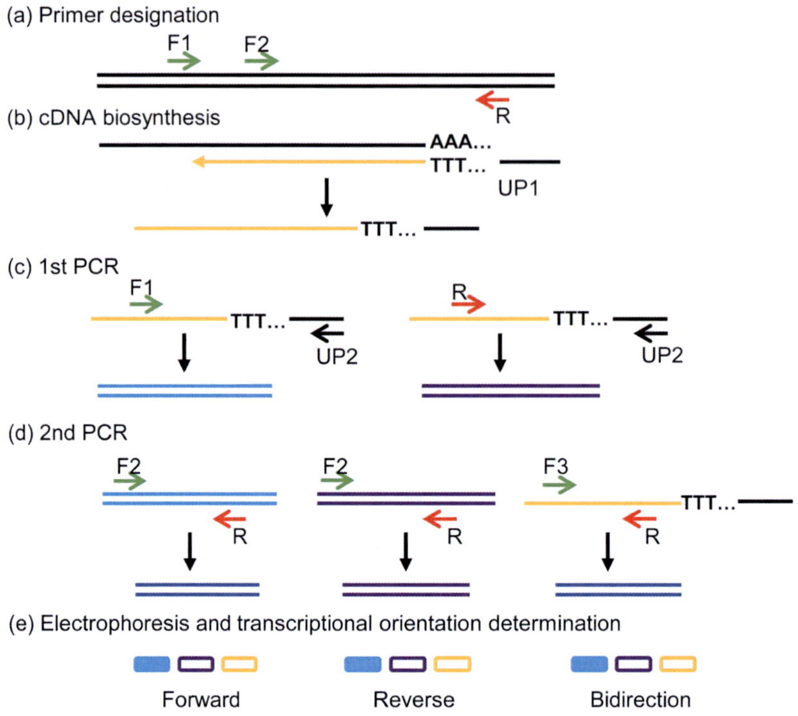

Fig. 7.2 Schematic representation of the modified 3' RACE method for transcriptional orientation determination. The method involves two universal primers (UP1 and UP2) and three gene-specific primers (F1, F2, and R). The entire procedure consists of one round of cDNA synthesis and two rounds of PCR amplification using particular pairs of primers. **a** Relative locations of primers F1, F2, and R are shown on a genomic locus. **b** The reverse transcription experiment intended to generate the first-strand cDNA using GeneRacer oligo dT primer (UP1). **c** The first round of PCR amplification using gene-specific primers F1 or R as the forward primer, GeneRacer 3' primer (UP2) as the reverse primer, and cDNA as the template. **d** The second round of PCR amplification using gene-specific primer F2 as the forward primer, gene-specific primer R as the reverse primer, and cDNA or PCR products from (**c**) as the template. **e** Electrophoresis of PCR products. The solid bars indicate the presence of the corresponding band. The hollow bars indicate the absence of the corresponding band. The yellow rounded circles were results from the negative control. For the negative control, no reverse transcription was performed. Also, genes with known transcription orientation were analyzed in the same process

lingzhi lincRNAs (Li et al. 2014). These lincRNAs may function as an enhancer of gene expression (Ørom et al. 2010; Kim et al. 2010; Wang et al. 2011).

Although current knowledge on lingzhi lincRNAs is very limited, functional characterization of the identified lingzhi lincRNAs may provide useful information for the mechanism of various biological processes. For example, circRNAs adjacent to triterpenoid biosynthesis-related genes may regulate triterpenoid production.

7.3 NATs in Lingzhi Mushroom

NATs, known as natural antisense RNAs, are a subset of lncRNAs derived from the opposite DNA strand (i.e., antisense strand) of protein-coding genes (i.e., the sense strand). It can be classified into two groups, including *cis*-NATs and *trans*-NATs. *Cis*-NATs are transcribed from the opposite DNA strand in the same genomic loci as the sense transcripts (STs) of protein-coding genes. The overlapping region of *cis*-

NATs and their corresponding STs has perfect complementarity. Differently, *trans*-NATs and their related STs are transcribed from separate loci and usually exhibit imperfect complementarity (Prasanth and Spector 2007). NATs widely exist in plants, animals, and fungi (Ansaldi et al. 2000; Katayama et al. 2005; Wang et al. 2005; Engström et al. 2006; Donaldson and Saville 2012; Ariel et al. 2015). They regulate the expression of protein-coding genes at the pre-transcriptional, transcriptional, and posttranscriptional levels through interaction with DNA, RNA, or proteins (Ariel et al. 2015; Villegas and Zaphiropoulos 2015). They inhibit the transcription of their STs by competing with transcription factors, hindering the transcriptional machinery, or using epigenetic processes. They may disrupt posttranscriptional modification and translation of their STs by forming RNA/RNA duplexes and RNAi-mediated gene silencing. They can also mask specific signals required for splicing and stability of mRNA transcribed from their STs (Munroe and Zhu 2006; Faghihi and Wahlestedt 2009; Li and Ramchandran 2010).

7.3.1 Identification of Lingzhi NATs

Using a bioinformatics pipeline (Fig. 7.3) modified from the pipeline used for lincRNA identification (Fig. 7.1) and strand-specific RNA-seq (ssRNA-seq) data generated from the mycelia, primordial, and fruiting bodies of lingzhi mushroom, Shao et al. (Shao et al. 2017) identified 1613 *cis*- and 244 *trans*-NAT, and ST pairs. The pipeline includes the following steps: assembling short reads from RNA-seq into long transcripts, merging with predicted genes into TUs, and applying a series of filtering processes, such as sequence length calculation, ORF length prediction, BLAST analysis against protein databases, Coding Potential Calculator analysis, and BLAST analysis against miRBase (Li et al. 2014). The lincRNA identification pipeline discarded TUs overlapping with the predicted genes. In contrast, the NAT identification pipeline checked the complementarity of TUs and predicted genes, and selected the TUs perfectly complementary to and coinciding with the predicted genes in the same genomic loci as *cis*-NAT candidates for further analysis. It also selected the TUs imperfectly complementary to and overlapping with the predicted genes in the separate genomic loci as *trans*-NAT candidates for further characterization (Shao et al. 2017).

7.3.2 Characterization of Lingzhi NATs

Lingzhi NATs demonstrated significant diversity in several aspects, including sequence length, NAT to ST ratio, and NAT to ST relative orientation. Sequence length distribution analysis showed that 1613 identified *cis*-NATs mostly ranged from 201 to 700nt, with the most abundant bin 201–300nt and an average of length 523nt. The identified 244 *trans*-NATs had the majority length from 201 to 800nt, with the most abundant bin 301–400nt and an average of length 580nt (Shao et al. 2017). Sequence comparison showed that the identified NATs had very complicated relationships with their STs. The pairing of STs and NATs could be 1:1 (one ST was mapped to one NAT), 1:n (one ST was mapped to at least two NATs), n:1 (at least two STs were mapped to one NAT), and n:n (an ST was mapped to at least two NATs and the NAT was mapped to at least two STs). Among the 1613 *cis*-ST and NAT pairs, 1175, 417, 26, and 12 formed 1:1, 1:n, n:1, and n:n relationships. Among the 244 *trans*-ST and NAT pairs, 156, 78, 29, and 61 formed 1:1, 1:n, n:1, and n:n relationships, respectively (Shao et al. 2017). Among the 1613 *cis*-ST and NAT pairs, 373 were tail-to-tail, 614 were head-to-head, 526 STs contained NATs, and 117 NATs contain STs. Analysis of NAT expression using RNA-seq data showed that 1477 NATs were expressed in all three development stages analyzed. 53, 12, and 54 were expressed only in the mycelium, primordial, and fruiting bodies, respectively (Shao et al. 2017).

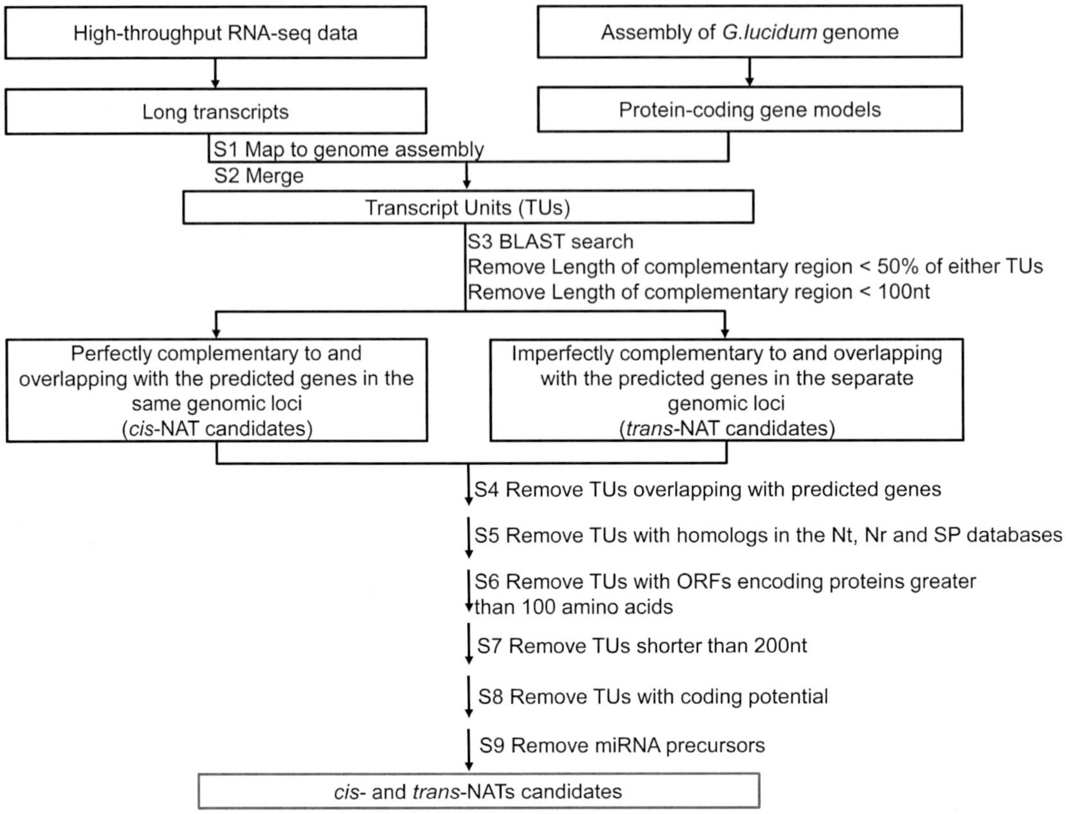

Fig. 7.3 Schematic representation of the bioinformatic pipeline for NAT identification

7.3.3 Possible Functions of Lingzhi NATs

NATs play significant roles in various organisms (Ariel et al. 2015; Villegas and Zaphiropoulos 2015; Munroe and Zhu 2006; Faghihi and Wahlestedt 2009; Li and Ramchandran 2010). Enrichment analysis of STs with NATs in lingzhi mushrooms showed that they belonged to different functional categories, such as transmembrane transport, monooxygenase activity, and DNA replication (Shao et al. 2017). Further functional analyses revealed that lingzhi NATs were non-randomly associated with particular functional groups of STs across all stages and in specific stages. It indicates that NATs play essential roles in regulating the expression of the corresponding STs, and NAT-mediated ST expression regulation is developmental stage-related and is associated with particular cellular functions.

Furthermore, NATs can be either stabilizers or repressors of STs in lingzhi mushrooms. Expression correlation analysis of 25 ST and NAT pairs showed that 15 pairs had NATs and STs significantly positively correlated and four pairs were significantly negatively correlated (Shao et al. 2017). Positive expression correlation could result from the increase of ST mRNA stability after binding with the corresponding NAT, which acted as a stabilizer. In contrast, negative expression correlation could result from the repression of ST expression by the corresponding NAT, which served as a repressor. So far, the underlying mechanisms remain to be elucidated. Preliminary evidence indicated that the interactions between lingzhi mushroom STs and NATs could be rather complicated. The expression regulation of lingzhi mushroom ST and NAT pairs might be mediated through a complex multi-layered regulatory network (Shao et al. 2017).

7.4 CircRNAs in Lingzhi Mushroom

CircRNAs are a class of npcRNAs found in eukaryotes. They are derived from the back-splicing of precursor mRNAs. CircRNAs can originate from different sources: exons (exonic circRNAs), introns (intronic circRNAs), both exons and introns (exon-intronic circRNAs), and intergenic regions (intergenic circRNAs). The 5' and 3' ends of a circRNA are linked together to form a covalent closed-loop structure. The circularization of linear RNA molecules enhances their stability. They become resistant to RNase R that preferentially degrades linear molecules. The biogenesis of circRNAs is considerably complicated. Several mechanisms have been proposed. These include (1) lariat-driven circularization or exon skipping; (2) intron-pairing-driven circularization or direct back-splicing; (3) intron circularization by tail trimming; and (4) RBP or trans-factor-driven circularization (Jeck et al. 2013).

CircRNAs were first reported in a virus in 1976 (Sanger et al. 1976). The presence of circRNAs was then tested in a wide range of species, with the maturation of high-throughput DNA sequencing technology and the development of many bioinformatics analysis methods. The results showed that circRNAs are present in various animals and plants, including humans (Julia et al. 2012), Archaea (Miri et al. 2012), *Arabidopsis thaliana* (Chu-Yu et al. 2015), rice (Lu et al. 2015), etc. However, to our best knowledge, to date, fungi circRNAs are described only in two species, namely Lingzhi mushroom (Shao et al. 2019) and *Cryptococcus neoformans* (Huo et al. 2018). Although many circRNAs have been identified, only a few are studied in detail (Kristensen et al. 2019). CircRNAs can regulate gene expression by acting as miRNA sponges to bind specific miRNAs, preventing them from regulating target genes (Piwecka et al. 2017). CircRNAs can also regulate the splicing of their parental genes through R-loop formation (Conn et al. 2017).

7.4.1 Strategies for the Identification of Lingzhi CircRNAs

Due to the specific structure of circRNAs, there are two commonly used methods for constructing a sequencing library. Firstly, circRNAs are detected in a sequencing library constructed with rRNA depletion but without mRNA enrichment. The library constructed using this method is called polyA(−) library. Secondly, total RNAs are treated with RNase that selectively degrades linear RNA molecules, followed by standard polyA(−) library construction procedure. This library construction method is known as polyA(−)/RNase R.

To construct Lingzhi circRNA sequencing libraries, the authors extracted RNA samples from three developmental stages of Lingzhi mushroom, which were mycelia, primordia, and fruiting bodies. Two biological replicates were prepared for each developmental stage above (Fig. 7.4) (Shao et al. 2019). Two types of libraries, including polyA(−)/RNase R and polyA(−), were constructed for each sample. Each library was then sequenced using the HiSeq 4000 platform. The sequence data were used to predict circRNAs by CIRCexplorer2 (Zhang et al. 2014). Analysis of the sequencing data obtained from the polyA(−)/RNase R libraries showed that 1273, 870, and 893 exonic and intronic circRNAs were identified in both replicate samples of mycelia, primordia, and fruiting bodies, respectively, using the read cutoff of 1 in at least one sample (Shao et al. 2019). Concerning the RNA-seq data generated from the polyA(−) libraries, 17, 123, and 144 exonic and intronic circRNAs were identified in both replicate samples from the three developmental stages using the read cutoff of 1 (Shao et al. 2019). The results suggest that the polyA(−)/RNase R libraries include significantly more circRNAs than the polyA(−) libraries.

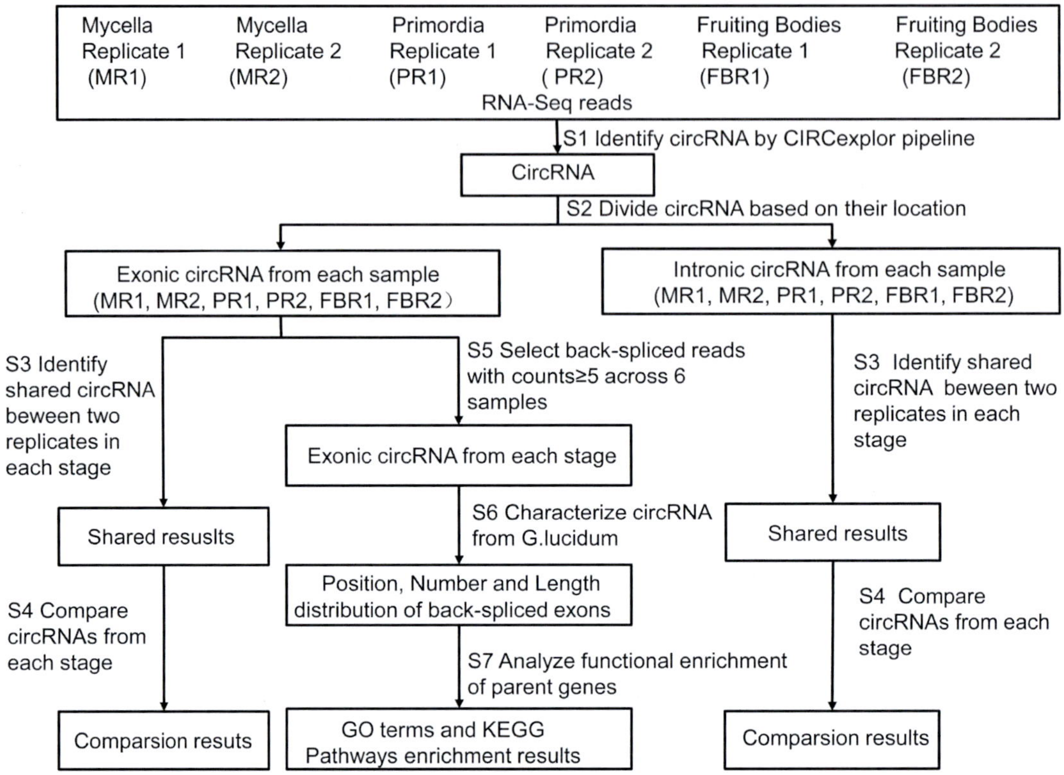

Fig. 7.4 Schematic representation of the bioinformatic pipeline for circRNA identification. The entire procedure consists of seven steps, which are labeled as S1 to S7 in the text. The results obtained from the analysis steps are shown in squares

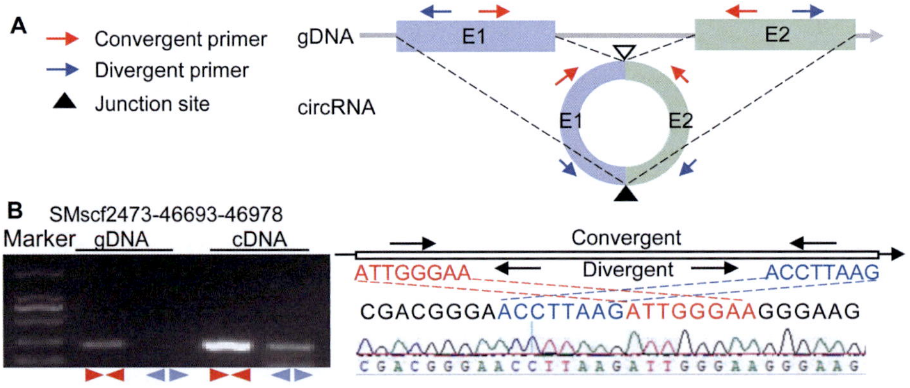

Fig. 7.5 Experimental validation of the circRNAs identified by high-throughput DNA sequencing. Panel **A** is the schematic representation of a circRNA derived from the back-splicing of two exons (E1) and (E2). The filled triangle indicates the back-splicing site. The open triangle indicates the regular splicing site. The positions and orientations of the convergent and divergent primers are shown. Panel **B** shows the experimental results of the validation. On the left is the gel electrophoresis result. The template for the PCR amplification is either genomic DNA (gDNA) or cDNA. The primers used for each PCR amplification are shown at the bottom. The PCR products were excised and subjected to Sanger sequencing. The corresponding results and their mapping to the genome are shown on the right side. The relevant location of the sequences on the genomic DNA sequence is shown

7.4.2 Validation of Lingzhi CircRNAs

The predicted circRNAs are usually validated by reverse transcription (RT)-PCR amplification and subsequent Sanger sequencing. The method of validation essentially utilizes two sets of primers, termed convergent primers and divergent primers. These primers are designed based on the genomic sequence generating circRNAs. The convergent primers face each other, while the divergent primers face away from the regular splicing site, and toward the back-splice site (Fig. 5a). Convergent primers generate PCR products from genomic DNA and cDNA. The products potentially have a different length (Fig. 5a).

In contrast, the divergent primers generate products only from cDNA. PCR products with expected size are further sequenced using the Sanger method (Fig. 5b). If the sequencing results are consistent with the expected back-splice sites, the verification is considered successful.

In Lingzhi, six predicted circRNAs have been experimentally confirmed. They are GaLu96scf_31_39627_39859 originated from GL24100 gene, GaLu96scf_18_226622_227010 originated from GL21985 gene, GaLu96scf_18_226875_227010 originated from GL21985 gene, GaLu96scf_2-1,057,062–1,057,450 originated from GL22329 gene, GaLu96scf_21_34416_34722 originated from GL22778 gene, and GaLu96scf_ 35_328572_328844 originated from GL24337 gene (Shao et al. 2019). Many other circRNAs were not validated using the methods described above. To explore the functions of these circRNAs, the authors calculated the correlation between the expression profiles of circRNAs and their parent genes (Shao et al. 2019). The results showed that the expression of 16.3% of exonic circRNAs and their parent genes was significantly positively correlated ($r \geq 0.9$, $q < 0.01$), whereas the expression of 30.9% was significantly negatively correlated ($r \leq -0.9$, $q < 0.01$).

7.4.3 Possible Functions of CircRNAs in Lingzhi Mushroom

CircRNAs have been proposed to be involved in diverse biological processes. In Lingzhi, circRNAs may participate in the expression regulation of their parent genes in a developmental-stage-specific manner. CircRNAs potentially regulate many genes involved in the growth, development, and secondary metabolism of Lingzhi mushrooms. For example, Lingzhi genes GL20553 and GL24505 encode proteins belonging to the CAZy families. They were mapped to the GO terms, including GO:0,003,824 (catalytic activity), GO:001,620 (membrane), GO:0,055,114 (oxidation–reduction), and GO:0,016,616 (oxidoreduction activity acting on the CH-OH group of donors NAD or NADP as acceptor). These terms were enriched in genes having circRNAs at all three stages. Parent gene GL24088 encodes a glutaryl-CoA synthase. This gene was mapped to the GO term GO:0,016,021 (integral to the membrane), which was only enriched in mycelia. Parent genes, GL17382, GL30772, and GL30807, encode proteins belonging to the CYP450 families. They were mapped to the GO terms GO:0,009,055 (electron carrier activity), GO:0,020,037 (heme-binding), and GO:0,004,497 (monooxygenase activity). These three genes were enriched in mycelia and fruiting bodies. The identification of circRNAs targeting these genes raised the possibility that circRNAs play a role in active compound biosynthesis in Lingzhi mushrooms.

7.5 MilRNAs in Lingzhi Mushroom

Small RNAs (sRNAs) are a group of RNA molecules. Their lengths are usually 19–28nt. Several types of sRNAs have been reported, including miRNAs (Lee and Ambros 2001; Lau et al. 2001; Lagosquintana et al. 2001), endogenous trans-acting siRNAs (ta-siRNAs), repeat-associated siRNAs (ra-siRNAs), piwi-interacting RNAs (piRNAs), and so on (Mello and Conte 2004; Aravin et al. 2007). Among them, miRNAs are endogenous sRNAs derived from primary transcripts (pri-miRNA) with a characteristic stem-loop structure (Bartel 2004). They function in regulating the expression of genes in the biological processes, such as cell differentiation, proliferation, and oxidative stress resistance (Moss et al. 1997), in all domains of lives, including plants (Mao et al. 2012; Jones-rhoades et al. 2006), animals (Lewis et al. 2003), and fungi (Lee et al. 2010).

In fungi, miRNAs are usually called miRNA-like RNAs (milRNAs) (Lee et al. 2010). Using a high-throughput small RNA sequencing method, Shao et al. (2020) systematically predicted and characterized milRNAs across three developmental stages of *G. lucidum*. A total of 168 unique milRNAs were predicted. Among them, 42 were selected and subjected to RT-PCR amplification. The corresponding PCR products were then sequenced. The results showed that five products had sequences similar to those predicted. The results confirm the presence of milRNAs in *G. lucidum*. However, the low success rate highlights the difficulty in validation. Among the 168 predicted milRNAs, 111 were found to be significantly differentially expressed ($q \leq 0.05$) across the three developmental stages of *G. lucidum*. The expression levels of 12 milRNAs were validated using stem-loop quantitative real-time polymerase chain reaction (qRT-PCR). The expression profiles of eight miRNAs showed a high correlation with the sequencing results ($r \geq 0.9$, $p \leq 0.05$). The results open new avenues for researches on the biosynthesis and function of milRNAs in basidiomycetes.

7.5.1 Identification of Lingzhi MilRNAs

Shao et al. (2020) used the small RNA sequencing and the degradome sequencing methods to generate high-throughput data for milRNA identification. The entire procedure is shown in Fig. 7.6. Raw sequences from each sample of mycelia, primordia, and fruiting bodies were pre-processed using FASTX-Toolkit (v0.0.14). The resulting reads were used to search against the Rfam (http://rfam.janelia.org/) and the Repeat databases (http://www.girinst.org/repbase/) using script rfam_scan.pl (Sarah et al. 2013). Any reads hitting entries of Rfam were filtered out. This process removed reads potentially derived from mRNA, rRNA, tRNA, snoRNA, snRNA, other noncoding RNAs, and repeat sequences. The resulting sequences were further compared against miRBase 21.0 (Griffithsjones et al. 2008). Sequences having three or few mismatches with mature miRNAs in miRBase were kept as potential miRNAs (Mu et al. 2015). The remaining sequences were analyzed using miRDeep2 software (FriedlanNder et al. 2012) with the default parameters.

7.5.2 Characterization of Lingzhi milRNAs

A total of 92, 149, and 134 milRNAs were identified in both replicate samples from mycelia, primordia, and fruiting bodies, respectively. Those milRNAs identified in both replicate samples were retrieved, resulting in 168 unique milRNAs across the three developmental stages. Seventy-seven out of 168 (45.83%) milRNAs were shared across all three developmental stages. By contrast, 2, 22, and 14 out of 168 milRNAs (1.19%, 13.10%, and 8.33%) were found only in the mycelia, primordia, and fruiting bodies, respectively. Twelve milRNAs whose putative target genes are involved in the lignin degradation pathway, MVA pathway, and polysaccharide biosynthesis pathway are as follows: milRNA008, milRNA020, milRNA042, milRNA061, milRNA063, milRNA120, milRNA139, milRNA143, milRNA152, milRNA154, milRNA159, and milRNA160. The hairpin structures of these 12 selected milRNAs sequences are shown in Fig. 7.7.

7.5.3 Validation of the Predicted milRNAs Using Stem-loop RT-PCR and Sanger Sequencing

Forty-two predicted milRNAs were selected for further validation. Primers were designed to amplify the expected milRNAs using cDNA or total RNAs as a template. The total RNAs were pre-treated with DNAse I to remove DNA

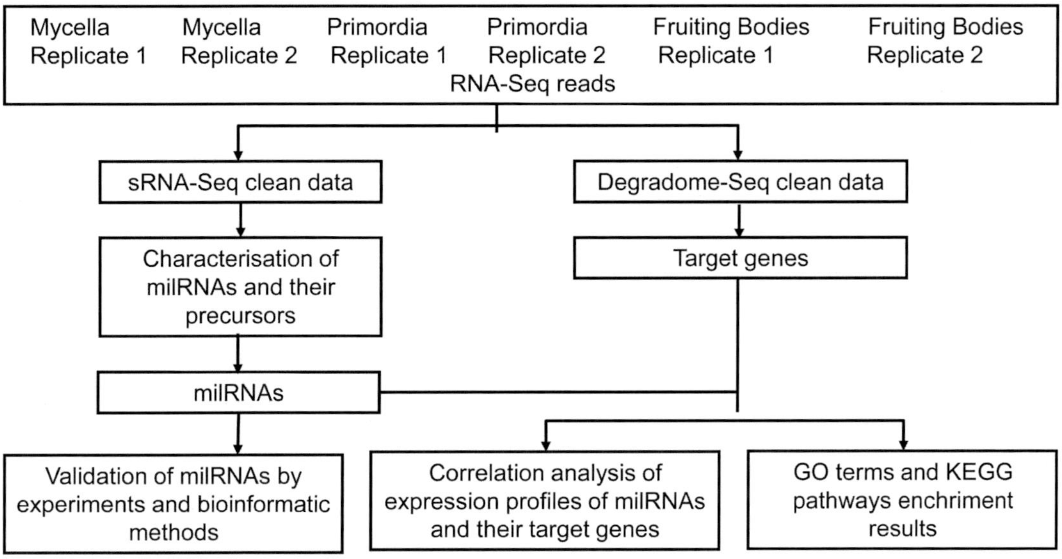

Fig. 7.6 Pipeline for the identification and validation of milRNAs

contaminations. The forward and reverse primers were designed according to those described previously (Chen et al. 2005) using software primer 6.0. The RNA samples used for RT-PCR validation were similar to those used for high-throughput DNA sequencing experiments.

Finally, 23 of the 42 predicted milRNAs had amplified products. Examination of the results showed that 18 of them contained only the primer sequences. The other five (milRNA024, milRNA033, milRNA038, milRNA067, and milRNA137) had sequences similar to those predicted. Primers specific for RPL4 (Xu et al. 2014) were also used to amplify the cDNAs. It served as a positive control to ensure that our PCR reactions worked adequately. The gel electrophoresis results of PCR products are shown in Fig. 7.8a. Lane 1 shows the results using cDNA as a template, lane 2 shows the results using water as a template, and lane 3 shows the results using total RNA treated with DNAase as a template. They used water as the template to ensure that the PCR products are specific to the input cDNA template. In contrast, they used total RNAs as the template to ensure that the PCR products were not amplified from contaminating genomic DNA. As expected, there are PCR products only in lane 1, where the cDNAs were used as the template. The PCR products amplified from the cDNA templates were then sent for Sanger sequencing.

Figure 7.8b shows the chromatograms of Sanger sequencing results. The five predicted sequences and those sequenced showed high sequence similarity (Fig. 7.8c). Three of the five sequences were identical to those predicted: milRNA024, milRNA033, and milRNA067. The predicted sequences of milRNA038 and milRNA137 had 4 and 9 base deletions compared with those obtained experimentally. The reasons remain to be identified.

7.5.4 Validation of the Predicted milRNAs Using Bioinformatic Analyses

The low validation rate obtained above suggests that experimental validation of milRNAs is difficult due to reasons currently unknown. As a result, Shao et al. (2020) performed an additional bioinformatic analysis to determine the likelihood of these predicted milRNAs being real.

Firstly, the copy numbers of each milRNA in the *G. lucidum* genome were determined by comparing the milRNA with the genome using

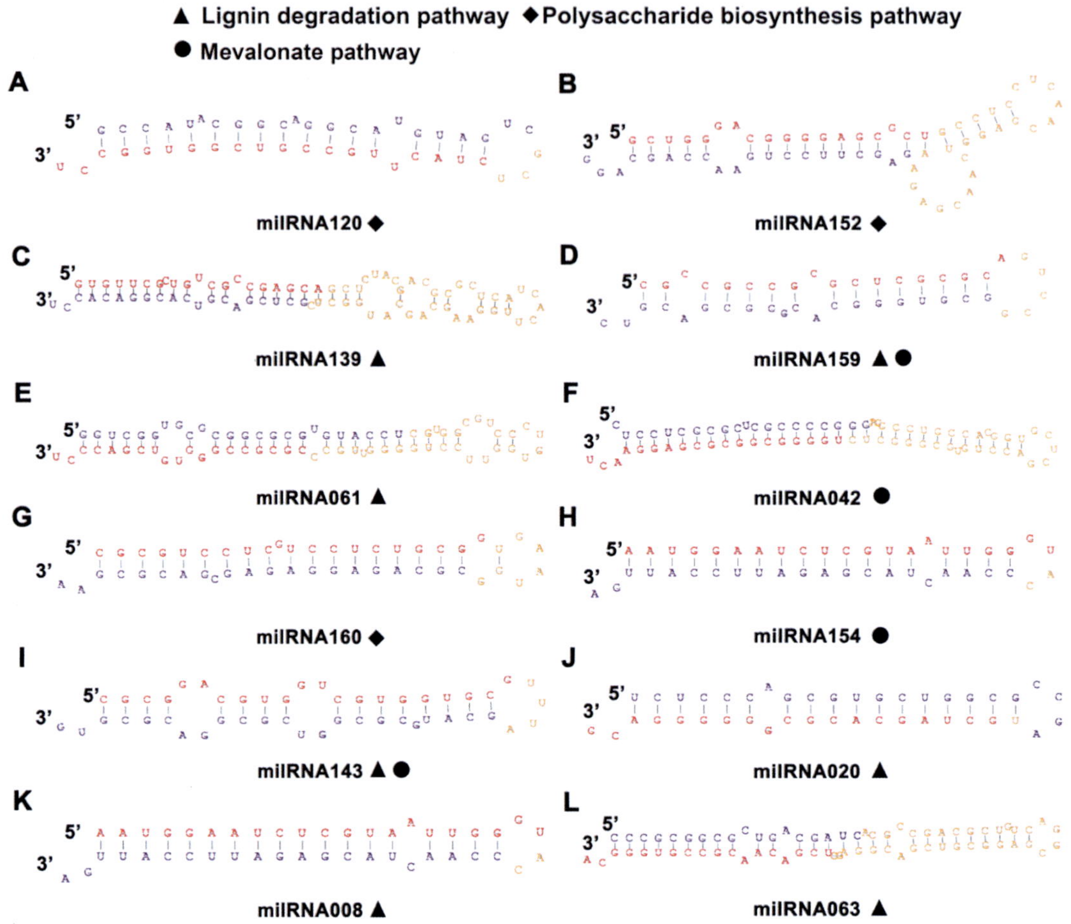

Fig. 7.7 Hairpin structure of 12 predicted milRNAs. The mature milRNAs sequence is highlighted in red, whereas the milRNAs* sequence is highlighted in purple, and the stem-loop region is shown in yellow. **A** milRNA120; **B** milRNA152; **C** milRNA139; **D** milRNA159; **E** milRNA061; **F** milRNA042; **G** milRNA160; **H** milRNA154; **I** milRNA143; **J** milRNA020; **K** milRNA008; and **L** milRNA063. '▲ ': Lignin degradation pathway; '◆ ': Polysaccharide biosynthesis pathway; '● ': Mevalonate pathway (reprinted from Shao et al. 2020 with permission).

the BLASTN program. All the 12 milRNA genes had only one copy in the genome. Secondly, the expression levels of 12 consensus milRNA precursor genes were determined using the lncRNA sequencing data reported before (Shao et al. 2017). Thirdly, the expression levels of the 12 milRNA were determined using the sRNA sequencing data. The majority of the 12 selected milRNAs exhibited tissue-specific expression. Stage-specific expression of milRNAs suggests that they might play species-specific functions (Xia et al. 2016). Fourthly, the correlation between the expression profiles of the 12 milRNAs and their target genes were analyzed. Fifthly, the cleaved transcripts of the 12 milRNA-target genes were analyzed. All the target genes had reads which were mapped to the cleaved positions. The target genes mainly belonged to categories 2, 3, and 4. Only one and two target genes of milRNA042 and milRNA152 belonged to category 0. These 12 milRNAs are likely to represent true milRNAs.

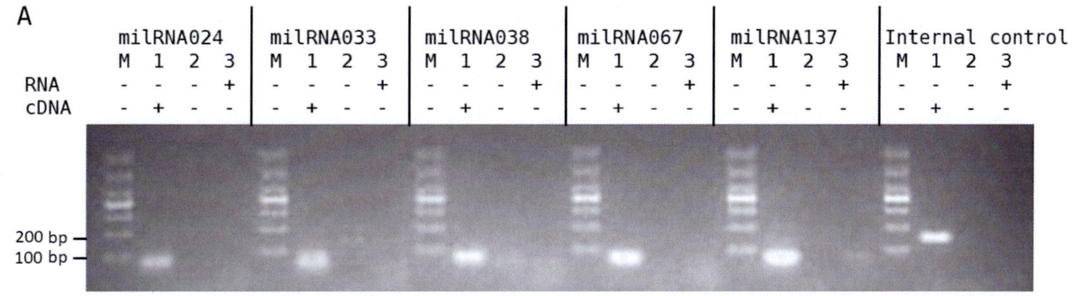

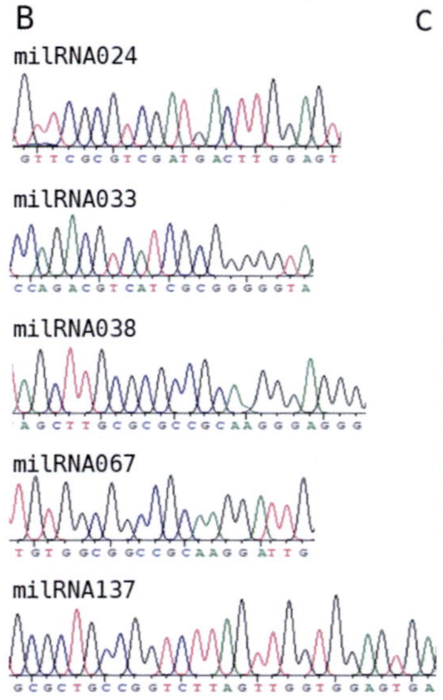

C

milRNA ID	Alignment of sequences predicted (top) and those obtained experimentally (bottom)(Length, nt)
milRNA024	GUUCGCGUCGAUGACUUGGAGU (22) GTTCGCGTCGATGACTTGGAGT (22)
milRNA033	CCAGACGUCAUCGCGGGGGUA (21) CCAGACGTCATCGCGGGGGTA (21)
milRNA038	AGCUUGCGCGCCGC----GAGGG (19) AGCTTGCGCGCCGCAAGGGAGGG (23)
milRNA067	UGUGGCGGCCGCAAGGAUUG (20) TGTGGCGGCCGCAAGGATTG (20)
milRNA137	GCGCUGCCGGUCUU---------AGUGA (19) GCGCTGCCGGTCTTAGTTGGTGGAGTGA (28)

Fig. 7.8 Validation of predicted milRNAs by stem-loop RT-PCR and Sanger sequencing methods. Five milRNAs (milRNA024, milRNA033, milRNA038, milRNA067, and mil-RNA137) were studied. **A** The gel electrophoresis results of PCR products using various templates. RPL4 gene was used as an internal control (Xu et al. 2014). M, marker; the lanes show the results for using cDNA as template (lane 1), using water as the template (lane 2), and using total RNA as template (lane 3). **B** The chromatograms of Sanger sequencing results of the PCR products were amplified from the cDNA template. The four colors represent different nucleotides: green, Adenine (A); blue, Cytidine (C); black, Guanine (G); red, Thymine (T). **C** Comparison of the predicted milRNA sequences and those obtained experimentally (reprinted from Shao et al. 2020 with permission)

7.5.5 Possible Functions of Lingzhi milRNA

Due to the lack of an efficient genetic transformation system for lingzhi mushroom, the functions of these predicted milRNAs have not been validated. Nevertheless, expression correlation analyses have been used to reveal potential interactions between the milRNAs and their targets. It is found that the correlation between milRNAs and their putative target genes can be either positive or negative. For example, eight milRNA-target pairs exhibited a negative relationship at the expression level. For instance, milRNA139-GL21192 and milRNA061-GL22248 were related to the copper radical oxidase and the benzoquinone reductase in the lignin degradation pathway (Kersten and Dan 2014; Mori et al. 2016), respectively. The milRNA154-GL23376 was associated with squalene monooxygenase in the MVA pathway (Shiao et al. 1994). In contrast, the milRNA152-GL23023 was linked to RHO1 GDP-GTP exchange protein in the polysaccharide biosynthesis pathway (Macías-Sánchez et al. 2011; Reyes-Medina and Macías-Sánchez 2015; Richthammer et al. 2012).

By contrast, the expression profiles of the other eight milRNA-target pairs were positively correlated. For instance, milRNA020-GL22248 was associated with benzoquinone reductase in the lignin degradation pathway (Mori et al. 2016). The milRNA159-GL24922 was related to glutaryl-CoA synthase 3-hydroxy-3-methyl in the MVA pathway (Ren et al. 2013). Lastly, and mlRNA160-GL20535 were associated with 1,3-beta-glucan synthase in the polysaccharide pathway (Moradali et al. 2007). Whether or not these observations represent novel mechanisms of milRNAs will be the subject of future study. These findings suggest that milRNAs may play crucial regulatory roles in the biological process of *G. lucidum* through complex networks.

7.6 Concluding Remarks

During the past decades, noncoding RNAs have been characterized extensively. It is well known that they played a diverse and vital role in the biological processes. As a result, it is no doubt that the discovery and characterization of noncoding RNAs will facilitate our understanding of the molecular mechanisms in *Ganoderma*, including but not limited to its growth, replication, and the biosynthesis of secondary metabolites. In this chapter, we summarize the recent progress of noncoding RNAs, particularly those for *G. lucidum*. The current study has focused mainly on the discovery and characterization of *Ganoderma* noncoding RNAs. In the future, work should focus on validating the function of these noncoding RNAs.

Acknowledgements This work was supported by the CAMS Innovation Fund for Medical Sciences (CIFMS) (2016-I2M-3-016) and National Science Foundation of China Funds (81872966).

References

Altschul SF, Madden TL, Schäffer AA, Zhang J, Zhang Z, Miller W, Lipman DJ (1997) Gapped BLAST and PSI-BLAST: a new generation of protein database search programs. Nucleic Acids Res 25(17):3389–3402

Ansaldi R, Chaboud A, Dumas C (2000) Multiple S gene family members including natural antisense transcripts are differentially expressed during development of maize flowers. J Biol Chem 275(31):24146–24155

Aravin AA, Hannon GJ, Brennecke J (2007) The Piwi-piRNA pathway provides an adaptive defense in the transposon arms race. Science 318(5851):761–764

Ariel F, Romero-Barrios N, Jégu T, Benhamed M, Crespi M (2015) Battles and hijacks: noncoding transcription in plants. Trends Plant Sci 20(6):362–371

Bartel DP (2004) MicroRNAs: genomics, biogenesis, mechanism, and function. Cell 116:281–297

Belousova EA, Filipenko ML, Kushlinskii NE (2018) Circular RNA: new regulatory molecules. Bull Exp Biol Med 164(6):803–815

Chen C, Ridzon DA, Broomer AJ, Zhou Z, Lee DH, Nguyen JT, Barbisin M, Xu NL, Mahuvakar VR, Andersen MR (2005) Real-time quantification of microRNAs by stem-loop RT-PCR. Nucleic Acids Res 33(20):e179

Chen S, Xu J, Liu C, Zhu Y, Nelson DR, Zhou S, Li C, Wang L, Guo X, Sun Y (2012) Genome sequence of the model medicinal mushroom *Ganoderma lucidum*. Nat Commun 3(2):913

Chu-Yu Y, Li C, Chen L, Qian-Hao Z, Longjiang F (2015) Widespread noncoding circular RNAs in plants. New Phytol 208(1):88

Conn VM, Hugouvieux V, Nayak A, Conos SA, Capovilla G, Cildir G, Jourdain A, Tergaonkar V, Schmid M, Zubieta C et al (2017) A circRNA from SEPALLATA3 regulates splicing of its cognate mRNA through R-loop formation. Nat Plants 3 (5):17053

Donaldson ME, Saville BJ (2012) Natural antisense transcripts in fungi. Mol Microbiol 85(3):405–417

Engström PG, Suzuki H, Ninomiya N, Akalin A, Sessa L, Lavorgna G, Brozzi A, Luzi L, Tan SL, Yang L et al (2006) Complex Loci in human and mouse genomes. PLoS Genet 2(4):e47

Faghihi MA, Wahlestedt C (2009) Regulatory roles of natural antisense transcripts. Nat Rev Mol Cell Biol 10 (9):637–643

FriedlanNder MRMS, Li N, Chen W, Rajewsky N (2012) miRDeep2 accurately indentifies known and hundreds of novel microRNA genes in seven animal clades. Nucleic Acids Res 40(1):37–52

Griffiths-Jones S (2007) Annotating noncoding RNA genes. Annu Rev Genomics Hum Genet 8(1):279–298

Griffithsjones S, Saini HK, Dongen SV, Enright AJ (2008) miRBase: tools for microRNA genomics. Nucleic Acids Res 36(Database issue):D154

Huo L, Zhang P, Li C, Rahim K, Hao X, Xiang B, Zhu X (2018) Genome-wide identification of circRNAs in pathogenic basidiomycetous yeast *Cryptococcus neoformans* suggests conserved circRNA host genes over kingdoms. Genes (basel) 9(3):118

Iseli C, Jongeneel CV, Bucher P (1999) ESTScan: a program for detecting, evaluating, and reconstructing potential coding regions in EST sequences. Proc Int Conf Intell Syst Mol Biol 1999:138–148

Jeck WR, Sorrentino JA, Wang K, Slevin MK, Burd CE, Liu J, Marzluff WF, Sharpless NE (2013) Circular RNAs are abundant, conserved, and associated with ALU repeats. RNA 19(2):141–157

Jonesrhoades MW, Bartel DP, Bartel B (2006) MicroRNAS and their regulatory roles in plants. Annu Rev Plant Biol 57(1):19–53

Julia S, Charles G, Peter Lincoln W, Norman L, Brown PO (2012) Circular RNAs are the predominant transcript isoform from hundreds of human genes in diverse cell types. Plos One 7(2):e30733

Kapranov P, Cheng J, Dike S, Nix DA, Duttagupta R, Willingham AT, Stadler PF, Hertel J, Hackermüller J, Hofacker IL et al (2007) RNA maps reveal new RNA classes and a possible function for pervasive transcription. Science 316(5830):1484–1488

Katayama S, Tomaru Y, Kasukawa T, Waki K, Nakanishi M, Nakamura M, Nishida H, Yap CC, Suzuki M, Kawai J et al (2005) Antisense transcription in the mammalian transcriptome. Science 309(5740):1564–1566

Kersten P, Dan C (2014) Copper radical oxidases and related extracellular oxidoreductases of wood-decay Agaricomycetes. Fungal Genet Biol 72:124–130

Kim TK, Hemberg M, Gray JM, Costa AM, Bear DM, Wu J, Harmin DA, Laptewicz M, Barbara-Haley K, Kuersten S et al (2010) Widespread transcription at neuronal activity-regulated enhancers. Nature 465 (7295):182–187

Kong L, Zhang Y, Ye ZQ, Liu XQ, Zhao SQ, Wei L, Gao G (2007) CPC: assess the protein-coding potential of transcripts using sequence features and support vector machine. Nucleic Acids Res **35**(Web Server issue):W345–349

Kozomara A, Griffiths-Jones S (2011) miRBase: integrating microRNA annotation and deep-sequencing data. Nucleic Acids Res 39(Database issue):D152–157

Kristensen LS, Andersen MS, Stagsted LVW, Ebbesen KK, Hansen TB, Kjems J (2019) The biogenesis, biology and characterization of circular RNAs. Nat Rev Genet 20(11):675–691

Lagosquintana M, Rauhut R, Lendeckel W, Tuschl T (2001) Identification of novel genes coding for small expressed RNAs. Science 294(5543):853–858

Lau NC, Lim LP, Weinstein EG, Bartel DP (2001) An abundant class of tiny RNAs with probable regulatory roles in *Caenorhabditis elegans*. Science 294(5543):858–862

Lee RC, Ambros V (2001) An extensive class of small RNAs in *Caenorhabditis elegans*. Science 294(5543):862

Lee HC, Li L, Gu W, Xue Z, Crosthwaite SK, Pertsemlidis A, Lewis ZA, Freitag M, Selker EU, Mello CC (2010) Diverse pathways generate MicroRNA-like RNAs and dicer-independent small interfering RNAs in fungi. Mol Cell 38(6):803–814

Lewis BP, Shih IH, Jonesrhoades MW, Bartel DP, Burge CB (2003) Prediction of mammalian microRNA targets. Cell 115(7):787–798

Li K, Ramchandran R (2010) Natural antisense transcript: a concomitant engagement with protein-coding transcript. Oncotarget 1(6):447–452

Li J, Wu B, Xu J, Liu C (2014) Genome-wide identification and characterization of long intergenic noncoding RNAs in *Ganoderma lucidum*. PLoS ONE 9 (6):e99442

Li C, Wang M, Qiu X, Zhou H, Lu S (2021) Noncoding RNAs in medicinal plants and their regulatory roles in bioactive compound production. Curr Pharm Biotechnol 22(3):341–359

Lu T, Cui L, Zhou Y, Zhu C, Fan D, Gong H, Zhao Q, Zhou C, Zhao Y, Lu D et al (2015) Transcriptome-wide investigation of circular RNAs in rice. RNA 21 (12):2076–2087

Macías-Sánchez K, García-Soto J, López-Ramírez A, Martínez-Cadena G (2011) Rho1 and other GTP-binding proteins are associated with vesicles carrying glucose oxidase activity from *Fusarium oxysporum* f. sp. *lycopersici*. *Antonie Van Leeuwenhoek* 99(3):671–680

Mao W, Li Z, Xia X, Li Y, Yu J (2012) A combined approach of high-throughput sequencing and degradome analysis reveals tissue-specific expression of microRNAs and their targets in cucumber. Plos One 7 (3):e33040

Ma L, Bajic VB, Zhang Z (2013) On the classification of long noncoding RNAs. RNA Biol 10(6):924–933

Mello CC, Conte D (2004) Revealing the world of RNA interference. Nature 431(7006):338–342

Miri D, Schraga S, Sarit E, Rotem S (2012) Transcriptome-wide discovery of circular RNAs in Archaea. Nucleic Acids Res 40(7):3131–3142

Moradali MF, Mostafavi H, Ghods S, Hedjaroude GA (2007) Immunomodulating and anticancer agents in the realm of macromycetes fungi (macrofungi). Int Immunopharmacol 7(6):701–724

Mori T, Koyama G, Kawagishi H, Hirai H (2016) Effects of homologous expression of 1,4-benzoquinone reductase and homogentisate 1,2-dioxygenase genes on wood decay in hyper-lignin-degrading fungus phanerochaete sordida YK-624. Curr Microbiol 73(4): 512–518

Moss EG, Lee RC, Ambros V (1997) The cold shock domain protein LIN-28 controls developmental timing in *C. elegans* and is regulated by the lin-4 RNA. Cell 88(5):637–646

Munroe SH, Zhu J (2006) Overlapping transcripts, double-stranded RNA and antisense regulation: a genomic perspective. Cell Mol Life Sci Cmls 63 (18):2102–2118

Mu DS, Li C, Shi L, Zhang X, Ren A, Zhao MW (2015) Bioinformatic identification of potential microRNAs and their targets in the Lingzhi or Reishi medicinal mushroom *Ganoderma lucidum* (higher Basidiomycetes). Int J Med Mushrooms 17(8):783–797

Ørom UA, Derrien T, Beringer M, Gumireddy K, Gardini A, Bussotti G, Lai F, Zytnicki M, Notredame C, Huang Q et al (2010) Long noncoding RNAs with enhancer-like function in human cells. Cell 143(1):46–58

Piwecka M, Glažar P, Hernandez-Miranda LR, Memczak S, Wolf SA, Rybak-Wolf A, Filipchyk A, Klironomos F, Cerda Jara CA, Fenske P et al (2017) Loss of a mammalian circular RNA locus causes miRNA deregulation and affects brain function. Science 357(6357):eaam8526

Prasanth KV, Spector DL (2007) Eukaryotic regulatory RNAs: an answer to the "genome complexity" conundrum. Genes Dev 21(1):11–42

Ransohoff JD, Wei Y, Khavari PA (2018) The functions and unique features of long intergenic noncoding RNA. Nat Rev Mol Cell Biol 19(3):143–157

Ren A, Ouyang X, Shi L, Jiang AL, Mu DS, Li MJ, Han Q, Zhao MW (2013) Molecular characterization and expression analysis of GlHMGS, a gene encoding hydroxymethylglutaryl-CoA synthase from *Ganoderma lucidum* (Ling-zhi) in ganoderic acid biosynthesis pathway. World J Microbiol Biotechnol 29 (3):523–531

Reyes-Medina MA, Macías-Sánchez KL (2015) GTPase *Rho*1 regulates the expression of *xyl*3 and laccase genes in *Fusarium oxysporum*. Biotech Lett 37 (3):679–683

Richthammer C, Enseleit M, Sanchez-Leon E, März S, Heilig Y, Riquelme M, Seiler S (2012) RHO1 and RHO2 share partially overlapping functions in the regulation of cell wall integrity and hyphal polarity in *Neurospora crassa*. Mol Microbiol 85(4):716–733

Rymarquis LA, Kastenmayer JP, Hüttenhofer AG, Green PJ (2008) Diamonds in the rough: mRNA-like non-coding RNAs. Trends Plant Sci 13(7):329–334

Sanger HL, Klotz G, Riesner D, Kleinschmidt GAK (1976) Viroids are single-stranded covalently closed circular RNA molecules existing as highly base-paired Rod-like structures. Proc Natl Acad Sci USA 73 (11):3852–3856

Sarah WB, Jennifer D, Ruth E, John T, Lars B, Eric P N, Sean R E, Paul P G, Alex B (2013) Rfam 11.0: 10 years of RNA families. Nucleic Acids Res 41 (Database issue):226–232

Shao J, Chen H, Yang D, Jiang M, Zhang H, Wu B, Li J, Yuan L, Liu C (2017) Genome-wide identification and characterization of natural antisense transcripts by strand-specific RNA sequencing in *Ganoderma lucidum*. Sci Rep 7(1):5711

Shao J, Wang L, Liu X, Yang M, Chen H, Wu B, Liu C (2019) Identification and characterization of circular RNAs in *Ganoderma lucidum*. Sci Rep 9(1):16522

Shao J, Wang L, Liu Y, Qi Q, Wang B, Lu S, Liu C (2020) Identification of milRNAs and their target genes in *Ganoderma lucidum* by high-throughput sequencing and degradome analysis. Fungal Genet Biol 136:103313

Shiao MS, Lee KR, Lin LJ, Wang CT (1994) Natural products and biological activities of the Chinese medicinal fungus *Ganoderma lucidum. ACS Symp Ser Am Chem Soc* 1994(547):342–354

Storz G (2002) An expanding universe of noncoding RNAs. Science 296(5571):1260–1263

Trapnell C, Roberts A, Goff L, Pertea G, Kim D, Kelley DR, Pimentel H, Salzberg SL, Rinn JL, Pachter L (2012) Differential gene and transcript expression analysis of RNA-seq experiments with TopHat and Cufflinks. Nat Protoc 7(3):562–578

Ulitsky I, Bartel DP (2013) lincRNAs: genomics, evolution, and mechanisms. Cell 154(1):26–46

Villegas V, Zaphiropoulos P (2015) Neighboring Gene regulation by antisense long non-coding RNAs. Int J Mol Sci 16(2):3251–3266

Wang XJ, Gaasterland T, Chua NH (2005) Genome-wide prediction and identification of *cis*-natural antisense transcripts in *Arabidopsis thaliana*. Genome Biol 6(4): R30

Wang D, Garcia-Bassets I, Benner C, Li W, Su X, Zhou Y, Qiu J, Liu W, Kaikkonen MU, Ohgi KA et al (2011) Reprogramming transcription by distinct classes of enhancers functionally defined by eRNA. Nature 474(7351):390–394

Wang M, Wu B, Chen C, Lu S (2015) Identification of mRNA-like noncoding RNAs and validation of a mighty one named MAR in *Panax ginseng*. J Integr Plant Biol 57(3):256–270

Wirth S, Crespi M (2009) Non-protein coding RNAs, a diverse class of gene regulators, and their action in plants. RNA Biol 6(2):161–164

Wu B, Li Y, Yan H, Ma Y, Luo H, Yuan L, Chen S, Lu S (2012) Comprehensive transcriptome analysis reveals novel genes involved in cardiac glycoside biosynthesis and mlncRNAs associated with secondary metabolism and stress response in *Digitalis purpurea*. BMC Genomics 13(1):15

Xia H, Zhang L, Wu G, Fu C, Long Y, Xiang J, Gan J, Zhou Y, Yu L, Li M (2016) Genome-wide identification and characterization of microRNAs and target genes in *Lonicera japonica*. PLoS One 11(10): e0164140

Xu J, Xu Z, Zhu Y, Luo H, Qian J, Ji A, Hu Y, Sun W, Wang B, Song J (2014) Identification and evaluation of reference genes for qRT-PCR normalization in *Ganoderma lucidum*. Curr Microbiol 68(1):120–126

Zhang XO, Wang HB, Zhang Y, Lu X, Chen LL, Yang L (2014) Complementary sequence-mediated exon circularization. Cell 159(1):134–147

Biosynthetic Pathway and Signal Transduction Regulation of Ganoderic Acid Biosynthesis in Lingzhi

8

Ang Ren, Liang Shi, Jing Zhu, Rui Liu, and Mingwen Zhao

Abstract

Ganoderic acid (GA) is an important bioactive metabolite of *G. lucidum*, whose biosynthesis pathway has been elucidated. We reviewed the studies on the biosynthetic pathway of GA and its regulation. The following are described: (1) Biosynthetic pathway of GA; (2) Regulation of GA biosynthesis by environmental factors; (3) Role of signaling in GA biosynthesis regulation; (4) Crosstalk of different signals for GA biosynthesis; and (5) Key genes involved in the signaling pathway regulation of GA biosynthesis. In particular, the mechanisms of different environmental factors that regulate GA biosynthesis are described in detail, including nutritional conditions, physical means, and single chemical or biological signals. More and more evidence indicates that various signaling molecules (such as ROS, Ca^{2+}, NO, H_2S, cAMP, and so on) are involved in GA biosynthesis regulation in response to environmental factors. Some key signaling molecules lead to crosstalk of different pathways. Also, under different environmental stress, these signaling molecules may interact differently, demonstrating the signaling regulatory networks' complexity. Lastly, the future direction are described in this area.

8.1 Biosynthetic Pathway of Ganoderic Acid

For a long time, the biosynthesis of terpenoids has been considered to derive from acetyl-CoA via mevalonate. It is believed that mevalonate is the only precursor of terpenoid biosynthesis. Therefore, this pathway is called the mevalonate pathway. The mevalonate pathway uses acetyl-coenzyme A (acetyl-CoA) as an initial substrate; acetyl-CoA condenses to the intermediate 3-hydroxy-3-methylglutaryl-CoA (HMG-CoA), which is subsequently reduced to the intermediate mevalonate, and then forms isopentenyl-diphosphate (IPP) and its double bond isomer dimethylallyl. Pyrophosphate (dimethylallyl diphosphate, DMAPP) further condenses to form sesquiterpenes, triterpenes, and steroids (Fig. 8.1). The mevalonate (MVA) pathway contains 11 steps, 16 chemical compounds, and nine enzymes (Hirotani et al. 1990). The pathway primarily exists in the cytosol and is also called the cytosolic pathway.

Previous isotope tracing experiments have proven that the MVA pathway in *G. lucidum* and ganoderic acids (GAs) are synthesized through the MVA pathway (Hirotani et al. 1990; Shiao 2003). Shiao et al. added 5-^{13}C-labeled mevalonate to the liquid culture of *G. lucidum* and

A. Ren · L. Shi · J. Zhu · R. Liu · M. Zhao (✉)
College of Life Sciences, Nanjing Agricultural University, Nanjing 210095, Jiangsu, People's Republic of China
e-mail: mwzhao@njau.edu.cn

C. Liu (ed.), *The Lingzhi Mushroom Genome*, Compendium of Plant Genomes,
https://doi.org/10.1007/978-3-030-75710-6_8

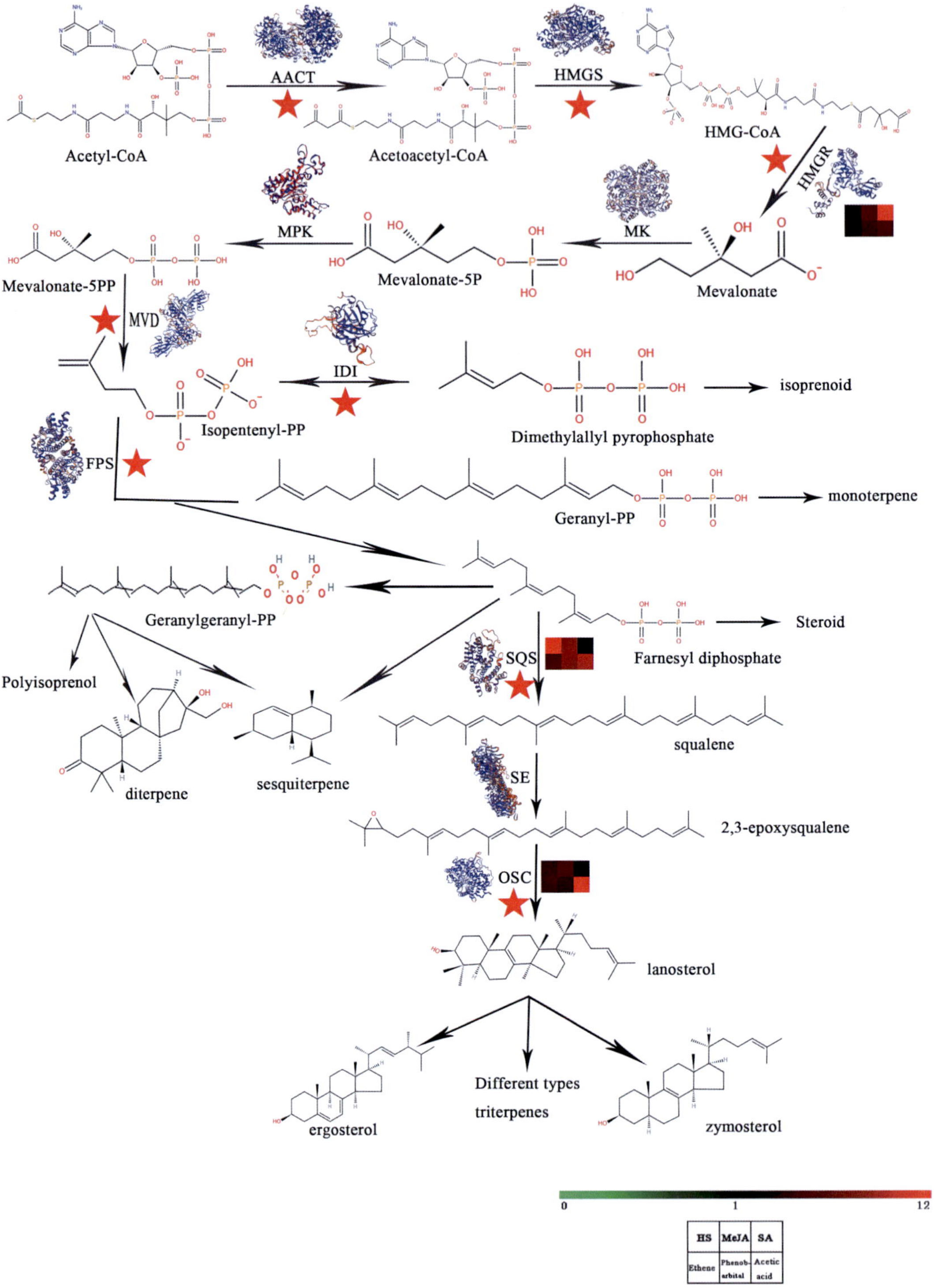
AACT
HMGS
Acetyl-CoA
Acetoacetyl-CoA
HMG-CoA
HMGR
MPK
MK
Mevalonate-5PP
Mevalonate-5P
Mevalonate
MVD
IDI
isoprenoid
Isopentenyl-PP
Dimethylallyl pyrophosphate
FPS
monoterpene
Geranyl-PP
Geranylgeranyl-PP
Steroid
SQS
Farnesyl diphosphate
Polyisoprenol
diterpene
sesquiterpene
squalene
SE
2,3-epoxysqualene
OSC
lanosterol
ergosterol
Different types triterpenes
zymosterol
0
1
12
HS
MeJA
SA
Ethene
Phenobarbital
Acetic acid

◀ **Fig. 8.1** The pathway of ganoderic acids biosynthesis, according to Shi et al (2010), with some modifications. The scheme shows the mevalonate pathway for the generation of the terpenoids. The protein 3D-structures of the key enzymes in the mevalonate pathway have been shown. The enzymes in the mevalonate pathway correlative with ganoderic acids biosynthesis are marked with the red star: 1. acetyl-CoA acetyltransferase, AACT (GenBank: AEB00648.1) (Fang et al. 2013); 2. 3-hydroxy-3-methylglutaryl-CoA synthase, HMGS (GenBank: AFM91095.1) (Wu et al. 2020); 3. 3-hydroxy-3-methylglutaryl-CoA reductase, HMGR (GenBank: ABY84848.1) (Xu et al. 2012); 4. mevalonate kinase, MK; 5. phosphomevalonate kinase, MPK; 6. pyrophosphomevalonate decarboxylase, MVD (GenBank: AEB00647.1) (Shi et al. 2012); 7. isopentenyl-diphosphate isomerase, IDI (GenBank: AGL94943.1) (Wu et al. 2013); 8. farnesyl diphosphate synthase, FPPs (GenBank: ACB37021.1) (Fei et al. 2019); 9. squalene synthase, SQS (GenBank: ABF57213.1) (Zhou et al. 2014); 10. squalene monooxygenase, SE (GenBank: ARU77555.1) (Zhang et al. 2017); 11. 2, 3-oxidosqualene-lanosterol cyclase, OSC (GenBank: ADD60469.1) (Zhang et al. 2017). SWISS-MODEL predicted the corresponding 3D structure of each enzyme, and the GenBank numbers for the sequences used to predict the 3D structure are given in parentheses. The effects of heat stress (HS) (Zhang et al. 2016), methyl jasmonate (MeJA) (Ren et al. 2010), salicylic acid (SA) (Cao et al. 2017), ethylene (Zhang et al. 2017), phenobarbital (Liang et al. 2010), and acetic acid (Ren et al. 2014) treatment on gene expression were presented as a heat map, which was drawn with GeneSpring GX 7.3.1 software (Agilent Technologies). The relative expression is shown as a mean value, from 0 to 12, in green to red (n = 3)

found that two labeled ^{13}C-oxidized triterpenes appeared in its metabolites, confirming that the two triterpenes were derived from mevalonate. Besides, the triterpenes were found stable in the labeling experiment. Furthermore, the signal intensity at positions C-2, C-6, C-11, C-12, C-16, and C-23 of these two triterpenes increases significantly as the detection number of 13C-NMR doubles in broadband. This result shows that there are many pairs of C-13 stereoisomers and C-3/C-15 positional isomers in *G. lucidum*. Through further experiments, it was found that many pairs of C-3 stereoisomers in mycelium were produced.

Meanwhile, ^{3}H-labeled sterols were converted into 3α and 3β series triterpenes with similar efficiency. When the labeled 3β triterpenes were added to the liquid culture regularly, 3α series metabolites were produced. The oxidative modification is observed on the lanosterol ring skeleton, and the oxidative cleavage is observed on the side chain. These results indicate that the 3α triterpenes are derived from the 3β triterpenes through an oxidation–reduction pathway. This result offered an in-depth understanding of steroid components' biotransformation ability and laid a solid foundation for examining the mechanism of GA biosynthesis and its regulation.

The squalene and lanosterol were produced from IPP. After a series of redox modifications, the triterpenes with different structures are formed. At present, studies have observed that there are many enzymes involved in the biosynthesis of *G. lucidum* triterpenes and that they catalyze numerous types of reactions. This complexity presents many difficulties to the study of biosynthesis regulation. However, with the progress made in molecular biology, some important enzymes have been identified. Increasing numbers of genes have been cloned, providing important tools for studying the regulation of biosynthetic pathways. Research on the GA biosynthesis pathway is primarily performed through key enzyme genes' cloning, isolation, characterization, and gene regulation. In total, 11 steps are thought to be involved in the MVA pathways. Below, we described three important enzymes in GA biosynthesis.

3-Hydroxy-3-methylglutaryl-CoA reductase (HMGR) was discovered in 1958 by Lynen et al. and Ferguson et al., working independently in their respective laboratories (Lynen et al. 1958). This enzyme catalyzes the formation of mevalonate (MVA) from HMG-CoA. Because the production of MVA is an irreversible process, this reaction is considered the rate-limiting step in the MVA pathway. Studies have shown that the increased expression of HMGR can accelerate the downstream synthesis of terpene compounds, thereby significantly increase its yield.

In 2008, Shang et al. cloned the HMGR gene from *G. lucidum* for the first time (Shang et al.

2008). The gene has a total length of 4262 bp, contains seven exons and six introns, and encodes 1226 amino acids. This author inserted the cDNA sequence of the cloned HMGR gene into the yeast expression vector pYF1845 and transferred it into the yeast mutant strain JRY1130, which is dependent on mevalonate for growth due to the lack of HMGR activity. Cells containing this construct were observed to grow on the culture medium, confirming the cloned gene's function. The HMGR gene of *G. lucidum* was also subjected to bioinformatics analysis. It was determined to be closely related to those from the basidiomycete smut fungus and *Cryptococcus neoformans*. Also, it has two conserved HMG-CoA binding sites, E (838) and T (867), and 2 NAD(H) binding sites, D (963) and G (1112) (Shang et al. 2008).

Farnesyl pyrophosphate synthase (FPPS) was first isolated by Lynen et al. in 1959 (Lynen et al. 1959). FPPS is at the first branch point in the isoprenoid biosynthetic pathway. In both prokaryotic and eukaryotic organisms, the activity of FPPS changes the output of downstream products. Therefore, FPPS might play a key role in the isoprenoid biosynthesis pathway. In 2008, Ding et al. cloned the FPPS gene from *G. lucidum* for the first time (Fei et al. 2019). The gene is 1388 bp in length, contains four exons and three introns, and encodes 360 amino acids. The author cloned the full-length cDNA sequence of the *G. lucidum* FPPS gene into the yeast expression vector pYF1845 and transferred it into the yeast mutant strain CC25, which depends on ergosterol due to the lack of FPPS activity. The transformants were observed to grow on an ergosterol-free medium at 36 °C, confirming the function of the cloned FPPS. The FPPS sequences of *G. lucidum* and another basidiomycete, *Lactarius*, showed the highest similarity (59%). Furthermore, *G. lucidum* FPPS contained four conserved regions with the FPPS sequences of different species. *G. lucidum* FPPS belongs to type I (eukaryotic state) farnesyl pyrophosphate synthase, consistent with the species classification of *G. lucidum*.

Squalene synthase (SQS) was first extracted from yeast by Agnew in 1978 as a soluble enzyme (Agnew and Popjak 1978). Due to its unstable form, it was not until 1988 that Sasiak et al. purified SQS with more than 95% purity for the first time (Sasiak and Rilling 1988). SQS is at the branch point of FPP to triterpenoids, sterols, cholesterol, and other terpenoids in the metabolic pathway. Therefore, its content and activity determine the yield of subsequent products. In 2007, Zhao et al. cloned the SQS gene from *G. lucidum* for the first time (Zhou et al. 2014). The gene is 1984 bp in length, contains four exons and three introns, and encodes 467 amino acids. The *G. lucidum* SQS gene cDNA sequence was cloned into the yeast expression vector pYF1845. The yeast mutant strain 2C1, which is dependent on ergosterol due to lack of SQS activity, was transformed. The transformant could grow on an ergosterol-free medium, confirming the function of the cloned SQS gene.

While the most important genes of the GA biosynthesis pathway have been identified, there are additional challenges in genetic engineering studies. These include identifying additional key enzyme genes, the small number of clones, and the speed limit of triterpene product synthesis. Future studies are needed to find solutions to these challenges.

8.2 Environmental Regulation of Ganoderic Acid Biosynthesis

Since the discovery of GAs and their beneficial effects on human health, efforts have been undertaken to improve the yield of GAs. The last decade has seen significant progress made in characterizing environmental factors affecting GA biosynthesis in *G. lucidum* (Ren et al. 2019). At present, three groups of parameters are adjusted to study GA biosynthesis's environmental regulation (Table 8.1). The first group of parameters includes nutritional conditions such as carbon sources, nitrogen sources, and the ratio of carbon and nitrogen sources. The second group of parameters includes physical means such as heat stress, solid culture, and pH change. The third group of parameters includes single chemical or biological signals such as methyl

Table 8.1 The content of ganoderic acids under different environmental conditions

The types of regulatory factors	Regulatory factors	References
Single chemical or biological signal	MeJA	Ren et al. (2010)
	SA	Cao et al. (2017)
	Ethylene	Zhang et al. (2017)
	Phenobarbital	Liang et al. (2010)
	Aspirin	You et al. (2013)
	Acetic acid	Ren et al. (2014)
	Na^+	Xu et al. (2013)
	Cu^{2+}	Tang and Zhu (2010)
	Mn^{2+}	Xu et al. (2014)
	Ca^{2+}	Xu and Zhong (2012)
Physical means	Heat stress	Zhang et al. (2016)
	Solid culture	You et al. (2012)
	Two-stage culture	Fang and Zhong (2002)
	Oxygen supply	Tang and Zhong (2003)
	Three-stage light irradiation	Zhang and Tang (2008)
	pH change	Wu et al. (2016)
Nutritional conditions	Carbon sources (microcrystalline cellulose)	Hu et al. (2017)
	Nitrogen sources	Zhu et al. (2019)
	LACTOSE	Tang and Zhong (2002)
	Complex media	Xu et al. (2008)
Other factors	Extract from the medicinal insect	Liu et al. (2011)
	Cellulase	Zhang et al. (2014)
	Protein elicitor	Zhu and Tang (2010)

jasmonate, salicylic acid, ethylene, phenobarbital, aspirin, acetic acid, Na^+, cellulose, and microcrystalline cellulose. These three types of parameters are related to the cultivation process of *G. lucidum*.

8.2.1 Nutritional Conditions Regulate GA Biosynthesis

8.2.1.1 Carbon Source

The effects of different carbon sources on GA biosynthesis were compared (Wei et al. 2016). The results showed that sucrose is the most suitable carbon source for GA biosynthesis. Liu investigated the effects of different glucose concentrations on GA-ME fermentation and subsequently optimized the culture conditions to obtain the individual GA-Me at a glucose concentration of 44.4 g/L (Liu et al. 2012). The results suggested that the type and availability of carbon sources affect GA production. Further research examining the mechanism of sensing and utilizing carbon sources in *G. lucidum* will provide new ideas for promoting GA biosynthesis through carbon source utilization.

8.2.1.2 Nitrogen Source

Various nitrogen sources have different effects on fungal secondary metabolism. Zhao et al. investigated the effects of four different nitrogen sources (ammonium sulfate, glycine, glutamine, and asparagine) on GA content (Zhao et al. 2011). The results showed that when ammonium sulfate, glutamine, or asparagine were used as nitrogen sources, limiting nitrogen sources significantly increased GA biosynthesis. Furthermore, different concentrations (0–60 mM) of glutamine were employed as nitrogen sources to determine their effects on GA content. The results showed that the expression levels of key genes of GA biosynthesis and GA content in 3 mM glutamine culture were significantly higher than those of the control group in 60 mM glutamine. Simultaneously, a putative nitrogen regulator gene, *areA,* was determined to be upregulated under glutamine limiting conditions. These findings provide a useful strategy for enhancing GA production by nitrogen limitation.

Also, Zhu et al. (2019) found that the GA content in *G. lucidum*, with nitrate serving as the nitrogen source, was significantly lower than that observed with ammonium salt as the nitrogen source. At the same time, the expression level of AreA was significantly increased. The transcription factor AreA can activate nitrate reductase (NR) to produce NO in the process of nitrate absorption and utilization, and NO can be utilized as a signal molecule to inhibit GA biosynthesis in *G. lucidum*.

Different nitrogen sources have different effects on the accumulation of triterpenoids in *G. lucidum* (Zhu et al. 2019). For example, the content of triterpenoids in *G. lucidum* decreased significantly under nitrate conditions. At the same time, the expression level of AreA was significantly increased. The transcription factor AreA can activate NR to produce NO in the process of nitrate absorption and utilization, and NO can be utilized as a signal molecule to inhibit the biosynthesis of triterpenes in *G. lucidum*.

8.2.1.3 Intermediate Metabolites

GA-T is a kind of GA and has significant antitumor activity (Chen et al. 2010). In *G. lucidum* fermentation, GA-T production is promoted by providing two key analogs (7-O-ethyl GA-O and GA-Mk). This method provides a new means of improving the specific GA molecule by feeding intermediate metabolites. The conversion involves three consecutive chemical steps, namely, hydrolysis–acetylation–hydrolysis (Wang et al. 2012).

8.2.2 Physical Means Regulate GA Biosynthesis

In a fungal growth environment, physical parameters (such as temperature, solid/liquid states, and pH) are important factors that affect fungal secondary metabolism (Calvo et al. 2002). Therefore, employing physical means to change these parameters has become a common method to improve GA biosynthesis.

8.2.2.1 Temperature

Environmental temperature can influence fungi's diverse physiological processes, including growth, secondary metabolism, and development (Yu and Keller 2005). In *G. lucidum*, heat stress at 42 °C induced the GA biosynthesis and the expression of heat shock protein (HSP) (Zhang et al. 2016). Besides, the change of GA content was detected in temperature shifting in the fermentation of *G. lucidum.* At the early stage of fermentation, a temperature of 32 °C induces the mycelial growth of *G. lucidum.* At the middle stage, the stepwise temperature decrease from 31 °C to 30 °C is conducive to the rapid biosynthesis of GA. At later stages, a temperature of 29 °C is conducive to reduce the inhibition of GA biosynthesis. Thus, an optimized temperature shifting strategy emerged to improve GA production. Clearly, different temperatures have different effects on GA biosynthesis (Feng et al. 2016). The specific mechanisms warrant further study.

8.2.2.2 Solid-Medium Culture Approach

Liquid culture and solid culture are the two types of culture methods commonly employed. You et al. showed that solid culture increased the biomass and GA biosynthesis compared with liquid culture under the same nutrient conditions (You et al. 2012). Surprisingly, GA extracted from the solid culture's mycelium has a higher activity than that from liquid against three tumor cell lines: human lung cancer cell CH27, melanoma cell M21, and oral cancer cell HSC-3. The successful production of enzymes or secondary metabolites by microorganisms through solid-state fermentation has been reported (Kumar et al. 2020). This work provides a method that enhances GA production (with higher biological activity) by mycelium culturing on a solid medium.

8.2.2.3 PH

In the fermentation process of *G. lucidum*, pH affects mycelium growth and GA biosynthesis (Fang and Zhong 2002). Fang and Zhong found that a maximum GA content was obtained at an initial pH of 6.5. However, since pH is a dynamic factor in fermentation, it is challenging to regulate GA biosynthesis by this single factor alone. Li et al. found that there are interactions between pH and other media factors (such as mineral ions) that affect GA biosynthesis (Li et al. 2006). Wu et al. reported a transcriptional factor PacC of *G. lucidum*, which can respond to environmental pH regulation (Wu et al. 2016). The GA content and levels of the SQS and LS transcripts increased significantly in the *PacC* gene silencing strain. Although the exact mechanism by which PacC regulates fungal secondary metabolism is not clear, it is found that PacC is involved in regulating the pH response, ion stress, and secondary metabolism (Luo et al. 2017). Follow-up mechanistic study may help to improve GA biosynthesis through the adjustment of pH and other culture conditions.

8.2.2.4 Light

Zhang et al. investigated the effects of different light qualities, i.e., blue light (390–500 nm, λ_{max} = 470 nm), red light (560–700 nm, λ_{max} = 625 nm), and white light (400–740 nm, λ_{max} = 550 nm), on GA biosynthesis (Zhang and Tang 2008). According to GA accumulation characteristics, a new three-stage light irradiation strategy was proposed for deep fermentation of *G. lucidum*. The first stage was two days of dark culture; the second stage was six days of 0.94 w/m^2 white light culture; and the third stage was 4.70 w/m^2 white light culture until the end of fermentation.

8.2.3 Single Chemical or Biological Signals Regulate GA Biosynthesis

In nature, *G. lucidum* obtains nutrients through wood decomposition. In this progress, fungus encounters single chemical or physical signals, which can induce GA biosynthesis. On the other hand, those studies (Shi et al. 2015; Ren et al. 2014) showed that a single chemical compound induces GA biosynthesis in *G. lucidum.* This opens up the possibility to regulate the expression of key enzymes at each step of the biosynthetic pathway under different environmental factors.

8.2.3.1 Methyl Jasmonate, Salicylic Acid, and Ethylene

Methyl jasmonate (MeJA), salicylic acid (SA), and ethylene are plant hormones that regulate various physiological processes in plants (Gunjegaonkar and Shanmugarajan 2019). In 1994, it was reported that MeJA induces steroid-derivative biosynthesis and the expression of 3-hydroxy-3-methylglutaryl-coenzyme A reductase (Ren et al. 2010). SA also induces the production of plant defensive metabolites, which are secondary metabolites (Schweiger et al. 2014). In *G. lucidum*, Ren et al. found that MeJA significantly induces GA biosynthesis, which was also reviewed in 2010 (Shi et al. 2010). In 2017, Cao et al. and Zhang et al. reported that SA and ethephon (2-chloroethyl phosphonic acid), an ethylene-releasing chemical, induce GA biosynthesis and upregulate the expression of genes associated with GA biosynthesis in *G. lucidum* (Zhang et al. 2017; Cao et al. 2017). However, it remains necessary to elucidate how SA or ethylene induces GA biosynthesis. In 2013, to further explore the mechanism of MeJA-induced GA biosynthesis, the differentially expressed genes in response to MeJA were identified using cDNA-amplified fragment length polymorphism (cDNA-AFLP). The transcriptional profiling analysis results showed that some MeJA-regulated genes are related to signal transduction, such as ROS, Ca^{2+}, MAPK, cAMP-dependent protein kinase, small G-proteins, and some transcription factors (Ren et al. 2013). The screening of these differentially expressed genes provides new target genes for further improvement of GA biosynthesis using genetic transformation and provides a new approach for studying the regulation mechanism of GA biosynthesis.

8.2.3.2 Phenobarbital, Aspirin, and Acetic Acid

Phenobarbital is a typical inducer for transcriptional activation of cytochrome P450 (CYP) genes (Czekaj 2000). The CYP superfamily plays a significant role in the biosynthesis of different GAs from lanosterol. Liang et al. found that phenobarbital enhanced GA biosynthesis and individual GA-Mk, -T, -S, and –Me (Liang et al. 2010). For a review of phenobarbital's effects on the expression of the genes involved in MVP, see reference Shi et al. (2010). In fact, there are 219 identified CYP sequences in the *G. lucidum* genome (Chen et al. 2012). Further study is needed to elucidate the roles played by each member of the CYP family in the biosynthesis and modification of GA or other terpenoids.

In *Saccharomyces cerevisiae* and HepG2 cells, aspirin is used as an inducer of apoptosis (Raza et al. 2011; Balzan et al. 2004). In *G. lucidum*, aspirin also induces cell apoptosis, and the induction of apoptosis coincides with ROS production, Hog1 phosphorylation, and lanosta-7,9(11), 24-trien-3alpha-01-26-oic acid (ganoderic acid 24, GA24) biosynthesis, and total GA biosynthesis (You et al. 2013). This work provides a novel approach for enhancing GA biosynthesis using aspirin induction. However, it remains unknown whether aspirin-induced apoptosis, ROS accumulation, and hog-1 phosphorylation are necessary for aspirin-induced GA biosynthesis. It is interesting to speculate what the relationship is between GA biosynthesis and the fungal apoptosis signaling cascade.

As a carbon source, acetic acid can be catalyzed by acetyl-coenzyme A synthase (ACS) to form acetyl-CoA. The addition of acetic acid to the medium stimulates intracellular metabolism and gene expression (Ren et al. 2014). In *G. lucidum*, 8.21 mM acetic acid with an induction time of 22.68 h was observed to induce a maximum GA content of 5.36 mg/100 mg mycelial dry weight, 2.05 times that of the control. Under acetic acid induction, a putative ACS gene was determined to be upregulated and the content of intracellular acetyl-CoA increased. These results show that acetic acid alters the expression of genes related to acetic acid assimilation. Considering that acetic acid is commonly employed in foods, such as vinegar, improving GA biosynthesis by acetic acid is a convenient, economic, and safe approach in the fermentation of *G. lucidum*.

8.2.3.3 Metal Ions (Na^+ and Cu^{2+})

Metal ions have significant effects on secondary metabolite biosynthesis in plants (Lajayer et al. 2017). The effects of metal ions on the secondary metabolism were also investigated. The GA content was increased by copper ion treatment in *G. lucidum* (Jain et al. 2020). Xu et al. found that the amount of sodium ions (100 mM Na^+) added to the medium can significantly increase GA biosynthesis (Xu et al. 2013). Further detection showed that sodium ions increased the intracellular calcium concentration and expression levels of the genes of Ca^{2+} sensors. The induction of GA biosynthesis by Na^+ was depressed by cyclosporine A, a calcineurin inhibitor. This work suggested an efficient induction strategy for improved GA production using the administration of medium with Na^+. Besides, three sensor genes (*cam*, *cna*, and *crz1*), *ena1* (a sodium transport ATPase 1), and Ca^{2+}-ATPase were upregulated by Na^+ addition but were significantly depressed by cyclosporine A. Although the work implied that calcineurin signaling transduction would be involved in the regulation of Na^+ on GA biosynthesis, it needs to clarify further the role of calcium signal transduction-related proteins in the regulation of GA biosynthesis.

8.2.3.4 Cellulase and Microcrystalline Cellulose

Cellulase is a common fungal elicitor to induce secondary metabolites' accumulation in plants (Thakur et al. 2019). Zhang et al. screened a series of exogenous compounds (cellulase, $CaCl_2$, $ZnCl_2$, malto-oligosaccharide, olive oil, and Chinese medicines (*Lycium chinense* and *Illicium verum*)) to induce GA biosynthesis (Zhang et al. 2014) significantly. The results showed that cellulase, $CaCl_2$, $ZnCl_2$, and malto-oligosaccharide could induce GA biosynthesis. Olive oil, *L. chinense*, and *I. verum* were observed to reduce GA biosynthesis. Under 5 mg/l cellulase supplementation on day 3 in the culture, total GA and extracellular GA increased rapidly on day 7, but intracellular GA increased gradually with fermentation time. The results suggest that cellulase enhancement effects result from its cell wall degradation feature (Zhang et al. 2014).

G. lucidum is a wood-decay fungus. In nature, this fungus obtains nutrients from wood degradation, including microcrystalline cellulose (MCC) and D-galactose. Therefore, MCC and D-galactose were used to test whether they can induce GA biosynthesis (Hu et al. 2017). The results showed that GA production increased 85.96% and 63.90% under treatment with 1.5% microcrystalline cellulose (MCC) and 0.5% D-galactose treatments. MCC and D-galactose significantly induced the transcription level of genes related to GA biosynthesis. Other wood degradation components (L-arabinose, D-xylose and D-rhamnose, D-mannose, glucan hyaluronic acid, and algin sodium mannuronic acid) could not contribute to GA biosynthesis. Further study is needed to explore the mechanism by which MCC and galactose induce GA biosynthesis.

8.3 Role of Signaling in GA Biosynthesis Regulation

Considerable progress in the study of *G. lucidum* has been made in the last decade, as the completed genome sequence (Chen et al. 2012), efficient and straightforward transformation systems (Shi et al. 2012; Yu et al. 2014), and reliable reverse genetic tools (Mu et al. 2012) become available for *G. lucidum*. *G. lucidum* is a nonmodel species, and it is difficult to perform research work at the genetic level. Therefore, clearly illustrating GA biosynthesis's regulatory mechanism may require a scheme that combines physiological and genetic research in *G. lucidum*. At present, it is worth learning from the idea of studying the environmental regulation of signal transduction.

Signals involved in regulating secondary metabolism are less studied in fungi. Therefore, plant physiology methods were introduced to study the functions of signals regulating secondary metabolism in *G. lucidum*. By examining the roles of ROS, Ca^{2+}, cAMP, and phospholipid signaling (Fig. 8.2), the regulatory mechanism of GA biosynthesis was gradually elucidated.

8.3.1 Reactive Oxygen Species

Reactive oxygen species (ROS), such as superoxide anion (O^{2-}), hydrogen peroxide (H_2O_2), and hydroxyl radical (OH^-), regulate various processes of cellular physiology, including growth, development, and metabolism (Sies and Jones 2020). The nicotinamide adenine dinucleotide phosphate oxidase (Nox) family produces intracellular ROS. Three Nox members, Nox-A, -B, and -R, are observed in *G. lucidum* (Mu et al. 2014). The silencing of the Nox isoforms and NoxR expression indicated a central role for these genes in ROS generation and GA biosynthesis regulation. A study found that MeJA could induce the production of intracellular ROS in *G. lucidum* (Shi et al. 2015). ROS scavengers could reverse the increase in ROS and GA content. It was further observed that the Nox family plays an important role in the induction of ROS production by MeJA (Shi et al. 2015). In the Nox silencing strain, MeJA could not induce GA biosynthesis. These results suggest that the ROS produced by Nox plays a vital role in MeJA-induced GA biosynthesis.

In addition to the ROS production system, the ROS scavenging system is also an important

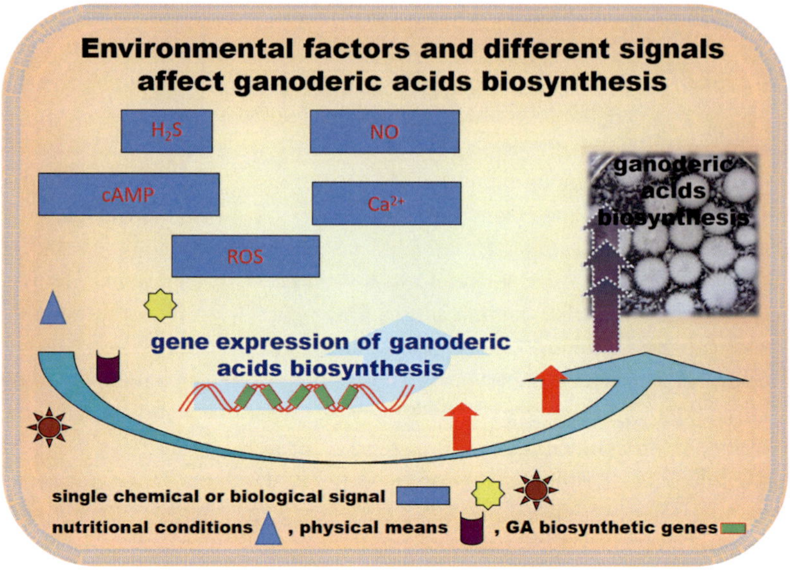

Fig. 8.2 Environmental factors and different signals affect GA biosynthesis, according to Ren et al. (2019) (Ren et al. 2019), with some modifications. ROS, Reactive oxygen species; NO, nitric oxide; H_2S, Hydrogen sulfide; cAMP, cyclic adenosine 3',5'-monophosphate

intracellular ROS source. The alternative oxidase (AOX), an enzyme at the branch of the mitochondrial respiratory electron transport pathway, alleviates reactive oxygen species (ROS) production by transferring electrons from ubiquinol to O_2 with the production of H_2O. When AOX was knocked down, the ROS content increased, followed by an increase in the GA content (Shi et al. 2017). Furthermore, NAC treatment decreased the GA content in AOX knockdown strains. Ornithine decarboxylase (ODC) is a key enzyme in the polyamine biosynthetic pathway (Wu et al. 2017). The ODC-mediated production of putrescine regulates intracellular ROS levels. It influenced GA biosynthesis and three key enzymes' expression levels. In contrast, glutathione peroxidase (GPx) is one of the most important antioxidant enzymes. Knockdown of *GPx* leads to ROS accumulation in *G. lucidum* (Li et al. 2015). An interesting phenomenon is that GA content is decreased, even though ROS is increased, in GPx-silenced strains. There are different mechanisms of ROS-regulated GA biosynthesis, as described below.

In 2012, You et al. reported that hydrogen peroxide or a pro-oxidant 1-chloro-2,4-dinitrobenzene (CDNB) in the solid medium increased total GA production, GA24, accompanied by the phosphorylation of MAPK-Hog1 (You et al. 2012). However, the relationship between MAPK-Hog1 and GA biosynthesis has not been elucidated to date. In another report, GlSlt2, a MAPK in the cell wall integrity pathway, was identified and characterized. Knockdown of *GlSlt2* resulted in decreased intracellular ROS in the mycelium and a decrease in GA biosynthesis (Zhang et al. 2017). Adding exogenous ROS may compensate for the reduction of GA in the GlSlt2 knockdown strains.

These results suggest that ROS plays a significant role in GA biosynthesis. However, at present, studies on the ROS regulation of GA biosynthesis are not complete. In the present study, these results showed that ROS mediate different functions in GA biosynthesis regulation. For example, ROS produced by NOX promoted GA biosynthesis. GPX is an enzyme for ROS scavenging. The accumulation of ROS due to a deficiency in GPX negatively correlated with GA biosynthesis. In other species, ROS have also been reported to play conflicting roles. For instance, in *Arabidopsis thaliana*, ROS can promote root growth under drought stress, but ROS can also inhibit root growth under copper stress. In addition, different locations of ROS can also lead to different functions. For instance, although ROS's increased production has long been considered a cause of aging, mitochondrial ROS produced via reverse electron transport extends the *Drosophila* lifespan (Scialo et al. 2016). Intracellular ROS is not only a substance but also serves as a signal. ROS from different sources and different production sites may have different mechanisms to regulate secondary metabolism in *G. lucidum* (Mu et al. 2014; Li et al. 2015; Zhang et al. 2017; Zhu et al. 2019). However, because of ROS and other signals' interaction, ROS's mechanisms that influence secondary metabolite production are also different. Further studies are needed to determine the different mechanisms by which ROS signaling, either by itself or with other signals, regulates GA biosynthesis.

8.3.2 Ca^{2+}

Ca^{2+} is a key second messenger for many various physiological changes and cellular processes, including growth, development, and secondary metabolism (Park et al. 2019). Some reports suggest that the change in intracellular calcium activates downstream calcium-related receptors and then regulates downstream genes. In *G. lucidum*, the addition of exogenous calcium ions (at 10 mM) to static liquid cultures enhances the production of antitumor GAs (total GAs, individual GA-Mk, -T, -S, and -Me) (Xu and Zhong 2012). The chlorpromazine is a calmodulin inhibitor, and cyclosporin A is a calcineurin inhibitor. They were used to confirm the role of calcium signaling in GA biosynthesis regulation. In addition, heat stress could also significantly elevate cytosolic Ca^{2+} concentration and the GA content.

Furthermore, treatment with the calcium ion chelator agent EGTA and the calcium channel inhibitor $LaCl_3$ partially prevented heat stress-induced GA biosynthesis (Zhang et al. 2016). The *Ca^{2+}*-permeable channel gene *cch and* phospholipase C gene *plc* were upregulated under heat stress conditions. Further results showed that the *cch* and *plc* gene silencing transformants caused a reduction in the HS-induced increase in GA biosynthesis and *hsp* expression in *G. lucidum* (Zhang et al. 2016).

8.3.3 CAMP

Cyclic adenosine monophosphate (cAMP) is also the second messenger participating in fungi's various physiological activities (Bores-Walmsley and Walmsley 2000; Kronstad et al. 1998). The intracellular cAMP dynamics primarily depend on two enzymes. Adenylated cyclase is the main enzyme for the synthesis of cAMP. Phosphodiesterase is involved in the degradation of cAMP. Caffeine is an inhibitor of phosphodiesterase, and NaF is an activator of adenylate cyclase. They have been employed to increase cytosolic cAMP levels in *G. lucidum* (You et al. 2017). The results showed that GA biosynthesis was induced by caffeine or NaF. At the same time, this process is accompanied by cell apoptosis.

8.3.4 H_2S

Hydrogen sulfide (H_2S), a small-molecule signaling agent, was recently shown to play a significant role in many physiological processes (Olas 2015). Tian et al. found that H_2S could reduce HS-induced GA biosynthesis (Tian et al. 2019). They found that H_2S could affect the physiological process of *G. lucidum* by interacting with multiple signals, including ROS, NO, AMPK, sphingolipid, mTOR, and phospholipase D. In particular, H_2S might alleviate the GA biosynthesis by inhibiting the intracellular calcium accumulation.

8.4 Membrane Fluidity and PLD-Phosphatidic Acid Signaling

The cell membrane plays an important role in sensing environmental change (Van Dijck et al. 2017). The cell membrane fluidity was observed to increase under heat stress (HS) in *G. lucidum* (Liu et al. 2017). Changing the membrane fluidity through the membrane rigidifier DMSO and the membrane fluidizer benzyl alcohol (BA) can alter HS-regulated GA biosynthesis.

A strain was constructed with a silenced delta 9 fatty acid desaturase (Δ9des). This strain's membrane fluidity was significantly reduced, resulting in the remission of heat stress-induced GA biosynthesis. Recently, HS-induced lipid remodeling was investigated using electrospray ionization–tandem mass spectrometry (ESI-MS/MS) analysis in *G. lucidum* under heat stress. The results showed that phosphatidic acid (PA) accumulated significantly under HS. Knocking out phospholipase D (PLD) showed that PA accumulation depended on HS-induced hydrolysis of phosphatidylethanolamine (Liu et al. 2017). In addition, 1-butanol, which decreased the PLD-mediated formation of PA, inhibited HS-induced GA accumulation.

Furthermore, knockdown of *PLD* partly blocked HS-induced GA biosynthesis, and this block can be reversed by adding PA to *PLD*-silenced strains. These results showed that PLD and PA mediated heat stress-induced GA biosynthesis. This study elucidated how *G. lucidum* responds to heat stress through phospholipid remodeling and subsequently accumulates secondary metabolites.

8.5 Crosstalk of Different Signals for GA Biosynthesis

Secondary metabolism usually involves the interaction between multiple signaling pathways. Under different environmental pressures, the interaction between these signals may be different. For example, in plants, ROS activates Ca^{2+} channels to regulate stomatal movement (Kong et al. 2016). However, Ca^{2+} enhances cellular ROS accumulation through phosphorylation of nicotinamide adenine dinucleotide phosphate oxidase RBOHB (Kobayashi et al. 2007). Current studies show that crosstalk between multiple signal transduction pathways also regulates secondary metabolism in *G. lucidum* (Gao et al. 2018). The study of signal interactions may help to elucidate the internal language transmission and mutual influence of *G. lucidum* and the influence of environmental factors on the regulatory network of GA biosynthesis.

Taking the ROS signal as an example, studies have shown that the interaction between ROS and Ca^{2+} in regulating GA biosynthesis is different under different conditions (Liu et al. 2018). In some cases, the ROS signal is positively correlated with GA biosynthesis (Mu et al. 2014). This work shows that ROS generated by NOx increases cytoplasmic Ca^{2+} by activating the plasma membrane Ca^{2+} influx pathway, thereby inducing the Ca^{2+} signaling pathway to regulate GA biosynthesis. In other cases, the ROS signal is negatively correlated with GA biosynthesis (Li et al. 2015). These results indicate that the repressed expression of GPx will lead to the accumulation of ROS but reduce the cytoplasmic Ca^{2+} concentration and GA content in *G. lucidum*. Further mechanistic studies also showed that the effect of GPx on GA biosynthesis via ROS is regulated by cytoplasmic Ca^{2+} content.

Another study showed that Cu^{2+}-induced GA content and cytoplasmic Ca^{2+} level depend on the increase in cytoplasmic ROS (Gao et al. 2018). The results also showed that the increased cytoplasmic Ca^{2+} could reduce cytoplasmic ROS by activating antioxidant enzymes, thereby reducing the increase in GA content under Cu^{2+} stress. This work demonstrates the presence of crosstalk between ROS and Ca^{2+} in regulating GA biosynthesis under Cu^{2+} stress. The result of this interaction is different from the mechanism observed under non-Cu^{2+} stress.

It has also been found that there is crosstalk between NO and Ca^{2+} under thermal stress (HS) (Liu et al. 2018). The results showed that heat stress induced the accumulation of NO and reduced GA biosynthesis induced by HS. At the same time, NO and Ca^{2+} promote each other in response to HS. However, the effect of Ca^{2+} on the increase in GA induced by HS is more direct and significant than that of NO (Fig. 8.3).

Although many signals have been involved in GA biosynthesis, few studies have investigated the regulation of signal interactions in GA biosynthesis. However, research may not

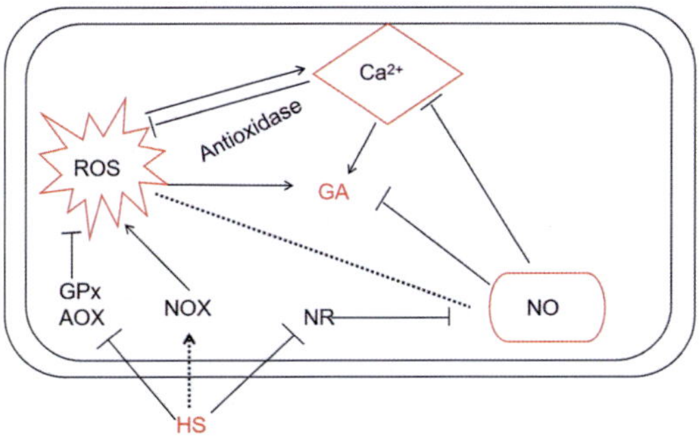

Fig. 8.3 The crosstalk of different signals (ROS, Ca^{2+}, NO) in the regulation progress of ganoderic acids biosynthesis according to Zhang et al. (Zhang et al. 2016; Shi et al. 2017; Liu et al. 2018), with some modifications. ROS, Reactive oxygen species; GPx, Glutathione peroxidase; AOX, Alternative oxidase; NOX, nicotinamide adenine dinucleotide phosphate oxidase; NR, Nitrate Reductase; GA, ganoderic acid; HS, heat stress. The solid black arrows indicate data supported by reported experiments, and the black dotted arrows indicate hypothetical steps based on the published papers in other species

determine that multiple signals, such as H_2S, CO, and CH_4, are involved in the regulation of GA biosynthesis (Ren et al. 2019). There may be interactions among these different signals. In addition, the signal interaction network may be different under different stress conditions. The study of these interactions needs further analysis. There are various models for regulating GA biosynthesis through different signals. Despite the interaction between the two signals, the functions of these signals in GA biosynthesis may be different under different conditions (such as Cu^{2+} stress, thermal stress, or Nox defects). Further research may still investigate the relationship between different signals under different conditions. It should also be investigated whether there are other signals (such as H_2S, CO, and CH_4) involved in the interaction regulation, which may lead to different results. However, research on these signals in fungi is still relatively limited. Studying these signals and their interactions may improve the understanding of GA biosynthesis's signal regulation network in *G. lucidum*.

8.6 Locus/Target of Signals in the Regulation of GA Biosynthesis

We found the mechanisms of GA biosynthesis involve a very complex network (Ren et al. 2019). Multiple signaling molecules are involved in the regulation of GA biosynthesis (Ren et al. 2019). In this process, there may be crosstalk among the different signaling molecules. However, the exact signaling molecules involved in regulating secondary metabolic biosynthesis have not been elucidated. Research on signal targets may provide a better explanation of the relationships between phenotypes and genes.

It was reported that treatment with hydrogen-rich water (HRW) could effectively reduce HAc-induced intracellular ROS production, inhibit mycelium growth, decrease mycelium biomass, and increase GA accumulation (Ren et al. 2017). Additional results indicate that HRW could reduce ROS levels by significantly improving the activities of the antioxidant enzyme glutathione peroxidase (gpx) under acetic acid (HAc)-induced oxidative stress. In addition, treatment with HRW did not significantly affect the gpx activities in the HRW and HAc co-treated samples in the gpx-silenced and gpx overexpression strains (Ren et al. 2017). The ROS levels, mycelium growth, and GA accumulation also demonstrated the ineffectiveness of HRW treatment in the absence of gpx. These results show that HRW was unable to alleviate the effects of HAc-induced phenotypic changes without gpx. The results indicate that GPX plays the most significant role in maintaining the ability of HRW to ameliorate ROS balance and regulate biochemical activity in *G. lucidum* under HAc-induced oxidative stress. Salicylic acid (SA) treatment induces GA biosynthesis by increasing ROS production in *G. lucidum*. Further research found that the ROS content was partially reduced in the NADPH oxidase-silenced strains with SA treatment. Mitochondria are a primary source of cellular ROS (Dan Dunn et al.

2015), and the localization of ROS shows that mitochondrial ROS increased in response to SA treatment. Further research also showed that SA treatment could inhibit mitochondrial complex III activity without causing obvious changes in the activities of mitochondrial complexes I and II. Treatment with SA in the mitochondria complex I and II silenced strains increased the mitochondrial ROS and GA content, but SA treatment did not significantly affect the GA content in the complex III strain (Liu et al. 2018). These results suggest that SA inhibits mitochondrial complex III activity to increase ROS generation and regulate GA biosynthesis in *G. lucidum*. In the case of nitrogen metabolism repression (NMR), the transcription factor AreA was activated and induced nitrogen assimilation or secondary metabolism. The areA could directly bind the promoter of the gene farnesyl diphosphate synthase (fps) and regulate the gene transcription level to enhance GA biosynthesis. The areA could also directly bind to the promoter of nitrate reductase (NR), a key enzyme for nitric oxide (NO) production which, in turn, enhanced NR activity. NO generated by NR could negatively regulate GA biosynthesis. The expression level of fps and the activity of NR were significantly decreased in the AreA-silenced strains. These results suggested that AreA regulated GA biosynthesis by directly binding fps and NR (Zhu et al. 2019). Glsnf1, a sucrose-nonfermenting serine–threonine protein kinase 1 (Snf1)/AMP-activated protein kinase homolog in *G. lucidum*, was activated when exposed to cellulose. Cellulase-related gene transcription was decreased in the Glsnf1-silenced strains, and related gene transcription was increased in the Glsnf1-overexpressing strains. The expression of carbon catabolite repressor (CreA) was the most significant among the cellulase-related genes (Hu et al. 2020). Further study on the yeast two-hybrid assay and immunoprecipitation showed that Snf1an directly interacts with CreA and that CreA is unable to localize in the nucleus in the presence of Glsnf1. In addition, Glsnf1 and GlCreA silencing inhibited the Glsnf1 silencing-induced decrease in cellulase activity. These results suggested that Glsnf1 inhibits the transfer of GlCreA to the nucleus and induces the expression of the cellulase-related gene GlCreA to increase cellulose (Hu et al. 2020).

The study of signal targets provides an important clue regarding directional induction of GA accumulation. The screening and determination of signal targets have a strong potential for the development of treatments to induce GA production. This approach may provide a more efficient and convenient method for further exploring the mechanism of GA biosynthesis.

8.7 Concluding Remarks

At present, there is no sophisticated research platform for *G. lucidum* comparable to those for model fungi (such as *Aspergillus* spp. and *Cryptococcus neoformans*). In the future, as genetic transformation and transcriptional gene silencing have been gradually developed in *G. lucidum*, the physiological research on *G. lucidum* should be strengthened. On the one hand, targets in the regulation network of GA biosynthesis have not been identified; therefore, further investigations should conduct more detailed analyses to clarify the molecular mechanisms by which environmental factors regulate GA biosynthesis. On the other hand, the elucidation of the GA's environmental regulation and the screening of specific GAs may provide better strategies for directed induction GA biosynthesis. This work will make *G. lucidum* a potential model species for studying the regulation of secondary metabolism.

References

Agnew WS, Popjak G (1978) Squalene synthetase. Solubilization from yeast microsomes of a phospholipid-requiring enzyme. J Biol Chem 253 (13):4574–83

Balzan R et al (2004) Aspirin commits yeast cells to apoptosis depending on carbon source. Microbiology 150(Pt 1):109–115

Bores-Walmsley MI, Walmsley AR (2000) cAMP signalling in pathogenic fungi: control of dimorphic switching and pathogenicity. Trends Microbiol 8 (3):133–141

Calvo AM et al (2002) Relationship between secondary metabolism and fungal development. Microbiol Mol Biol Rev 66(3):447–459

Cao PF et al (2017) Effects of exogenous salicylic acid on ganoderic acid biosynthesis and the expression of key genes in the ganoderic acid biosynthesis pathway in the Lingzhi or Reishi medicinal mushroom, *Ganoderma lucidum* (Agaricomycetes). Int J Med Mushrooms 19(1):65–73

Chen NH, Liu JW, Zhong JJ (2010) Ganoderic acid T inhibits tumor invasion in vitro and in vivo through inhibition of MMP expression. Pharmacological Reports 62(1):150–163

Chen S et al (2012) Genome sequence of the model medicinal mushroom *Ganoderma lucidum*. Nat Commun 3:913

Czekaj P (2000) Phenobarbital-induced expression of cytochrome P450 genes. Acta Biochim Pol 47 (4):1093–1105

Dan Dunn J et al (2015) Reactive oxygen species and mitochondria: a nexus of cellular homeostasis. Redox Biol 6:472–485

Fang QH, Zhong JJ (2002a) Effect of initial pH on production of ganoderic acid and polysaccharide by submerged fermentation of *Ganoderma lucidum*. Process Biochem 37(7):769–774

Fang QH, Zhong JJ (2002b) Two-stage culture process for improved production of ganoderic acid by liquid fermentation of higher fungus *Ganoderma lucidum*. Biotechnol Prog 18(1):51–54

Fang X et al (2013) The cloning, characterization and functional analysis of a gene encoding an acetyl-CoA acetyltransferase involved in triterpene biosynthesis in *Ganoderma lucidum*. Mycoscience 54(2):100–105

Fei Y et al (2019) Increased production of ganoderic acids by overexpression of homologous farnesyl diphosphate synthase and kinetic modeling of ganoderic acid production in *Ganoderma lucidum*. Microb Cell Fact 18(1):115

Feng J et al (2016) A new temperature control shifting strategy for enhanced triterpene production by *Ganoderma lucidum* G0119 based on submerged liquid fermentation. Appl Biochem Biotechnol 180(4):740–752

Gao T et al (2018) Cross Talk between calcium and reactive oxygen species regulates hyphal branching and ganoderic acid biosynthesis in *Ganoderma lucidum* under copper stress. Appl Environ Microbiol 84(13)

Gunjegaonkar SM, Shanmugarajan TS (2019) Molecular mechanism of plant stress hormone methyl jasmonate for its anti-inflammatory activity. Plant Signal Behav 14(10)

Hirotani M, Asaka I, Furuya T (1990) Studies on the metabolites of higher Fungi.9. Investigation of the biosynthesis of 3-alpha-hydroxy triterpenoids, ganoderic acid-T and acid-S by application of a feeding experiment using [1,2-(C2)-C-13]Acetate. J Chem Soc-Perkin Trans 1(10):2751–2754

Hu YR et al (2017) Improved ganoderic acids production in *Ganoderma lucidum* by wood decaying components. Sci Rep 7:46623

Hu Y et al (2020) In *Ganoderma lucidum*, Glsnf1 regulates cellulose degradation by inhibiting GlCreA during the utilization of cellulose. Environ Microbiol 22(1):107–121

Jain KK et al (2020) De novo transcriptome assembly and protein profiling of copper-induced lignocellulolytic fungus Ganoderma lucidum MDU-7 reveals genes involved in lignocellulose degradation and terpenoid biosynthetic pathways. Genomics 112(1):184–198

Kobayashi M et al (2007) Calcium-dependent protein kinases regulate the production of reactive oxygen species by potato NADPH oxidase. Plant Cell 19 (3):1065–1080

Kong D et al (2016) L-Met activates Arabidopsis GLR Ca (2+) channels upstream of ROS production and regulates stomatal movement. Cell Rep 17(10):2553–2561

Kronstad J et al (1998) Signaling via cAMP in fungi: interconnections with mitogen-activated protein kinase pathways. Arch Microbiol 170(6):395–404

Kumar V et al (2020) Recent developments on solid-state fermentation for production of microbial secondary metabolites: challenges and solutions. Bioresour Technol 323:124566

Lajayer BA, Ghorbanpour M, Nikabadi S (2017) Heavy metals in contaminated environment: destiny of secondary metabolite biosynthesis, oxidative status and phytoextraction in medicinal plants. Ecotoxicol Environ Saf 145:377–390

Li N et al (2006) Analysis of influence of environmental conditions on ganoderic acid content in *Ganoderma lucidum* using orthogonal design. J Microbiol Biotechnol 16(12):1940–1946

Li C et al (2015) Functional analysis of the role of glutathione peroxidase (GPx) in the ROS signaling pathway, hyphal branching and the regulation of ganoderic acid biosynthesis in *Ganoderma lucidum*. Fungal Genet Biol 82:168–180

Liang CX et al (2010) Enhanced biosynthetic gene expressions and production of ganoderic acids in static liquid culture of *Ganoderma lucidum* under phenobarbital induction. Appl Microbiol Biotechnol 86(5):1367–1374

Liu R et al (2018) Cross talk between nitric oxide and calcium-calmodulin regulates ganoderic acid biosynthesis in *Ganoderma lucidum* under heat stress. Appl Environ Microbiol 84(10)

Liu GQ et al (2011) Stimulated production of triterpenoids of *Ganoderma lucidum* by an ether extract from the medicinal insect, Catharsius molossus, and identification of the key stimulating active components. Appl Biochem Biotechnol 165(1):87–97

Liu GQ et al (2012) Improving the fermentation production of the individual key triterpene ganoderic acid me by the medicinal fungus *Ganoderma lucidum* in submerged culture. Molecules 17(11):12575–12586

Liu YN et al (2017a) Membrane fluidity is involved in the regulation of heat stress induced secondary metabolism in *Ganoderma lucidum*. Environ Microbiol 19 (4):1653–1668

Liu YN et al (2017b) Phospholipase D and phosphatidic acid mediate heat stress induced secondary metabolism in *Ganoderma lucidum*. Environ Microbiol 19 (11):4657–4669

Liu R et al (2018) SA inhibits complex III activity to generate reactive oxygen species and thereby induces GA overproduction in *Ganoderma lucidum*. Redox Biol 16:388–400

Luo Z et al (2017) The PacC transcription factor regulates secondary metabolite production and stress response, but has only minor effects on virulence in the insect pathogenic fungus *Beauveria bassiana*. Environ Microbiol 19(2):788–802

Lynen F et al (1958) The chemical mechanism of acetic acid formation in the liver. Biochem Z 330(4):269–295

Lynen F et al (1959) γ, γ-Dimethyl-allyl-pyrophosphat und Geranyl-pyrophosphat, biologische Vorstufen des Squalens Zur Biosynthese der Terpene, VI1). Angew Chem 71(21):657–663

Mu D et al (2012) The development and application of a multiple gene co-silencing system using endogenous URA3 as a reporter gene in *Ganoderma lucidum*. PLoS ONE 7(8):

Mu D et al (2014) Functions of the nicotinamide adenine dinucleotide phosphate oxidase family in *Ganoderma lucidum*: an essential role in ganoderic acid biosynthesis regulation, hyphal branching, fruiting body development, and oxidative-stress resistance. Environ Microbiol 16(6):1709–1728

Olas B (2015) Hydrogen sulfide in signaling pathways. Clin Chim Acta 439:212–218

Park HS et al (2019) Calcium-calmodulin-calcineurin signaling: a globally conserved virulence cascade in eukaryotic microbial pathogens. Cell Host Microbe 26 (4):453–462

Raza H, John A, Benedict S (2011) Acetylsalicylic acid-induced oxidative stress, cell cycle arrest, apoptosis and mitochondrial dysfunction in human hepatoma HepG2 cells. Eur J Pharmacol 668(1–2):15–24

Ren A et al (2010) Methyl jasmonate induces ganoderic acid biosynthesis in the basidiomycetous fungus *Ganoderma lucidum*. Bioresour Technol 101 (17):6785–6790

Ren A et al (2013) Profiling and quantifying differential gene transcription provide insights into ganoderic acid biosynthesis in *Ganoderma lucidum* in response to methyl jasmonate. PLoS ONE 8(6):

Ren A et al (2014) Transcript and metabolite alterations increase ganoderic acid content in *Ganoderma lucidum* using acetic acid as an inducer. Biotech Lett 36(12):2529–2536

Ren A et al (2017) Hydrogen-rich water regulates effects of ROS balance on morphology, growth and secondary metabolism via glutathione peroxidase in *Ganoderma lucidum*. Environ Microbiol 19(2):566–583

Ren A et al (2019) Shedding light on the mechanisms underlying the environmental regulation of secondary metabolite ganoderic acid in *Ganoderma lucidum* using physiological and genetic methods. Fungal Genet Biol 128:43–48

Sasiak K, Rilling HC (1988) Purification to homogeneity and some properties of squalene synthetase. Arch Biochem Biophys 260(2):622–627

Schweiger R et al (2014) Interactions between the jasmonic and salicylic acid pathway modulate the plant metabolome and affect herbivores of different feeding types. Plant Cell Environ 37(7):1574–1585

Scialo F et al (2016) Mitochondrial ROS produced via reverse electron transport extend animal lifespan. Cell Metab 23(4):725–734

Shang CH et al (2008) Cloning and characterization of a gene encoding HMG-CoA reductase from *Ganoderma lucidum* and its functional identification in yeast. Biosci Biotechnol Biochem 72(5):1333–1339

Shi L et al (2010) Current progress in the study on biosynthesis and regulation of ganoderic acids. Appl Microbiol Biotechnol 88(6):1243–1251

Shi L et al (2012a) Development of a simple and efficient transformation system for the basidiomycetous medicinal fungus *Ganoderma lucidum*. World J Microbiol Biotechnol 28(1):283–291

Shi L et al (2012b) Molecular cloning, characterization, and function analysis of a mevalonate pyrophosphate decarboxylase gene from *Ganoderma lucidum*. Mol Biol Rep 39(5):6149–6159

Shi L et al (2015) The regulation of methyl jasmonate on hyphal branching and GA biosynthesis in *Ganoderma lucidum* partly via ROS generated by NADPH oxidase. Fungal Genet Biol 81:201–211

Shi DK et al (2017) Alternative oxidase impacts ganoderic acid biosynthesis by regulating intracellular ROS levels in *Ganoderma lucidum*. Microbiology (Reading) 163(10):1466–1476

Shiao MS (2003) Natural products of the medicinal fungus *Ganoderma lucidum*: occurrence, biological activities, and pharmacological functions. Chem Rec 3 (3):172–180

Sies H, Jones DP (2020) Reactive oxygen species (ROS) as pleiotropic physiological signalling agents. Nat Rev Mol Cell Biol 21(7):363–383

Tang YJ, Zhong JJ (2002) Fed-batch fermentation of *Ganoderma lucidum* for hyperproduction of polysaccharide and ganoderic acid. Enzym Microb Technol 31(1–2):20–28

Tang YJ, Zhong JJ (2003) Role of oxygen supply in submerged fermentation of *Ganoderma lucidum* for production of Ganoderma polysaccharide and ganoderic acid. Enzym Microb Technol 32(3–4):478–484

Tang YJ, Zhu LW (2010) Improvement of ganoderic acid and Ganoderma polysaccharide biosynthesis by *Ganoderma lucidum* fermentation under the inducement of Cu^{2+}. Biotechnol Prog 26(2):417–423

Thakur M et al (2019) Improving production of plant secondary metabolites through biotic and abiotic elicitation. J Appl Res Med Aromat Plants 12:1–12

Tian JL et al (2019) Hydrogen sulfide, a novel small molecule signalling agent, participates in the regulation of ganoderic acids biosynthesis induced by heat stress in *Ganoderma lucidum*. Fungal Genet Biol 130:19–30

Van Dijck P et al (2017) Nutrient sensing at the plasma membrane of fungal cells. Microbiol Spectr 5(2)

Wang JL, Gu T, Zhong JJ (2012) Enhanced recovery of antitumor ganoderic acid T from *Ganoderma lucidum* mycelia by novel chemical conversion strategy. Biotechnol Bioeng 109(3):754–762

Wei ZH et al (2016) Sucrose fed-batch strategy enhanced biomass, polysaccharide, and ganoderic acids production in fermentation of *Ganoderma lucidum* 5.26. Bioprocess Biosyst Eng 39(1):37–44

Wu CG et al (2017) Ornithine decarboxylase-mediated production of putrescine influences ganoderic acid biosynthesis by regulating reactive oxygen species in *Ganoderma lucidum*. Appl Environ Microbiol 83(20)

Wu FL et al (2013) The cloning, characterization, and functional analysis of a gene encoding an isopentenyl diphosphate isomerase involved in triterpene biosynthesis in the Lingzhi or reishi medicinal mushroom *Ganoderma lucidum* (higher Basidiomycetes). Int J Med Mushrooms 15(3):223–232

Wu FL et al (2016) The pH-responsive transcription factor PacC regulates mycelial growth, fruiting body development, and ganoderic acid biosynthesis in *Ganoderma lucidum*. Mycologia 108(6):1104–1113

Wu L et al (2020) Overexpression of the 3-hydroxy-3-methylglutaryl-CoA synthase gene LcHMGS effectively increases the yield of monoterpenes and sesquiterpenes. Tree Physiol 40(8):1095–1107

Xu YN, Zhong JJ (2012) Impacts of calcium signal transduction on the fermentation production of antitumor ganoderic acids by medicinal mushroom *Ganoderma lucidum*. Biotechnol Adv 30(6):1301–1308

Xu P et al (2008) Improved production of mycelial biomass and ganoderic acid by submerged culture of *Ganoderma lucidum* SB97 using complex media. Enzym Microb Technol 42(4):325–331

Xu JW, Xu YN, Zhong JJ (2012) Enhancement of ganoderic acid accumulation by overexpression of an N-terminally truncated 3-hydroxy-3-methylglutaryl coenzyme a reductase gene in the basidiomycete *Ganoderma lucidum*. Appl Environ Microbiol 78 (22):7968–7976

Xu YN, Xia XX, Zhong JJ (2013) Induced effect of Na plus on ganoderic acid biosynthesis in static liquid culture of *Ganoderma lucidum* via calcineurin signal transduction. Biotechnol Bioeng 110(7):1913–1923

Xu YN, Xia XX, Zhong JJ (2014) Induction of ganoderic acid biosynthesis by Mn^{2+} in static liquid cultivation of *Ganoderma lucidum*. Biotechnol Bioeng 111 (11):2358–2365

You B-J et al (2017) Induction of apoptosis and ganoderic acid biosynthesis by cAMP signaling in *Ganoderma lucidum*. Sci Rep 7(1)

You BJ et al (2018) A novel approach to enhancing ganoderic acid production by *Ganoderma lucidum* using apoptosis induction. Plos One 8(1)

You BJ et al (2012) Enhanced production of ganoderic acids and cytotoxicity of *Ganoderma lucidum* using solid-medium culture. Biosci Biotechnol Biochem 76 (8):1529–1534

Yu JH, Keller N (2005) Regulation of secondary metabolism in filamentous fungi. Annu Rev Phytopathol 43:437–458

Yu X et al (2014) Development of an expression plasmid and its use in genetic manipulation of Lingzhi or Reishi medicinal mushroom, *Ganoderma lucidum* (higher Basidiomycetes). Int J Med Mushrooms 16 (2):161–168

Zhang W, Tang YJ (2008) A novel three-stage light irradiation strategy in the submerged fermentation of medicinal mushroom *Ganoderma lucidum* for the efficient production of ganoderic acid and ganoderma polysaccharides. Biotechnol Prog 24(6):1249–1261

Zhang JM, Zhong JJ, Geng AL (2014) Improvement of ganoderic acid production by fermentation of *Ganoderma lucidum* with cellulase as an elicitor. Process Biochem 49(10):1580–1586

Zhang X et al (2016) Heat stress modulates mycelium growth, heat shock protein expression, Ganoderic acid biosynthesis, and hyphal branching of G*anoderma lucidum* via cytosolic Ca^{2}. Appl Environ Microbiol 82 (14):4112–4125

Zhang G et al (2017a) Ethylene promotes mycelial growth and ganoderic acid biosynthesis in *Ganoderma lucidum*. Biotechnol Lett 39(2):269–275

Zhang G et al (2017b) The mitogen-activated protein kinase GlSlt2 regulates fungal growth, fruiting body development, cell wall integrity, oxidative stress and ganoderic acid biosynthesis in *Ganoderma lucidum*. Fungal Genet Biol 104:6–15

Zhang DH et al (2017c) Overexpression of the squalene epoxidase gene alone and in combination with the 3-hydroxy-3-methylglutaryl coenzyme a gene increases ganoderic acid production in *Ganoderma lingzhi*. J Agric Food Chem 65(23):4683–4690

Zhang DH et al (2017d) Overexpression of the homologous lanosterol synthase gene in ganoderic acid biosynthesis in *Ganoderma lingzhi*. Phytochemistry 134:46–53

Zhao W, Xu JW, Zhong JJ (2011) Enhanced production of ganoderic acids in static liquid culture of *Ganoderma lucidum* under nitrogen-limiting conditions. Bioresour Technol 102(17):8185–8190

Zhou JS et al (2014) Enhanced accumulation of individual ganoderic acids in a submerged culture of *Ganoderma lucidum* by the overexpression of squalene synthase gene. Biochem Eng J 90:178–183

Zhu J et al (2019) Functions of reactive oxygen species in apoptosis and ganoderic acid biosynthesis in *Ganoderma lucidum*. FEMS Microbiol Lett 366(23): fnaa015

Zhu LW, Tang YJ (2010) Significance of protein elicitor isolated from Tuber melanosporum on the production of ganoderic acid and Ganoderma polysaccharides during the fermentation of *Ganoderma lucidum*. Bioprocess Biosyst Eng 33(8):999–1005

Zhu J et al (2019) Dual functions of AreA, a GATA transcription factor, on influencing ganoderic acid biosynthesis in *Ganoderma lucidum*. Environ Microbiol 21(11):4166–4179

9 Genetic Transformation System

Jun-Wei Xu

Abstract

Lingzhi (*Ganoderma lucidum*) contains various bioactive constituents, such as polysaccharides and triterpenoids. The development of genetic transformation systems of Lingzhi is necessary for the characterization of the function of the target genes and the efficient production of bioactive compounds by genetically engineered strains. This chapter describes major genetic transformation methods for Lingzhi, namely, polyethylene glycol-mediated transformation (PMT), restriction enzyme-mediated integration (REMI), electroporation, and *Agrobacterium tumefaciens*-mediated transformation (ATMT). These transformation methods support the genetic engineering of Lingzhi to enhance polysaccharide and triterpenoid production. Gene silencing of Lingzhi through transformation is also described in this chapter. Additionally, recent advances in disrupting and deleting target genes using the CRISPR/Cas9 technology are discussed.

9.1 Introduction

Lingzhi (*Ganoderma lucidum*) is a well-known mushroom in China used in Chinese medicine to treat and prevent different diseases for more than 2000 years. It was listed as an effective drug without toxicity in ancient Chinese medicine books "*Shen Nong's Ben Cao Jing*" and "*Ben Cao Gang Mu*." Modern scientific research has demonstrated that *G. lucidum* possesses diverse pharmacological activities, such as immunomodulation, anti-tumor, anti-aging, anti-inflammatory, anti-diabetic, antibacterial, anti-viral (including anti-HIV), cholesterol synthesis inhibitory, hepatoprotective, and chemopreventive activities (Baby et al. 2015; Pan and Lin 2019). The annual global turnover of *Ganoderma* products is worth over USD 2.5 billion (Bishop et al. 2015; Hsu and Cheng 2018). Polysaccharides and triterpenes are two major types of bioactive compounds of *G. lucidum* (Boh et al. 2007). To date, more than 200 different polysaccharides and 150 triterpenes have been identified from the mycelia, spores, and fruiting bodies of *G. lucidum* (Ahmad 2018; Cor et al. 2018). Cultures of fruiting bodies and mycelia have been widely used to produce *Ganoderma* polysaccharides and triterpenes (Hsu and Cheng 2018; Wagner et al. 2003; Zhong and Tang 2004). With the release of the *G. lucidum* genome sequence (Chen et al. 2012; Kues et al. 2015; Liu et al. 2012), elucidation of biosynthetic pathways and regulatory mechanisms of these

J.-W. Xu (✉)
Faculty of Life Science and Technology, Kunming University of Science and Technology, Kunming 650500, China
e-mail: xjuwei@163.com; jwxu@kust.edu.cn

C. Liu (ed.), *The Lingzhi Mushroom Genome*, Compendium of Plant Genomes,
https://doi.org/10.1007/978-3-030-75710-6_9

bioactive compounds has recently become an important topic in *G. lucidum*. These research studies will pave the way for the efficient production of *Ganoderma* bioactive compounds. The development of genetic transformation approaches is a precondition for the functional characterization of the genes of interest and genetic improvement of *G. lucidum* for biotechnological applications. Here, we summarize major methods of genetic transformation of *G. lucidum*, application of genetic transformation to improve its production of polysaccharides and triterpenoids, and gene silencing methods for this species. Moreover, recent progress on the disruption and deletion of *Ganoderma* target genes is also presented.

9.2 Genetic Transformation Methods

Here, major genetic transformation methods for *G. lucidum*, including PMT, REMI, electroporation, and ATMT are described (Table 9.1).

9.2.1 PMT

PMT is the most commonly used genetic transformation methods for *G. lucidum* and relies on a sufficient amount of competent protoplasts. The principle is to use lytic enzymes to disrupt cell wall components for the preparation of protoplasts. Subsequently, PEG and calcium chloride ($CaCl_2$) are used to promote the fusion of protoplasts and DNA. PEG increases membrane permeability, and $CaCl_2$ participates in forming channels in the cytomembrane (Poyedinok and Blume 2018).

Li et al. (2006) reported the development of an efficient PMT method for *G. lucidum*. To increase the PMT efficiency of *G. lucidum*, they included heparin, aurintricarboxylic acid, and spermidine in the PMT method. This approach yielded 120–150 transformants per μg plasmid pAN7-1 per 10^7 viable protoplasts with the selection marker hygromycin B resistance gene (*hph*). This result indicates that the PMT is a useful tool for *G. lucidum*. Yu et al. (2014) transferred plasmid pJW-EXP and pJW-EXP-HMGR into *G. lucidum* by PMT. The carboxin-resistance gene (*cbx*) was used as a dominant selection marker in genetic transformation. The transformation efficiency was 15–20 per μg plasmid per 10^7 viable protoplasts. Southern blotting showed that multiple copies of *cbx* were stably integrated into the transformant genomes.

The PMT method is convenient and effective, and it does not require specialized equipments. However, its use has been limited by complicated procedures and the low yield and regeneration rate of protoplasts. Optimization of each step is thus necessary to facilitate its application.

9.2.2 REMI

Based on PMT, REMI is a common method for transferring linearized vectors and a specific restriction enzyme into host cells. The transfer of the linearized vector into host cells is accomplished by enzyme-mediated integration. The mechanism involves cutting the host genome at

Table 9.1 Methods for genetic transformation of *Ganoderma lucidum*

Method	Selection marker gene	Efficiency	References
PMT	hph	120–150 per μg DNA per 10^7 protoplasts	Li et al. (2006)
PMT	cbx	15–20 per μg DNA per 10^7 protoplasts	Yu et al. (2014)
REMI	kan	4–17 per μg DNA per 10^7 protoplasts	Kim et al. (2004)
Electroporation	bar	15 per μg DNA per 10^7 protoplasts	Sun et al. (2001, 2002)
ATMT	cbx	10–15 per 10^7 protoplasts	Xu et al. (2012)
ATMT	hph	200 per 10^5 protoplasts	Shi et al. (2012)

specific restriction sites and transforming linearized vectors using the same sites (Qin et al. 2016; Turgeon et al. 2010).

Kim et al. (2004) transformed plasmid pJS205-1, a geneticin resistance gene (*kan*), into protoplasts of *G. lucidum* via REMI. When the transformation was conducted using three different types of restriction enzymes such as *Eco*RV, *Not*I, and *Xho*I, the obtained transformation efficiency was 4–17 per μg plasmid per 10^7 protoplasts. Several *G. lucidum* mutants were generated during REMI. Genome PCR analysis confirmed the integration of plasmid pJS205-1 into the chromosomes of *G. lucidum*.

REMI has opened up prospects for integrating labeled genes into host cells' chromosomes (He et al. 2017; Poyedinok and Blume 2018). One of the advantages of REMI is the formation of many mutants that can be used to study the function of genes related to mutant characteristics (Kim et al. 2004; Qin et al. 2016). However, REMI also involves many steps, and restriction enzymes are influenced by protoplasts' status and require careful testing of transformation enzymes and vectors.

9.2.3 Electroporation

Electroporation is a common physical technique for the genetic transformation of fungi (He et al. 2017). Strong electrical fields increase the permeability of cytomembrane reversibly. Electric pulses lead to the formation of micropores in the cytomembrane, and these micropores are large enough to allow the transformation of exogenous DNA into the host cells. Treatment of cells and physical parameters such as electric field intensity, capacitance, pulse duration, and frequency are key factors that influence electroporation-mediated transformation efficiency (Li et al. 2017; Poyedinok and Blume 2018).

Sun et al. (2001) developed an electroporation procedure for the genetic transformation of *G. lucidum* protoplasts. The optimal condition for electroporation was obtained at 2.5 kV cm^{-1}, 25 μF, and 400 Ω. The transformation efficiency was about 15 per μg plasmid per 10^7 protoplasts, with the selection marker bialaphos resistance gene (*bar*). Dot blot results showed that the exogenous gene was integrated into the transformant genome. When plasmid p301-bG containing the CaMV 35S promoter and β-glucuronidase (GUS) gene was transferred into the protoplasts of *G. lucidum* by electroporation, Sun et al. (2002) obtained some transformants after four days of cultivation under selective CYM media with bialaphos. The transformation efficiency was 15 per μg plasmid per 10^7 protoplasts. Moreover, GUS activity was observed in five randomly picked *G. lucidum* transformants, conferring bialaphos resistance.

Compared with PMT where complicated steps are involved, electroporation is a convenient and straightforward method. However, the transformation of intact fungal mycelia and spores has only been reported in a few species (He et al. 2017; Meyer 2008). Furthermore, the adjustment of physical parameters is sometimes correlated with high transformation efficiency (He et al. 2017).

9.2.4 ATMT

ATMT of filamentous fungi was first reported by de Groot et al. in 1998. The tumor-inducing (Ti) plasmid in *A. tumefaciens* can transfer a piece of its DNA (T-DNA) into mycelial fungi, where it is integrated into the fungal genome and expressed. The transfer depends on the induction of virulence genes, which results in T-DNA processing and the establishment of pili/pores that can mediate T-DNA transfer into host cells (de Groot et al 1998). ATMT efficiency depends on several factors, including the type of fungal material, the ratio of fungus to *Agrobacterium* cells, acetosyringone concentration, and co-cultivation conditions (Li et al. 2017).

ATMT has been successfully applied to *G. lucidum*. Protoplasts of *G. lucidum* were transformed into carboxin resistant using the selection marker *cbx*, under the control of a homologous *sdhB* promoter. The transformation efficiency was 10–15 per 10^7 protoplasts. However, no stable transformants were produced when mycelia were used as starting material. *Cbx* was inserted into the transformant genome as a single copy, and the transformants exhibited 100% stability for the *cbx* (Xu et al. 2012). Shi et al. (2012) also reported the transformation of *G. lucidum* with *A. tumefaciens* LBA4404 for the integration of the *hph*. Co-cultivation of *A. tumefaciens* and protoplasts of *G. lucidum* in the presence of 0.2 mM acetosyringone resulted in the formation of hygromycin B-resistant colonies with high transformation efficiency, which is 200 transformants per 10^5 protoplasts. The results showed that over 70% of transformants exhibited mitotic stability for the *hph*.

The ATMT method has many advantages. Different types of fungal materials, such as protoplasts, mycelia, spores, and fruiting bodies, can be transformed by ATMT (He et al. 2017). Several fungal species subjected to this method exhibit a higher degree of mitotic stability (Li et al. 2017; Poyedinok and Blume 2018). Moreover, it can achieve high homologous recombination (HR) frequency in targeted gene deletion and disruption experiments (Li et al. 2017). However, multiple parameters during co-cultivation need to be considered during optimization of the transformation process. Besides, this method is more time-consuming than the other methods described above.

9.3 Application of Genetic Transformation of *G. Lucidum*

The establishment of genetic transformation for *G. lucidum* may facilitate the enhancement of the production of bioactive compounds. This section summarizes the efficient production of polysaccharides and ganoderic acids in *G. lucidum* by genetic engineering.

9.3.1 Efficient Production of Polysaccharides by Genetic Engineering of *G. Lucidum*

Genetic engineering has been successfully applied to enhance polysaccharide production in *G. lucidum*. Phosphoglucomutase (PGM) and UDP-glucose pyrophosphorylase (UGP) are critical enzymes in the nucleotide sugar biosynthesis. Xu et al. (2015) overexpressed the homologous PGM gene (*pgm*) in *G. lucidum*. The content of IPS in the transgenic strain was 23.67%, which was 1.4-fold higher than the control. EPS yield in the transgenic strain was 1.76 g/L, which was 1.44 higher than that of the wild-type strain. The transcription levels of downstream genes, such as UGP and β-1, 3-glucan synthase (GLS), were also upregulated in the engineered strain. Hu et al. (2018) also reported that the production of EPS decreased in the PGM gene knockdown strain to 20–40% of the control (Hu et al. 2018). Ji et al. (2015) improved *G. lucidum* polysaccharides' yield by overexpressing the homologous gene of UGP (*ugp*). UGP is an enzyme catalyzing the synthesis of UDP-glucose. Overexpression of the *ugp* did not significantly affect cell growth. The transgenic strain's EPS production and IPS content reached 1.66 g/L and 24.32%, respectively, 36% and 42% higher than the control. Li et al. (2015) also found that the UGP gene knockdown in *G. lucidum* reduced EPS production by approximately 30–56%. These results showed that manipulating the biosynthetic gene expression helps polysaccharide production in *G. lucidum* (Table 9.2).

The heterologous expression of *the Vitreoscilla* hemoglobin (VHb) gene is another way to improve polysaccharides production in *G. lucidum*. Li et al. (2016b) transferred the VHb gene (*vgb*) into *G. lucidum*. CO-difference spectrum results confirmed the bioactivity of expressed VHb in the engineered strain. The IPS content and EPS yield of the engineered strain were 1.3-

Table 9.2 Genetic engineering of *Ganoderma lucidum* for the efficient production of polysaccharides

Overexpressed genes	Fermentation conditions	Results	References
pgm	Submerged cultivation	Increased the IPS content and EPS production by 40% and 44%, respectively	Xu et al. (2015)
ugp	Submerged cultivation	Increased the IPS content and EPS production by 42% and 36%, respectively	Ji et al. (2015)
vhb	Submerged cultivation	Increased the IPS content and EPS production by 30% and 88%, respectively	Li et al. (2016b)

and 1.9-fold higher than that of the control, respectively (Fig. 9.1 and Table 9.2). The expression of VHb may lead to more efficient energy production in *G. lucidum*, which is favorable to polysaccharide biosynthesis from simple sugars. This study highlights VHb technology's potential in improving the production of *G. lucidum* polysaccharides by large-scale fermentation.

9.3.2 Efficient Production of Ganoderic Acid by Genetically Modified *G. Lucidum*

Ganoderic acids are synthesized via the mevalonate/isoprenoid (MVA) pathway in *G. lucidum*. The key enzymes in this pathway included hydroxyl-methyl-glutaryl coenzyme A

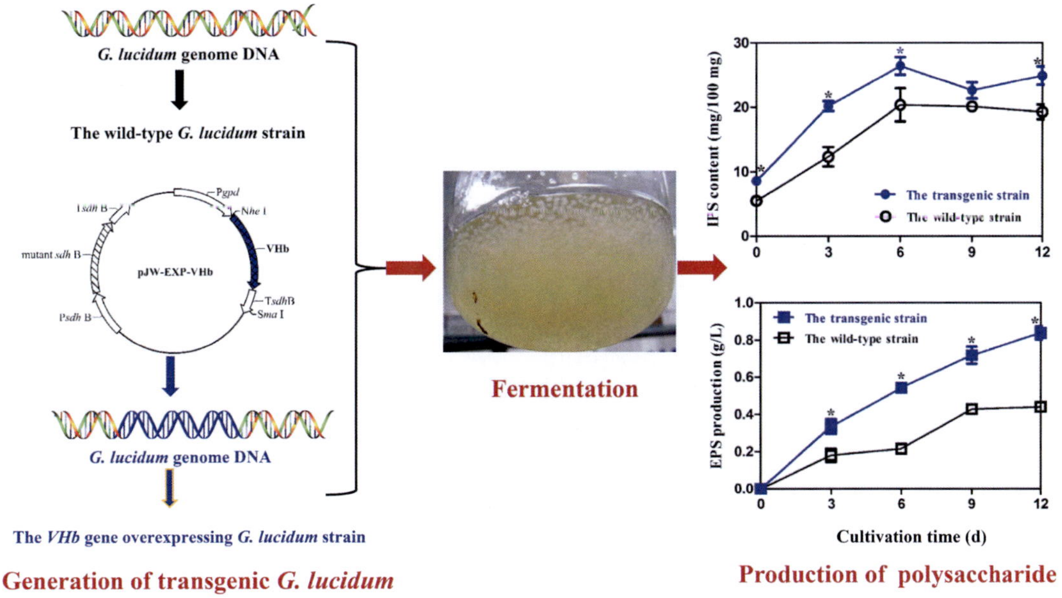

Fig. 9.1 Improved polysaccharide production in Ganoderma lucidum by the heterologous expression of the Vitreoscilla hemoglobin (VHb) gene

synthase (HMGS), 3-hydroxy-3-methyl-glutaryl coenzyme A reductase (HMGR), farnesyl diphosphate synthase (FPS), squalene synthase (SQS), squalene epoxidase (SE), and lanosterol synthase (LS). They have all been cloned in *G. lucidum* (Fei et al. 2019; Ren et al. 2013; Shi et al. 2010; Zhang et al. 2017a, b). Recently, gene overexpression is used to efficiently modify *G. lucidum* to produce ganoderic acids (Table 9.3).

Ren et al. (2013) investigated the effect of overexpression of the HMGS gene on ganoderic acid content. The transcription level of the HMGS gene in the engineered strain was 2.3-fold higher than the control, which resulted in approximately 15.1–24.2% improvement in total ganoderic acid content (Ren et al. 2013). Xu et al. overexpressed the homologous HMGR gene (*hmgr*) in *G. lucidum*. The content of total ganoderic acids increased by two-fold in the *hmgr* overexpressed strain compared to that of the control. The content of intermediates (lanosterol and squalene) and expression levels of downstream genes also increased in the engineered strain (Xu et al. 2012; Yu et al. 2014). However, the individual ganoderic acids' contents were not altered in the *hmgr* overexpressed strain, which suggested that there are other flux control points. Zhou et al. (2014) improved the accumulation of individual ganoderic acids through the SQS gene's overexpression (*sqs*) in *G. lucidum*. The contents of individual ganoderic acids, GA-Mk, GA-T, GA-Me, and GA-S in the *sqs* overexpressed strain were 16, 40, 43, and 53 μg/100 mg dry cell weight (DCW), respectively, which were 2.9, 2.7, 2.0, and 1.3 times higher than that of the control. Attempts have also been made to increase the contents of total and individual ganoderic acids by genetic manipulation of FPS gene (*fps*) and LS gene (*ls*) in *G. lucidum*. LS catalyzes the conversion of 2, 3-oxidoqualene to lanosterol. Zhang et al. (2017b) demonstrated

Table 9.3 Genetic engineering of *Ganoderma lucidum* for the efficient production of ganoderic acids

Overexpressed genes	Fermentation conditions	Results	References
hmgs	Submerged cultivation	Increased content of total GA by 15.1–24.2%	Ren et al. (2013)
hmgr	Submerged cultivation	Increased content of total GA by 2-fold	Xu et al. (2012)
sqs	Submerged cultivation	Increased contents of GA-Mk, -T, -Me, and -S by 2.9, 2.7, 2.0, and 1.3 times, respectively	Zhou et al. (2014)
ls	Submerged cultivation	Increased contents of GA-O, -Mk, -T, -S, and Me by 6.1-, 2.2-, 3.2-, 4.8-, and 1.9-times, respectively	Zhang et al. (2017b)
fps	Submerged cultivation	Increased contents of GA-Mk, -T, -S, and -Me by 2.0-, 2.2-, 2.7-, and 2.9-fold, respectively	Fei et al. (2019)
se	Submerged cultivation	Increased contents of GA-T, -S, -Mk, and -Me by 3.2, 2.4, 1.8, and 2.9 times, respectively	Zhang et al. (2017a)
hmgr and se	Submerged cultivation	Increased contents of GA-T, -S, -Mk, and -Me by 5.9-, 4.5-, 2.4-, and 5.8-times, respectively	Zhang et al. (2017a)
vhb	Submerged cultivation	Increased contents of GA-S, -T, -Mk, and -Me by 1.4-, 2.2-, 1.9-, and 2.0 times, respectively	Li et al. (2016a)
vhb	Liquid static cultivation with Ca^{2+} elicitor	Increased contents of GA-O, -Mk, -T, and –S and Me by 2.6-, 2.6-, 2.5-, 3.5, and 6.0 times, respectively	Xu et al. (2019)

that overexpression of the homologous LS increased lanosterol content and enhanced the accumulations of ganoderic acids and ergosterol in *G. lucidum*. The contents of individual ganoderic acids, GA-O, GA-Mk, GA-T, GA-S, GA-Mf, and GA-Me, increased by 6.1-, 2.2-, 3.2-, 4.8-, 2.0-, and 1.9-fold in the *ls* overexpressed strain (Zhang et al. 2017b). Fei et al. (2019) studied the influence of fps' overexpression on ganoderic acid production by developing mathematical models. They found that the contents of individual ganoderic acids, GA-Mk, GA-T, GA-S, and GA-Me in the *fps* overexpressed strain, increased by 2.0-, 2.2-, 2.7-, and 2.9-fold compared with the control. FPS gene overexpression resulted in higher non-growth-associated-content over the growth-associated-constant in the developed mathematical models. These results showed the potential of genetic manipulation of key genes in the ganoderic acid biosynthesis via genetic transformation.

Due to the complex regulation systems of ganoderic acid biosynthesis, a single gene's overexpression has modestly enhanced ganoderic acid production in *G. lucidum*. Zhang et al. (2017a) studied the influence of simultaneously overexpressing the SE gene (*se*) and *hmgr* on the production of ganoderic acids in *G. lucidum*. The amounts of individual ganoderic acids, GA-T, GA-S, GA-Mk, and GA-Me, produced by *G. lucidum* transformed with *se* and *hmgr* were 90.4, 35.9, 6.2, and 61.8 μg/100 mg DCW respectively, which were 5.9-, 4.5-, 2.4-, and 5.8-fold higher than the control strain. Moreover, the individual GA contents in *G. lucidum* transformed with two biosynthetic genes were approximately two-fold higher than strains overexpressing only one gene (Fig. 9.2). The simultaneous overexpression of the *se* and *hmgr* increased metabolic flux towards specific ganoderic acid pathway in *G. lucidum*.

The *vgb* was also overexpressed heterologously in *G. lucidum* to increase ganoderic acid production. Considering several oxidation steps in ganoderic acid biosynthesis, Li et al. (2016a) introduced the *vgb* into *G. lucidum* protoplasts using plasmid pJW-EXP-VGB. The heterologous expression of the *vgb* led to enhanced contents of total and individual ganoderic acids. The contents of GA-S, GA-T, GA-Mk, and GA-Me in the engineered strain were 1.4-, 2.2-, 1.9-, and 2.0-fold higher than the control strain. Expression of the *vgb* may increase oxygen utilization and accelerate oxygen-requiring steps in the ganoderic acid biosynthetic pathway. Xu et al. (2019) demonstrated that the *vgb* and calcium ion elicitor's overexpression could efficiently increase ganoderic acid yield in liquid static cultures of *G. lucidum*. The contents of GA-O, GA-Mk, GA-T, GA-S, and GA-Me with the integrated approach were 1,451, 180, 1,321, 1,431, and 1,284 μg/100 mg DCW, respectively. The obtained contents of GA-O, GA-S, and GA-Me using the integrated approach were the highest ever reported in *Ganoderma*. These results showed that the combination of genetic engineering and fermentation engineering might be a more promising approach in producing ganoderic acids.

9.4 Gene Silencing

Gene silencing inhibits the expression of target genes in host cells. RNA-mediated silencing is triggered by double-stranded RNA molecules, which leads either to mRNA degradation or translational repression (Espino et al. 2014). It is a reliable reverse genetics tool for functional gene analysis in filamentous fungi (Nakayashiki and Nguyen 2008).

Mu et al. (2012) analyzed four different approaches to silence the endogenous orotidine 5'-monophosphate decarboxylase gene (*ura3*) in *G. lucidum*. Four different plasmids (i.e., pAN7-ura3-dual, pAN7-ura3-s, pAN7-ura3-as, and pAN7-ura3-hp) that expressed an *ura3* fragment in a dual promoter, sense, antisense, and hairpin constructs, respectively, were generated. They transformed these plasmids into the protoplasts of *G. lucidum* by electroporation. Analysis of silencing efficiency illustrated that the four different methods partially suppressed the *ura3* expression in *G. lucidum*. The highest silencing rate (up to 81.9%) was obtained by the dual promoter plasmid (Table 9.4). This method was

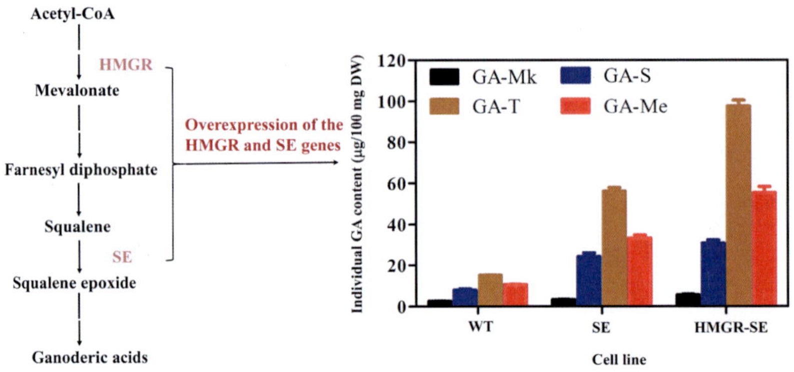

Fig. 9.2 Enhanced production of ganoderic acids in *G. lucidum* by simultaneously overexpressing the 3-hydroxy-3-methyl-glutaryl coenzyme A reductase (HMGR) and squalene epoxidase (SE) genes. WT: The wild-type strains; SE: The *G. lucidum* strains overexpressing the SE gene; HMGR-SE: The *G. lucidum* strains overexpressing the HMGR and SE genes

also used for functional gene analysis of three *G. lucidum* genes coding for nicotinamide adenine dinucleotide phosphate oxidases, 14-3-3 proteins, and AreA. They were silenced by the dual promoter plasmid system (Mu et al. 2014; Zhang et al. 2018; Zhu et al. 2019).

Gene silencing does not usually result in complete suppression of the target genes. Besides, silencing efficiency usually varies among genes, which may be related to the transgene's integrated site, the copy number, and the activity of the RNA polymerase. However, gene silencing can be applicable to study the function of lethal genes and multigene families.

9.5 Recent Advances in the Disruption and Deletion of Target Genes in *G. Lucidum*

CRISPR/Cas9 is a useful system for editing target genes in the *G. lucidum* genome. The Cas9-single-guide RNA (sgRNA) complex can catalyze double-stranded breaks (DSBs) in the target gene, which induces HR or non-homologous end joining (NHEJ) for repairing the formed DSBs. Qin et al. (2017) reported the CRISPR-Cas9-assisted disruption of *ura3* by the Cas9 and in vitro transcribed sgRNA in *G. lucidum*. The obtained disruption efficiency was 0.2–1.78 mutants in 10^7 protoplasts. Wang et al. (2020) developed a CRISPR/Cas9 system enabling the transcription of sgRNA in vivo using the *G. lucidum* u6 promoter of small nuclear RNA. The disruption efficiency of *ura3* (5.3 mutants in 10^7 protoplasts) was higher than that reported by Qin et al. The method may avoid the lower delivery efficiency and degradation of sgRNA during the transformation of *G. lucidum* protoplasts. Apart from optimizing sgRNA expression, Liu et al. (2020) improved the efficiency of gene disruption in *G. lucidum* by introducing an intron into the codon-optimized Cas9. The disruption efficiency of *ura3* (16 mutants in 10^7 protoplasts) improved by 10.6-fold by using this powerful strategy reported by Qin et al. (Figure 9.3). The introduction of an intron led to increased expression of the Cas9

Table 9.4 Strategies for silencing of *ura3* in *Ganoderma lucidum*

Strategy	Transforming plasmid	Number of transformants	Efficiency(%)
Expression of hairpin RNA	pAN7-ura3-hp	40	60.4
Expression of sense RNA	pAN7-ura3-s	41	56.5
Expression of antisense RNA	pAN7-ura3-as	43	51.5
Dual-promoter	pAN7-ura3-dual	58	81.9

gene, resulting in enhanced gene disruption efficiency. Target gene deletion is a valuable tool for *Ganoderma* research in the post-genomic era. Compared with gene disruption, it is a more useful approach to study the function of regulatory elements or non-coding region sequences in the genome. Recently, dual sgRNA-directed gene deletion has been reported in *G. lucidum* using the CRISPR/Cas9 system (Liu et al. 2020). Genomic fragments in the *ura3* and GL 17624 genes were successfully deleted in the genome of *G. lucidum* using this technology, and the average deletion efficiency for *ura3* was 36.7%. The application of dual sgRNA results in two DSBs in the gene of interest, leading to the knockout of the intervening fragment and ligation of two breakpoints by the NHEJ repair mechanism (Fig. 9.3). The CRISPR/Cas9-based gene deletion method provides an efficient platform for studying gene function in *G. lucidum*. This method will also help in the molecular breeding of *G. lucidum* and metabolic engineering of its bioactive compounds.

9.6 Concluding Remarks

Over the past decades, genetic transformation systems have been developed to enhance polysaccharide and triterpene production. With the release of the Lingzhi genome and transcriptome sequences, genetic transformation systems have played a vital role in characterizing gene function in Lingzhi. Besides, the genome editing of Lingzhi will facilitate functional genomics research in the post-omics era. However, genome editing technology applications to Lingzhi are still at their early stage, and thus additional studies are warranted.

Acknowledgements The National Natural Science Foundation of China (No. 81860668) and the Yunnan Applied Basic Research Project (No. 2018FB065) sup-

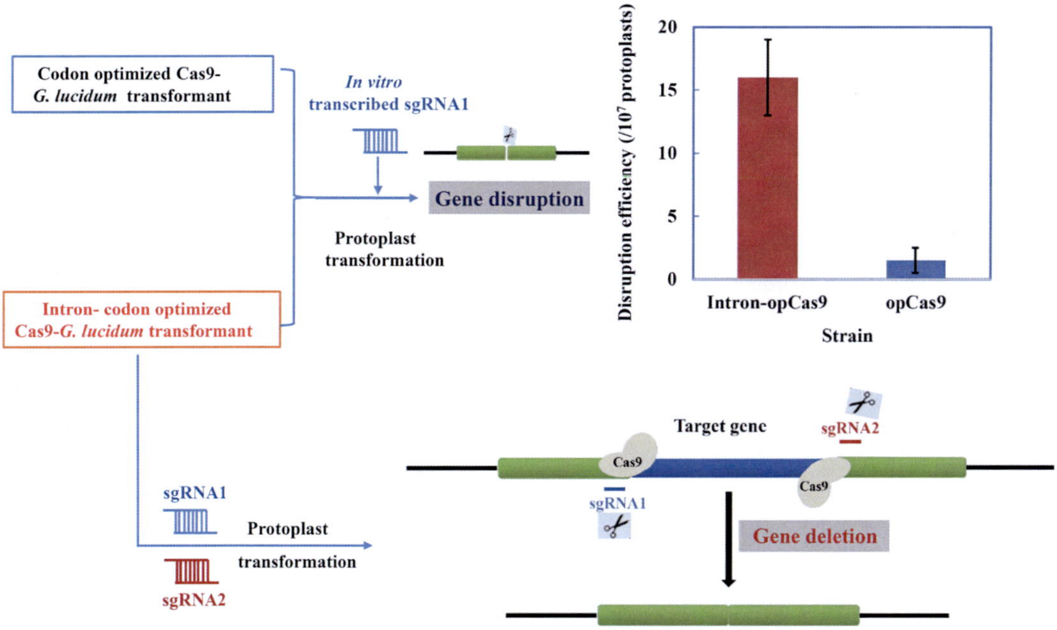

Fig. 9.3 An improved CRISPR/Cas9 system that contained the intron from the glyceraldehyde-3-phosphate dehydrogenase gene (gpd) for gene disruption is described in *G. lucidum*. An effective platform for target gene deletion was is illustrated using a dual sgRNA-directed CRISPR/Cas9 system in *G. lucidum*

ported this study. J.-W. Xu also thanks the Yunnan Ten Thousand Talents Plan-Young and Elite Talents Project.

References

Ahmad MF (2018) *Ganoderma lucidum*: Persuasive biologically active constituents and their health endorsement. Biomed Pharmacother 107:507–519. https://doi.org/10.1016/j.biopha.2018.08.036

Baby S, Johnson AJ, Govindan B (2015) Secondary metabolites from *Ganoderma*. Phytochemistry 114:66–101. https://doi.org/10.1016/j.phytochem.2015.03.010

Bishop KS, Kao CHJ, Xu Y, Glucina MP, Paterson RRM, Ferguson LR (2015) From 2000 years of *Ganoderma lucidum* to recent developments in nutraceuticals. Phytochemistry 114:56–65. https://doi.org/10.1016/j.phytochem.2015.02.015

Boh B, Berovic M, Zhang J, Lin ZB (2007) *Ganoderma lucidum* and its pharmaceutically active compounds. Biotechnol Annu Rev 13:265–301. https://doi.org/10.1016/s1387-2656(07)13010-6

Chen S, Xu J, Liu C, Zhu Y, Nelson DR, Zhou S, Li C, Wang L, Guo X, Sun Y, Luo H, Li Y, Song J, Henrissat B, Levasseur A, Qian J, Li J, Luo X, Shi L, He L, Xiang L, Xu X, Niu Y, Li Q, Han MV, Yan H, Zhang J, Chen H, Lv A, Wang Z, Liu M, Schwartz DC, Sun C (2012) Genome sequence of the model medicinal mushroom *Ganoderma lucidum*. Nat Commun 3:913. https://doi.org/10.1038/ncomms1923

Cor D, Knez Z, Knez Hrncic M (2018) Antitumour, antimicrobial, antioxidant and antiacetylcholinesterase effect of *Ganoderma lucidum* terpenoids and polysaccharides: a review. Molecules 23(3):649. https://doi.org/10.3390/molecules23030649

de Groot MJA, Bundock P, Hooykaas PJJ, Beijersbergen AGM (1998) *Agrobacterium* tumefaciens-mediated transformation of filamentous fungi. Nat Biotechnol 16(9):839–842. https://doi.org/10.1038/nbt0998-839

Espino J, Gonzalez M, Gonzalez C, Brito N (2014) Efficiency of different strategies for gene silencing in *Botrytis cinerea*. Appl Microbiol Biotechnol 98(22):9413–9424. https://doi.org/10.1007/s00253-014-6087-7

Fei Y, Li N, Zhang DH, Xu JW (2019) Increased production of ganoderic acids by overexpression of homologous farnesyl diphosphate synthase and kinetic modeling of ganoderic acid production in *Ganoderma lucidum*. Microb Cell Fact 18:115. https://doi.org/10.1186/s12934-019-1164-3

He L, Feng J, Lu S, Chen Z, Chen C, He Y, Yi X, Xi L (2017) Genetic transformation of fungi. Int J Dev Biol 61(6–7):375–381. https://doi.org/10.1387/ijdb.160026lh

Hsu KD, Cheng KC (2018) From nutraceutical to clinical trial: frontiers in *Ganoderma development*. Appl Microbiol Biotechnol 102(21):9037–9051. https://doi.org/10.1007/s00253-018-9326-5

Hu Y, Li M, Wang S, Yue S, Shi L, Ren A, Zhao M (2018) *Ganoderma lucidum* phosphoglucomutase is required for hyphal growth, polysaccharide production, and cell wall integrity. Appl Microbiol Biotechnol 102(4):1911–1922. https://doi.org/10.1007/s00253-017-8730-6

Ji SL, Liu R, Ren MF, Li HJ, Xu JW (2015) Enhanced production of polysaccharide through the overexpression of homologous rridine diphosphate glucose pyrophosphorylase gene in a submerged culture of Lingzhi or Reishi medicinal mushroom, *Ganoderma lucidum* (higher basidiomycetes). Int J Med Mushrooms 17(5):435–442. https://doi.org/10.1615/IntJMedMushrooms.v17.i5.30

Kim S, Song J, Choi HT (2004) Genetic transformation and mutant isolation in *Ganoderma lucidum* by restriction enzyme-mediated integration. FEMS Microbiol Lett 233(2):201–204. https://doi.org/10.1016/j.femsle.2004.02.010

Kues U, Nelson DR, Liu C, Yu GJ, Zhang J, Li J, Wang XC, Sun H (2015) Genome analysis of medicinal *Ganoderma* spp. with plant-pathogenic and saprotrophic life-styles. Phytochemistry 114:18–37. https://doi.org/10.1016/j.phytochem.2014.11.019

Li D, Tang Y, Lin J, Cai W (2017) Methods for genetic transformation of filamentous fungi. Microb Cell Fact 16(1):168. https://doi.org/10.1186/s12934-017-0785-7

Li G, Li RX, Liu QY, Wang Q, Chen M, Li BJ (2006) A highly efficient polyethylene glycol-mediated transformation method for mushrooms. FEMS Microbiol Lett 256(2):203–208. https://doi.org/10.1111/j.1574-6968.2006.00110.x

Li HJ, He YL, Zhang DH, Yue TH, Jiang LX, Li N, Xu JW (2016a) enhancement of ganoderic acid production by constitutively expressing *Vitreoscilla* hemoglobin gene in *Ganoderma lucidum*. J Biotechnol 227:35–40. https://doi.org/10.1016/j.jbiotec.2016.04.017

Li HJ, Zhang DH, Yue TH, Jiang LX, Yu X, Zhao P, Li T, Xu JW (2016b) Improved polysaccharide production in a submerged culture of *Ganoderma lucidum* by the heterologous expression of *Vitreoscilla* hemoglobin gene. J Biotechnol 217:132–137. https://doi.org/10.1016/j.jbiotec.2015.11.011

Li M, Chen T, Gao T, Miao Z, Jiang A, Shi L, Ren A, Zhao M (2015) UDP-glucose pyrophosphorylase influences polysaccharide synthesis, cell wall components, and hyphal branching in *Ganoderma lucidum* via regulation of the balance between glucose-1-phosphate and UDP-glucose. Fungal Genet Biol 82:251–263. https://doi.org/10.1016/j.fgb.2015.07.012

Liu D, Gong J, Dai W, Kang X, Huang Z, Zhang HM, Liu W, Liu L, Ma J, Xia Z, Chen Y, Chen Y, Wang D, Ni P, Guo AY, Xiong X (2012) The genome of *Ganderma lucidum* provide insights into triterpense biosynthesis and wood degradation. Plos One 7(5). https://doi.org/10.1371/journal.pone.0036146

Liu K, Sun B, You H, Tu J-L, Yu X, Zhao P, Xu JW (2020) Dual sgRNA-directed gene deletion in basidiomycete *Ganoderma lucidum* using the CRISPR/Cas9 system. Microb Biotechnol 13 (2):386–396. https://doi.org/10.1111/1751-7915.13534

Meyer V (2008) Genetic engineering of filamentous fungi–progress, obstacles and future trends. Biotechnol Adv 26(2):177–185. https://doi.org/10.1016/j.biotechadv.2007.12.001

Mu D, Li C, Zhang X, Li X, Shi L, Ren A, Zhao M (2014) Functions of the nicotinamide adenine dinucleotide phosphate oxidase family in *Ganoderma lucidum*: an essential role in ganoderic acid biosynthesis regulation, hyphal branching, fruiting body development, and oxidative-stress resistance. Environ Microbiol 16 (6):1709–1728. https://doi.org/10.1111/1462-2920.12326

Mu D, Shi L, Ren A, Li M, Wu F, Jiang A, Zhao M (2012) The development and application of a multiple gene co-silencing system using endogenous URA3 as a reporter gene in *Ganoderma lucidum*. Plos One 7(8). https://doi.org/10.1371/journal.pone.0043737

Nakayashiki H, Nguyen QB (2008) RNA interference: roles in fungal biology. Curr Opin Microbiol 11 (6):494–502. https://doi.org/10.1016/j.mib.2008.10.001

Pan Y, Lin Z (2019) Anti-aging effect of *Ganoderma* (Lingzhi) with health and fitness. In: Lin Z, Yang B (eds) *Ganoderma* and health: pharmacology and clinical application. Adv Exp Med Biol 1182:299–309

Poyedinok NL, Blume YB (2018) Advances, problems, and prospects of genetic transformation of fungi. Cytol Genet 52(2):139–154. https://doi.org/10.3103/s009545271802007x

Qin H, Xiao H, Zou G, Zhou Z, Zhong JJ (2017) CRISPR-Cas9 assisted gene disruption in the higher fungus *Ganoderma species*. Process Biochem 56:57–61. https://doi.org/10.1016/j.procbio.2017.02.012

Qin H, Xu JW, Xiao JH, Tang YJ, Xiao H, Zhong JJ (2016) Cell factories of higher fungi for useful metabolite production. In: Ye Q, Bao J, Zhong JJ (eds) Bioreactor engineering research and industrial applications i: cell factories. Adv Biochem Eng Biotechnol 155:199–235

Ren A, Ouyang X, Shi L, Jiang AL, Mu DS, Li MJ, Han Q, Zhao MW (2013) Molecular characterization and expression analysis of GlHMGS, a gene encoding hydroxymethylglutaryl-CoA synthase from *Ganoderma lucidum* (Ling-zhi) in ganoderic acid biosynthesis pathway. World J Microbiol Biotechnol 29 (3):523–531. https://doi.org/10.1007/s11274-012-1206-z

Shi L, Fang X, Li M, Mu D, Ren A, Tan Q, Zhao M (2012) Development of a simple and efficient transformation system for the basidiomycetous medicinal fungus *Ganoderma lucidum*. World J Microbiol Biotechnol 28(1):283–291. https://doi.org/10.1007/s11274-011-0818-z

Shi L, Ren A, Mu D, Zhao M (2010) Current progress in the study on biosynthesis and regulation of ganoderic acids. Appl Microbiol Biotechnol 88(6):1243–1251. https://doi.org/10.1007/s00253-010-2871-1

Sun L, Cai H, Xu W, Hu Y, Gao Y, Lin Z (2001) Efficient transformation of the medicinal mushroom ganoderma lucidum. Plant Mol Biol Rep 19(4):383a–383j

Sun L, Cai H, Xu W, Hu Y, Lin Z (2002) CaMV 35S promoter directs β-glucuronidase expression in *Ganoderma lucidum* and Pleurotus citrinopileatus. Appl Biochem Biotechnol-Part B Mol Biotechnol 20 (3):239–244. https://doi.org/10.1385/MB:20:3:239

Turgeon BG, Condon B, Liu J, Zhang N (2010) Protoplast transformation of filamentous fungi. Methods Mol Biol 638:3–19. https://doi.org/10.1007/978-1-60761-611-5_1

Wagner R, Mitchell DA, Sassaki GL, Amazonas M, Berovic M (2003) Current techniques for the cultivation of *Ganoderma lucidum* for the production of biomass, ganoderic acid and polysaccharides. Food Technol Biotechnol 41(4):371–382

Wang PA, Xiao H, Zhong JJ (2020) CRISPR-Cas9 assisted functional gene editing in the mushroom *Ganoderma lucidum*. Appl Microbiol Biotechnol 104(4):1661–1671. https://doi.org/10.1007/s00253-019-10298-z

Xu JW, Ji SL, Li HJ, Zhou JS, Duan YQ, Dang LZ, Mo MH (2015) Increased polysaccharide production and biosynthetic gene expressions in a submerged culture of *Ganoderma lucidum* by the overexpression of the homologous alpha-phosphoglucomutase gene. Bioproc Biosyst Eng 38(2):399–405. https://doi.org/10.1007/s00449-014-1279-1

Xu JW, Xu N, Zhong JJ (2012) Enhancement of ganoderic acid accumulation by overexpression of an N-Terminally truncated 3-hydroxy-3-methylglutaryl coenzyme A reductase gene in the Basidiomycete *Ganoderma lucidum*. Appl Environ Microbiol 78(22):7968–7976. https://doi.org/10.1128/aem.01263-12

Xu JW, Yue TH, Yu X, Zhao P, Li T, Li N (2019) Enhanced production of individual ganoderic acids by integrating Vitreoscilla haemoglobin expression and calcium ion induction in liquid static cultures of *Ganoderma lingzhi*. Microb Biotechnol 12(6):1180–1187. https://doi.org/10.1111/1751-7915.13381

Yu X, Ji SL, He YL, Ren MF, Xu JW (2014) Development of an expression plasmid and its use in genetic manipulation of Lingzhi or Reishi medicinal mushroom, *Ganoderma lucidum* (higher Basidiomycetes). Int J Med Mushrooms 16(2):161–168. https://doi.org/10.1615/IntJMedMushr.v16.i2.60

Zhang DH, Jiang LX, Li N, Yu X, Zhao P, Li T, Xu JW (2017a) Overexpression of the squalene epoxidase gene alone and in combination with the 3-hydroxy-3-methylglutaryl coenzyme A gene increases ganoderic acid production in *Ganoderma lingzhi*. J Agr Food Chem 65(23):4683–4690. https://doi.org/10.1021/acs.jafc.7b00629

Zhang DH, Li N, Yu X, Zhao P, Li T, Xu JW (2017b) Overexpression of the homologous lanosterol synthase gene in ganoderic acid biosynthesis in *Ganoderma lingzhi*. Phytochemistry 134:46–53. https://doi.org/10.1016/j.phytochem.2016.11.006

Zhang TJ, Shi L, Chen D-D, Liu R, Shi DK, Wu CG, Sun ZH, Ren A, Zhao MW (2018) 14-3-3 proteins are involved in growth, hyphal branching, ganoderic acid biosynthesis, and response to abiotic stress in *Ganoderma lucidum*. Appl Microbiol Biotechnol 102(4):1769–1782. https://doi.org/10.1007/s00253-017-8711-9

Zhong JJ, Tang YJ (2004) Submerged cultivation of medicinal mushrooms for production of valuable bioactive metabolites. Adv Biochem Eng Biotechnol 87:25–59. https://doi.org/10.1007/b94367

Zhou JS, Ji SL, Ren MF, He YL, Jing XR, Xu JW (2014) Enhanced accumulation of individual ganoderic acids in a submerged culture of *Ganoderma lucidum* by the over-expression of squalene synthase gene. Biochem Eng J 90:178–183. https://doi.org/10.1016/j.bej.2014.06.008

Zhu J, Sun Z, Shi D, Song S, Lian L, Shi L, Ren A, Yu H, Zhao M (2019) Dual functions of AreA, a GATA transcription factor, on influencing ganoderic acid biosynthesis in *Ganoderma lucidum*. Environ Microbiol 21(11):4166–4179. https://doi.org/10.1111/1462-2920.14769

Chemical Components and Cancer Immunotherapy of *Ganoderma* 10

Linfang Huang and Yu Cao

Abstract

Ganoderma (also known as Lingzhi) is an important source of natural medicines. It contains many well-known chemical components such as triterpenes, terpenes, alkaloids, steroids, nucleotides, nucleobases, polysaccharides, and so on. Lingzhi exhibits broad-spectrum biological activities and has been utilized as a traditional drug and functional food for the treatment of various diseases for over two thousand years in China. Particularly, Lingzhi has shown significant efficacy as an antitumor agent, which has immunomodulatory, anti-inflammatory, anticancer, and antioxidant activities both in preclinical and clinical studies. However, few studies have focused on the usage of Lingzhi in cancer immunotherapy. Here, we reported a comprehensive view of the active chemical compositions and immunomodulatory actions of the Lingzhi for treating various cancers and the main signaling pathways of immune cells in response to Lingzhi treatment. Additionally, we demonstrated that polysaccharides and immunomodulatory proteins of Lingzhi represent the core chemical compositions underlying the cancer immunotherapeutic activities. In the meantime, the NF-κB and MAPK pathways are the primary pathways related to the effects of Lingzhi. Moreover, the toxicology and clinical studies of Lingzhi are also summarized in this chapter. The results imply that Lingzhi have a wide range of applications for cancer treatment through regulating the immune system. The literature review offers valuable and informative references that warrant further investigation of *Ganoderma*'s potential cancer immunotherapy applications.

10.1 Chemical Components

Ganoderma Karst. is a fungal genus belonging to the family Polyporaceae. More than six hundred compounds have been purified and isolated from the *Ganoderma* species. The most studied 20 species of *Ganoderma* are—*G. sinense, G. lucidum, G. japonicum, G. australe, G. applanatum, G. capense, G. tropicus, G. tsugae, G. resinaceum, G. boniense, G. theaecolum, G. cochlear, G. formosanum, G. amboinense, G. boninense, G. duropora, G. colossum, G. concinna, G. pfeifferi, G. atrum, and G. orbiforme*. These chemical

L. Huang (✉) · Y. Cao
Institute of Medicinal Plant Development, Chinese Academy of Medical Sciences and Peking Union Medical College, Beijing, China
e-mail: lfhuang@implad.ac.cn

Y. Cao
Chengdu Institute of Biology, Chinese Academy of Sciences, Chengdu, China

C. Liu (ed.), *The Lingzhi Mushroom Genome*, Compendium of Plant Genomes,
https://doi.org/10.1007/978-3-030-75710-6_10

components mainly contained steroids, nucleosides, polysaccharides, meroterpenoids, alkaloids, and triterpenes. Among them, triterpenes and polysaccharides were the main active compounds.

10.1.1 Triterpenoids

Triterpenoids have always been a research hotspot in medicinal mushrooms. They have been considered the main active compounds in *Ganoderma* species. At present, over three hundred triterpenoids have been isolated from *Ganoderma* species. The structures of *Ganoderma* triterpenoids, such as pentacyclic and tetracyclic lanostane carbon skeleton, are complicated. The triterpenoids can be divided into three groups C30, C27, and C24 based on the number of carbons in the compound skeleton. Furthermore, they can be classified into triterpenoid acids, triterpenoid lactones, and triterpenoid alcohols, according to the substituting groups. The relative molecular weights (MW) of *Ganoderma* triterpenoids usually range from 400 to 600 Da. *Ganoderma* triterpenoids are generally regarded as having antiplatelet aggregation, antidiabetic, antioxidant, antiaging, and anti-inflammatory activities.

10.1.2 Meroterpenoids

To date, the advancement in instrument analysis technology has promoted the analysis of meroterpenoids structure. About two hundred meroterpenoids have been isolated from the *Ganoderma* species. Meroterpenoids are natural hybrid products derived from the biological, genetic pathway of shikimic acid and valeric acid. *Ganoderma* meroterpenoids (GMs) possessed diverse biological activities, including antioxidant, antimicrobial, anti-allergic, anti-AChE, anti-fibrotic, cytotoxic activities, inhibition of NO, etc.

10.1.3 Polysaccharides

Polysaccharide is a long-chain, polymeric carbohydrate molecule composed of monosaccharide units connected by glycosidic bonds. Research shows that most *Ganoderma lucidum* polysaccharides (GLPS) were heteropolysaccharides formed by different combinations of galactose, arabinose, glucose, xylose, fucose, and mannose. And $(1 \rightarrow 3)$-α, $(1 \rightarrow 4)$-α, $(1 \rightarrow 6)$-α, and β-glucans were also identified from polysaccharides. The MW of polysaccharides and glycoprotein usually ranges from 103 to 106 Da. GLPS is considered the key contributor to the antitumor effect. Besides, it is found that polysaccharides have immunomodulatory, antioxidant, antidiabetic, antiaging, and anti-inflammatory activities.

10.1.4 Others

To date, over 30 alkaloids have been isolated and identified in *Ganoderma* species. These alkaloids contain adenine nucleoside, urinary purine nucleoside, adenine, urine purine, ganoder purine, betaine, and ganoine. Moreover, more than 30 steroids have been found, including ergosterol, sitosterol, stigmastol, and carotenosterol. In particular, the most common compounds found in the *Ganoderma* species are ergosterols.

10.2 Cancer Immunotherapy

Ganoderma lucidum exerts its antitumor effects through regulating the immune system (Boh et al. 2007), and its therapeutic effects can be attributed to triterpenes, polysaccharides, and fungal immunomodulatory proteins (FIPs). Below, we provide an overview of the Lingzhi chemical compositions, their potential pharmacological effects, and the underlying molecular mechanism for the effects. In particular, we focus on the immunoregulatory activities of Lingzhi that are most likely to be responsible for its therapeutic potential on cancer.

10.2.1 Bioactive Compounds of *Ganoderma* and Their Immunomodulatory Effect on Cancer Treatment

Boh et al. showed that *Ganoderma* species is rich in fungal immunoregulatory proteins (FIPs), polysaccharides, triterpenes, etc. These components can regulate the immune system, through which they exert their antitumor effects.

10.2.1.1 Fungal Immunomodulation Proteins

Fungal immunomodulation proteins (FIPs) are a group of low molecular-weight protein molecules isolated from various fungi, including Lingzhi. Lingzhi FIPs are part of the functional Lingzhi compositions having anticancer properties (Table 10.1). At present, four FIPs have been isolated and purified from Lingzhi: Lingzhi 8 (LZ-8), family of immunoprotein from *Ganoderma tsugae* (Fip-gts), *Ganoderma Microsporum* immunoprotein (GMI), and Family of immunoprotein from *Ganoderma atrum* (Fip-gat).

Lz-8 was first isolated and cloned in 1989 from *G. Lucidum*. It is 110 amino acids long. Lz-8 folds into a unique structure similar to immunoglobulin and can form a bioactive non-covalent homodimer (Kino et al. 1989). The homodimer has a significant effect on the specific type of lung cancer. Liang et al. reported that the recombinant Lz-8 (rLz-8) induces endoplasmic reticulum (ER) aggregation and autophagic cell death, which in turn, triggers the ER stress and results in the activation of the ATF4 and CHOP pathway in a human gastric cancer cell line (Hsin et al. 2015a). Furthermore, Lin and Hsu found that rLz-8 represents a promising chemotherapy agent for lung cancer because it can target FAK, a protein implicated in cancer metastasis. Additionally, Lin et al. investigated the rLz-8's anticancer effect in lung cancer cells by targeting EGFR-dependent processes, EGFR mutation, and EGFR overexpression (Liang et al. 2012).

Fip-gts is the first FIP isolated and cloned from *Ganoderma tsugae*. Lin et al. isolated the DNA fragment encoding the protein by screening a cDNA library constructed with reverse-transcribed mRNAs (Lin et al. 1997). Liao et al. reported that recombinant FIP-GTs (rFip-gts) could inhibit the cellular telomerase activity dose dependently. The effect might be mediated by down-regulating the telomerase catalytic subunit expression (Lin et al. 2017). In vitro, rFip-gts inhibits telomerase activity in lung cancer cells by a nuclear export mechanism. The process might be mediated by intracellular calcium levels induced by ER stress (Liao et al. 2007). In addition, the rFIP-gts can potentially inhibit the growth of lung tumors. For example, in in vivo studies, after treating nude mice with rFIP-gts, the A549 cells isolated from the nude mice grew dramatically slower than the mock control (Liao et al. 2008). Lastly, Fip-gts have significant inhibitory effects on cervical cancer cells.

GMI is a FIP identified from *G. Microsporum*. Chiu et al. showed that 83% of the protein sequence is similar to FIP-gts (Chiu et al. 2015). GMI could inhibit EGF-induced protein phosphorylation in vitro. It then activates AKT and EGFR pathways in a dose-dependent manner (Lin et al. 2010). Hsin et al. reported that significant autophagosome accumulation in the cell model treated with GMI was observed, further inducing autophagy death. Besides, the combination of cisplatin and GMI could induce apoptosis. The effects were likely to be mediated by autophagy. The autophagy can be initiated through the survivin and ERCC1 independent pathways or the caspase-7-dependent pathway (Hsin et al. 2012). Oral intake of GMI could inhibit the growth of tumor cells. In contrast, subcutaneous injection of GMI into nude mice induced autophagy of mouse A549 cells (Hsin et al. 2011).

Fip-gat is another FIP isolated from *G. atrum*. The protein contains one hundred and eleven amino acids. In vitro, treating MDA-MB-231 cells with various concentrations of recombinant

Table 10.1 List of immunomodulatory proteins and their pharmacological effects (modified from Cao et al. 2018 with permission)

Source	Protein	Cell lines/mice	Optimum treatment concentration/dose	Duration	Pharmacological effect(s)	References
G. lucidum	recombinant (r) Lz-8	MBT-2 cell, C57BL/6, C3H/HeN, C3H/HeJ mice	10 μg/ml	90 days	Improved the therapeutic effect of DNA vaccine on mice MBT-2 tumor	Lin et al. (2011)
		Human primary and Jurkat T cells	1 μg/ml	24 h	Inducement of IL-2 gene expression through the Src-family protein tyrosine kinase Stress-mediated autophagic cell death	Hsu et al. (2008)
		LLC1 cell, C57BL/6 mice	10 μg/ml, 7.5 mg/kg	48 h, 18 days	Promoting degradation of epidermal growth factor receptor (EGFR), inhibited growth, and induced lung	Lin et al. (2017)
G. tsugae	(r)Fip-gts	HeLa, SiHa, and Caski cells	0.15 μM	24 h	Suppressed cervical cancer cell migration and enhanced the inhibition of FIP-gts upon migration	Wang et al. (2007)
		A549, H1299, A549-p53, H1299-p53 stable cells	1.2 μM	48 h	Induced suppression of telomerase activity in lung cancer cells via post-translational modifications on the protein of hTERT	Liao et al. (2007)
		A549, CaLu-1 cells, nude mice	1.2 μM, 12.8 mg/kg	48 h, 33 days	Inhibited A549 cell growth. A549 cells treated with reFIP-gts grew slower than cells treated with PBS alone in vivo	Liao et al. (2008)
G. Microsporum	GMI	A549, CaLu-1 cells	1.2 μM (GMI) + 5 μM (Cisplatin)	48 h	Induced apoptosis via autophagy and might be a potential cisplatin adjuvant agent against lung	Hsin et al. (2015)
		A549, CaLu-1 cells	1.2 μM	48 h	Inhibited lysosome degradation on autophagosome formation, migration, and invasion	Hsin et al. (2012)
G. atrum	(r)Fip-gat	MDA-MB-231 cell	9.96 μg/ml	48 h	Triggered cell cycle arrest during the transition of G1 to S phases and significant increased in cell apoptosis	Xu et al. (2016)

Fip-gat reduced the cell viability in a dose-dependent manner (Xu et al. 2016). In particular, the treatment with recombinant FIP-gat triggered the block of the G1/S transition in the cell cycle to a remarkable degree. This led to a significant increase in cell apoptosis and possible tumor cell death.

10.2.1.2 Effects of Polysaccharides

GPLS also play an important role in treating cancer because of the immunoregulatory effects (Table 10.2) (Meng et al. 2014). Below we will describe the effect on different types of cancer in detail.

Leukemia

Several lines of evidence suggest that GPLS possess anti-leukemia activities. Firstly, GPLS could increase the levels of IL-6 and IL-1 indirectly, potentiating antitumor immunity in vivo (Wang et al. 1997). Secondly, it is reported that GLPS can potentiate the specific cytotoxic T-lymphocytes (CTL) that were activated by dendritic cells (Cao and Lin 2003). These lymphocytes targeted the P815 tumor antigens, activating pathways related to cytotoxicity, including the IFN-γ and the granzyme B pathways. Thirdly, GLPS in 400 or 100 mg/mL concentrations enhanced the proliferation and cytotoxicity of CIK cells and increased the protein and mRNA expression of granzyme B and perforin in CIK cells. And, it also increased the production of TNF and IL-2 through interaction with other cytokines. Fourthly, GLPS can decrease 50% expression of anti-CD3 and 75% expression of IL-2, respectively. The effects were not likely to relate to nitric oxide (NO) (Zhu and Lin 2006; Chan et al. 2007). It was demonstrated that GLPS could differentiate selected monocytic leukemic cell populations (Chan et al. 2007). Lastly, Sun et al. used *G. lucidum* water extract and investigated its effects on NK cells. GPLS increased NK cell cytotoxicity. The underlying mechanism might relate to the stimulation of perforin and granulysin secretion (Sun et al. 2011a).

Melanoma

Several lines of evidence suggest that GPLS possess anti-melanoma activities. Firstly, GLPS could promote the growth of the B16F10 melanoma cells. The effects are accompanied by CD69 and FasL expression, IFN-γ production, and lymphocyte proliferation (Sun et al. 2011a). The melanoma cells, in turn, can activate lymphocytes. However, some controversial results were observed. For example, the supernatant of cultured melanoma cell line B16F10 was found to be bioactive. It can inhibit lymphocyte proliferation (Sun et al. 2011b). Secondly, Sun et al. reported that GLPS could enhance the activity of MHC class I molecules, leading to the increased cytotoxicity against melanoma cells (Sun et al. 2012). Thirdly, the *G. lucidum* ethanolic extracts also have bioactivity. They can dramatically inhibit the release of MMP-2/MMP-9, IL-6, and IL-8 in cancer cells under specific pro-inflammatory conditions (Barbieri et al. 2017). Lastly, Wang et al. reported that *G. formosanum* polysaccharide PS-F2 could strengthen the host immune responses against continuous tumor growth under the condition that PS-F2 was used continuously (Wang et al. 2011, 2014). Overall, the results are somewhat controversial, and future investigations are needed.

Lung Cancer

Several lines of evidence suggest that GPLS possess anticancer activities in the lung. Firstly, Feng et al. studied whether *G. lucidum* triterpenoids can inhibit cell proliferation and tumor growth. It was shown that the IC50 of triterpenes was 24.63 g/mL in A549 cells (Liang et al. 2013). Triterpenoids could inhibit tumor growth significantly in tumor-bearing mice at a dose ranging from 30 to 120 mg/kg. The treatment significantly improves the indexes of immune organs such as the spleen and thymus. Secondly, Reishi polysaccharide fraction (FMS) of *G. lucidum* inhibited cancer cell growth by increasing cytotoxicity mediated by antibodies and reducing the production of inflammatory mediators associated with tumor growth, especially the production of monocyte chemotactic protein-1. In vivo, FMS mediated the production of anti-glycan IgM, leading to a significant increase in the cell population of the peritoneal B1 B-cells. Sun et al. recently found that plasma of lung cancer patients inhibits the CD69

Table 10.2 List of bioactive compounds and proteins of Lingzhi preparations and their Pharmacological effects (modified from Cao et al. 2018 with permission)

Source	Components	Cell lines/mice	Optimum treatment concentration/dose	Duration	Pharmacological effect(s)	References
G. lucidum	Water extract	γ-ray-irradiated mice	400 mg/kg	35 days	Enhanced the recovery of cellular immunocompetence from γ-ray-irradiation	Chen et al. (1995)
		RAW 264.7 cell	100 μg/ml	24 h	Inhibited LPS-induced NO production in RAW 264.7 macrophages	Song et al. (2004)
	Ethanolic extract	MDA-MB 231, B16-F10 cells	250 μg/ml	48 h	Decreased the feasibility of cancer cells in a time- and concentration-dependent manner	Barbieri et al. (2017)
	Polysaccharide	HL-60 and U937 cells	100 μg/ml	5 days	Enhanced IL-1 and IL-6 and might have an indirect role in potentiating against tumor Immunity in vitro	Wang et al. (1997)
		C57BL/6j, BALB/c mice	12.8 mg/L	5 days	Promoted the cytotoxicity of specific cytotoxic T-lymphocytes induced by dendritic cells (DC), which were pulsed with P815 tumor antigen during the stage of antigen presentation	Cao and Lin (2003)
		LAK cells, C57BL/6j mice	100 or 400 mg/L	8 days	Mediated the antitumor activity through complement receptor type 3	Zhu and Lin (2005)
		L929, P815, YAC-1 cells, C57BL/6 mice	100 or 400 mg/L	15 days	Promoted cytokine-induced killer (CIK) cell Proliferation and cytotoxicity were closely relevant to intensifying the production of IL-2, TNF	Zhu and Lin (2006)
		S180, Heps, EAC cells, ICR species mice	300 mg/kg	8 days	Restrained the growth of inoculated Heps, S180, and EAC tumor cells in ICR mice	Pang et al. (2007)
		S180 cell, BALB/c mice	200 mg/kg	14 days	Stimulated the immune response of the host organism via the activation of T cells, NK cells, and macrophages	Wang et al. (2012)
		rats of Wistar strain	2.6 mg/ml	48 h	Enhanced the antioxidant enzyme activities, And lowered levels of IL-6, IL-1b, and TNF-α in Wistar rats with cervical cancer	Chen et al. (2009)
		B16F10 cell, C57BL/6 and BABL/c mice	12.8 μg/ml	72 h	Had antagonistic effects on the immunosuppression	Sun et al.

(continued)

Table 10.2 (continued)

Source	Components	Cell lines/mice	Optimum treatment concentration/dose	Duration	Pharmacological effect(s)	References
					induced by B16F10 culture supernatant	
		B16F10 cell, BALB/c mice	400 μg/ml	5 days	Suppressed lymphocyte proliferation and granzyme B and perforin production in lymphocytes after introduction with Phytohemagglutinin	Sun et al.
		B16F10 cell	400 μg/ml	48 h, 21 days	Enhanced main histocompatibility complex (MHC) class I, more efficient immune cell-mediated cytotoxicity as opposed to these B16F10 cells might be induced	Sun et al. (2012)
		Lymphocytes of cancer patients	12.8 μg/ml	48 h	Antagonized lung cancer patient, plasma-induced suppression of lymphocyte, Stimulation via phytohemagglutinin	Sun et al. (2014)
	β-glucan	Neutrophils	100 μg/ml	24 h	Induced antiapoptotic effects on neutrophils Relying on stimulation of Akt-regulated signaling pathways	Hsu et al. (2002)
G. lucidum	Polysaccharide	THP-1, U937 cells	10 μg/ml	24 h	Boosted the stimulation and maturation of immature DC	Lin et al. (2005)
			100 μg/ml	72 h	Induced selected monocytic leukemic cell divergence into DCs with immuno-stimulatory function	Chan et al. (2007)
	A fucose-containing glycoprotein F3	Con A-stimulated mouse spleen cells	0.01–0.1 μg/ml	72 h	Stimulated the expression of cytokines, particularly IL-2, IL-1, and INF-g	Wang et al. (2002)
		BALB/c mice spleen cells	100 μg/ml	48 h	Stimulated the expression of TNF-a, GM-CSF, G-CSF, IL-1, IL-6, IL-12, IFN-c, and M-CSF	Chen et al. (2004)
	Ganoderic acid Me	YAC-1, LLC cells, C57BL/6 mice	28 mg/kg	20 days	Up-regulated expression of Nuclear Factor-κB after the treatment of GA-Me, which might be Involved in the production of IL-2	Wang et al. (2007)
		2LL cells, C57BL/6 mice	10 μg/ml	48 h	Increased the proportion of Treg cells and induced the apoptosis of competent T cells	Que et al. (2014)

(continued)

Table 10.2 (continued)

Source	Components	Cell lines/mice	Optimum treatment concentration/dose	Duration	Pharmacological effect(s)	References
G. sinensis	Lipid extract	U937, HepG2 cells	12.8 μg/ml	72 h	Re-establish the anticancer activity of the immunosuppressive tumor-related macrophages	Sun et al.
G. applanatum	Polysaccharide	S180 Transplanted Mice	20 mg/kg	10 days	Restored the NK activity and the IFNy and IL-2 producing activity of the spleen cells, which were restrained by the tumor	Gao and Yang (1991)
	Exo-biopolymer (EXP)	S180 cell, BALB/c mice	80 mg/kg	16 days	Suppressed the growth of solid tumor and enhanced the natural killer (NK) cell activity	Jeong et al. (2008)
	unknown	Breast cancer cells	Unknown	Unknown	Stimulated macrophages in immunosuppressive breast cancer microenvironment	Payne et al. (2016)
G. atrum	Polysaccharide	S180 cell, Kunming mice	100 mg/kg	18 days	Induced against tumor activity through the Mitochondria-mediated apoptotic pathway related to the stimulation of host immune response	Li et al. (2011)
		CT26 cell, BALB/c mice	200 mg/kg	14 days	Activated macrophages by TLR4-dependent signaling pathways, improved immunity, and thus inhibited tumor growth	Zhang et al. (2013)
		RAW264.7 cell, C3H/HeN, C3H/HeJ mice	160 μg/ml	48 h	Induced TNF-a secretion through TLR4, ROS, PI3K, Akt, MAPKs, and NF-κ, pathways during macrophage activation	Yu et al. (2014)
		CT26 cell, BALB/c mice	200 mg/kg	15 days	Exerted anticancer activity in vivo by inducing apoptosis through mitochondria-mediated apoptotic pathway and increased host immune system function	Zhang et al. (2014)
			100 mg/kg	18 days	Activated spleen lymphocytes and peritoneal macrophages in mice treated with cyclophosphamide	Yu et al. (2015)
G. formosanum	PS-F2	S180, B16, C26 cells C57BL/6, BALB/c mice	50 mg/kg	24 days	Activated the host immune responses against tumor growth	Wang et al. (2014)

expression, perforin and granzyme B production, and lymphocyte proliferation (Sun et al. 2014). Thirdly, administration of GLPS can completely or partially reverse the above effects. Besides, Que et al. reported that purified lanostane triterpenes, *Ganoderma* acid ME, contributed to indoleamine 2, 3-dioxygenase. And, ganoderic acid ME could directly inhibit the activation of CD8 + T cells, promote T cell apoptosis, improve Treg-mediated immunosuppression, and create a tolerant environment in lung tumors (Que et al. 2014). Lastly, the results are controversial regarding the effects of Lingzhi composition on lung cancer. Additional investigations are needed to solve the discrepancies in the experimental results.

Liver Cancer

Several lines of evidence suggest that GPLS possess anticancer activities in the liver. Firstly, Zhang et al. believed that the lipid extract of *G. sinensis* spores has direct tumoricidal activity. Besides, the extract played an anticancer role by stimulating the activation of human macrophages/monocytes (Zhang et al. 2010). Secondly, Shen et al. reported that GLPS had anticancer activities. Specifically, GLPS can be used to study the human miRNA differential expression in hepatoma cells, as treating cancer cells with polysaccharides led to the differential expression of miRNAs (Zhang et al. 2014). Lastly, Li et al. showed that GLPS could potently inhibit tumor growth in mice. These effects were associated with increased proportions of the Tregs (regulatory T cells) and the Teffs (effector T cells). Furthermore, GLPS could increase IL-2 secretion. While Treg can inhibit the proliferation of Teffs, treatment with GLPS can eliminate Treg's inhibitory effects.

Colon Cancer

In colon cancer, *G. atrum* polysaccharides (GAPS) could potentiate macrophages through activation of the TLR4-dependent signaling pathway. As a result, GAPS can improve the body's immunity and inhibit tumor growth (Zhang et al. 2013). PS-F2 is a fraction of chrysanthemum polysaccharide. Wang et al. reported that continuous treatment of PS-F2 activates the host immune response, which, in turn, inhibits tumor growth (Wang et al. 2014). Moreover, Yu et al. noted that the GAPS had a chemoprotective effect against cyclophosphamide treatment in mice. The effects might result from GAPS's ability to activate splenic lymphocytes and intraperitoneal macrophages (Yu et al. 2015).

10.2.2 Pathways in Immune Cells Implicated in the Cancer Immunotherapy by *Ganoderma*

10.2.2.1 T-Lymphocytes and Dendritic Cell

Toll-like receptor (TLR)-4 is a well-known polysaccharide receptor. GLPS could activate TLR4, which, in turn, inhibited the IL-10 and IL-12 production. This observation suggested that dendritic cell (DC) signaling played an important role in this process after treatment with GLPS. In-depth research demonstrated that GLPS could enhance the activity of nuclear factor (NF)-κB inhibitors (IκB), as well as p38 mitogen-activated protein kinase (MAPK) and the phosphorylation of IκBα (Fig. 10.1a) (Lin et al. 2005). GLPS could increase the function of four proteins, H-2Kb, H-2Db, and B7-1 and B7-2, on the B16F10 cells. And, the expression upregulation of these molecules could further enhance the antitumor cytotoxicity of the cells after GLPS treatment (Sun et al. 2012). Please note that B7-1 and B7-2 are MHC class I molecules playing important roles in cellular immune responses. Furthermore, GLPS could bind to extracellular receptors, activating T cells via the protein kinase C (PKC) and inositol triphosphate/Ca^{2+} (IP3/Ca^{2+}) pathways (Fig. 10.1a) (Li et al. 2013,2015).

10.2.2.2 Natural Killer Cells and B Lymphocytes

F3 is the main fraction of the GPLS. F3 could interact with TLR4 and TLR2 receptors and then transducted intracellular signaling via p38 MAPK and the NF-κB pathway, leading to Blimp-1 mRNA expression (Fig. 10.1b) (Lin et al. 2006). F3 induced the cytotoxicity of

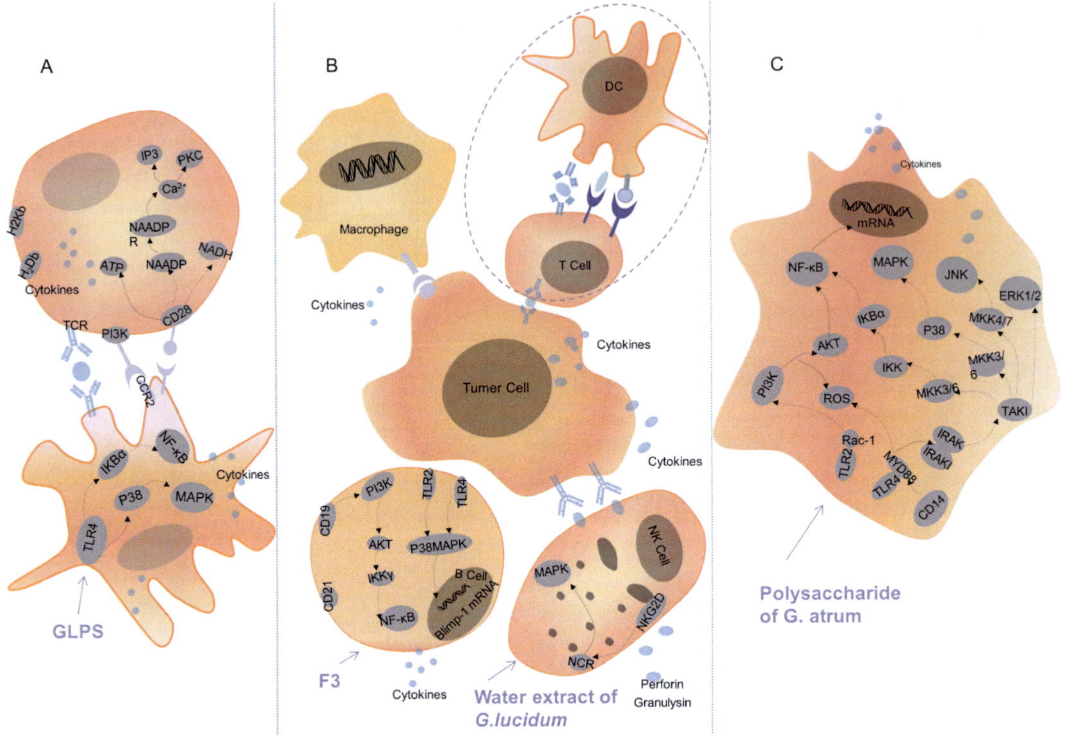

Fig. 10.1 Main pathways involved in the cancer immunotherapy in immune cells response to Ganoderma (Lingzhi) treatment. **a** GLPS stimulates Dendritic Cells and then T cells. **b** The *Ganoderma* water extract (F3) stimulates Macorphage, Dendritic Cells, T Cells, and NK cells. **c** GAPS induces macrophage stimulation. The arrows indicate the direction of activation. The figure was modified from Wang et al. (2011) with permission

various cancer cells through activating NKG2D and NCR receptors and then the MAPK signaling pathways, resulting in the exocytosis of perforin and granulosin (Fig. 10.1b) (Sun et al. 2011a).

10.2.2.3 Macrophage

The dried mycelia of *G. lucidum* were found to play an important role in the activation of NF-κB in murine macrophages (Kuo et al. 2006). The receptor binds to pro-inflammatory cytokines such as IL-1β, TNF-α, and IFN-γ. At last, it activates NF-κB to increase the expression of iNOS. The activation of the signal transduction mechanism may be mediated by TLR4 pathways, followed by the MAPK and NF-κB signal transduction pathways (Qiang et al. 2015). The studies showed that the GAPS could be used as an immunomodulatory antitumor agent to exert antitumor activity and to improve the immune system function (Fig. 10.1c). The GAPS promoted TNF-α secretion by activating several pathways, including the TLR4 pathway, ROS pathway, PI3K pathway, Akt pathway, MAPKs pathway, and the NF-κB pathways when it activates the macrophage (Yu and Nie 2014). In another study, Huang et al. explored the potential signaling pathways associated with macrophage activation in S180 mice bearing tumors by GAPS. Increased phosphorylation of proteins from the Akt, NF-κB, and MAPK families was observed, indicating the activation of the NF-κB pathway (Huang et al. 2016). In conclusion, these results further demonstrated that the NF-κB signaling pathway might be involved in mRNA expression and TNF-α secretion (Fig. 10.1c).

10.3 Clinical Studies, Toxicology, and Discussion

10.3.1 Clinical Studies

Lingzhi preparations have been tested in several clinical studies. Firstly, Gao et al. studied the bioactivity of Lingzhi preparation called Ganopoly on 34 patients with various stages of cancers. They found that it improved the quality of life for patients experiencing advanced-stage malignant growth. The likely mechanism is the increase in CD^{3+} cell numbers (Gao et al. 2003). Secondly, Shing et al. described that a six-month treatment of Lingzhi preparations expanded mitogen-stimulated proliferative responses of lymphocytes in youngsters who had tumors and were immunocompromised (Shing et al. 2008). Thirdly, a pilot study conducted in 2012 showed that the *G. lucidum* spore powder (GLSP) had beneficial effects on malignancy-related exhaustion. The administration of GLSP improved personal satisfaction in patients having breast cancer and going through endocrine treatment, with no emotional antagonistic effects. The treatment significantly improved the patients' physical well-being and fatigue sub-scale (Zhao et al. 2012). Lastly, another study focused on five patients diagnosed with gynecological cancer. The tumors reached a stage of stability after the patients had taken water extracts of Lingzhi fruiting bodies. Significant advantages were observed when the Lingzhi preparations were administrated in combination with standard chemotherapy (Chen and Alpert 2016).

10.3.2 Toxicity

The toxicity of Lingzhi preparations has also been studied in the past. Firstly, Wan et al. described an instance of lethal fulminant hepatitis after the patient started to consume Lingzhi powder for about two months (Wanmuang et al. 2007). Secondly, a patient was determined to have non-Hodgkins lymphoma. He was given Lingzhi preparations and experienced continuous watery loose bowels afterward (Zhang et al. 2014). Thirdly, no strange clinical side effects and no significant difference in food consumption rate and body weight were seen in Wistar rodents during the 30 days administration (Cheng et al. 2008). Lastly, no increase of mutagenicity in mice was noticed after administering Lingzhi preparations, as shown by negative results from polychromatic erythrocyte, chromosome variation test, the Ames test, and micronucleus trial of sperm irregularity test (Zhang et al. 2016).

10.3.3 Discussion

Cancer is a disease currently having the highest mortality in the world. Chemotherapy is the first-line therapeutic strategy. However, chemotherapy usually comes with significant side effects. Furthermore, alternative treatments, particularly those stimulating the patient's immunity, are needed to boost the effectiveness of chemotherapy.

Ganoderma is a medicinal mushroom genus that contains many species used as traditional medicines. It can treat a wide variety of diseases through mostly immunoregulatory activities. Among the many *Ganoderma* species, *G. lucidum* is the most studied species. Recent studies have shown that it can be used as an adjunct treatment to stimulate host immunity and enhance the response against the tumor.

Concerning the active components responsible for its antitumor activity, polysaccharides and FIPs have caught the most attention. Among the FIPs, Lz-8 isolated from *G. lucidum* is the most examined. However, the exact mechanisms can be multiple. For example, Lingzhi preparations can activate NK cells, macrophages, T lymphocytes, B lymphocytes, and other immune cells. It can also promote the in vitro expansion of the spleen cells that have not differentiated. Lastly, it can increase antibody and cytokine production.

The MAPK and NF-κB pathways are well-studied signal-transduction pathways. Previous studies showed that Lingzhi preparations could activate this pathway, leading to the production of cytokines that stifle the development of tumor cells. The likely target of Lingzhi preparations is the TLR-4, which is involved in the host immune response to polysaccharides.

While the Lingzhi preparations can be used alone, Lingzhi *preparations* can also be used together with other drugs for cancer therapies. For example, GMI and cisplatin were combined in one study, and *G. atrum* polysaccharides were combined with cyclophosphamide. Both of them were shown to decrease the adverse effects of standard therapy.

Immunotherapy by Lingzhi preparations on several cancers, including lung, melanoma, liver, colon cancer, and leukemia, have been studied extensively. For the most part, the effects can be explained by the theory of "channel tropism" in the traditional medicine system. Furthermore, we surveyed the safety of Lingzhi preparations through the investigation of toxicological studies. Overall, no serious adverse effects were found to be associated with Lingzhi preparations. However, chronic watery diarrhea and liver toxicity have been reported for patients treated with Lingzhi preparations. As a result, careful monitoring of patients after treatment is warranted.

Ganoderma is one of the most widely used medicinal fungi in the world. The finding from this study suggests that it is a potential immunotherapeutic anticancer agent. It possesses a low level of toxicity and high efficacy as an adjunct treatment. This review also identified several problems in the past studies that need to be taken into consideration in the future development of Lingzhi preparations. Firstly, the reviewed studies did not identify the particular cellular targets of Lingzhi preparations. As a result, the observed effects lack specificity. Secondly, most of the effects were observed in vitro. Thirdly, the potential adverse effects and interactions between Lingzhi preparations and other drugs or chemotherapy medications need further investigation.

To realize the potential of Lingzhi preparations, the main bioactive compounds should be further explored. The in vivo pharmacokinetic research needs to be conducted. The underlying mechanisms of the immunoregulatory activities ought to be determined adequately.

References

Barbieri A, Quagliariello V, Vecchio Vd, Falco M, Luciano A, Amruthraj NJ, et al. (2017) Anticancer and anti-inflammatory properties of *Ganoderma lucidum* extract effects on melanoma and triple-negative breast cancer treatment. Nutrients 9(3):210

Boh B, Berovic M, Zhang J, Zhi-Bin L (2007) *Ganoderma Lucidum* and its pharmaceutically active compounds. Biotechnol Annu Rev 13:265–301

Cao L, Lin Z (2003) Regulatory effect of *Ganoderma lucidum* polysaccharides on cytotoxic T-lymphocytes induced by dendritic cells in vitro. Acta Pharmacol Sin 24(4):321–326

Cao Y, Xu XW, Liu SJ, Huang LF, Gu J (2018) *Ganoderma*: a cancer immunotherapy review. Front Pharmacol 9:1217. https://doi.org/10.3389/fphar.2018.01217

Chan W, Cheung C, Law H, Lau Y, Chan G (2007) *Ganoderma lucidum* polysaccharides can induce human monocytic leukemia cells into dendritic cells with immunotolerogenic function. Blood 110:4906–4906

Chen Q, Alpert J (2016) Nutraceuticals: evidence of benefit in clinical practice? Am J Med 129(9):897–898

Chen WC, Hau DM, Lee SS (1995) Effects of *Ganoderma lucidum* and krestin on cellular immunocompetence in gamma-ray-irradiated mice. Am J Chin Med 23(1):71–80

Chen HS, Tsai YF, Lin S, Lin CC, Khoo KH, Lin CH, et al. (2004) Studies on the immuno-modulating and antitumor activities of *Ganoderma lucidum* (Reishi) polysaccharides. Bioorg Med Chem 12(21)595–601

Chen X-P, Yan C, Shui-bing L, You-Guo C, JianYun L, Lan-ping L (2009) Free radical scavenging of *Ganoderma lucidum* polysaccharides and its effect on antioxidant enzymes and immunity activities in cervical carcinoma rats. Carbohydr Polym 77:389–393

Cheng PC, Hsu CY, Chen CC, Lee KM (2008) In vivo immunomodulatory effects of *Antrodia camphorata* polysaccharides in a T1/T2 doubly transgenic mouse model for inhibiting infection of *Schistosoma mansoni*. Toxicol Appl Pharmacol 227(2):291–298

Chiu L-Y, Hu M, Yang T-Y, Hsin I-L, Ko J, Tsai K, Sheu G-T (2015) Immunomodulatory protein from *Ganoderma microsporum* induces pro-death autophagy through Akt-mTOR-p70S6K pathway inhibition in multidrug resistant lung cancer cells. PLoS ONE 10 (5):e0125774

Gao B, Yang GZ (1991) Effects of *Ganoderma applanatum* polysaccharide on cellular and humoral immunity in normal and sarcoma 180 transplanted mice. Phytother Res 5(3) 134-138

Gao Y, Zhou SW, Huang M, Dai X (2003) Effects of ganopoly (a *Ganoderma lucidum* polysaccharide extract) on the immune functions in advanced-stage cancer patients. Immunol Invest 32(3):201–215

Hsin IL, Ou CC, Wu TC, Jan MS, Wu MF, Chiu LY et al (2011) GMI, an immunomodulatory protein from *Ganoderma microsporum*, induces autophagy in non-small cell lung cancer cells. Autophagy 7(8):873–882

Hsin IL, Sheu GT, Jan MS, Sun HL, Wu TC, Chiu LY et al (2012) Inhibition of lysosome degradation on autophagosome formation and responses to GMI, an immunomodulatory protein from *Ganoderma microsporum*. Br J Pharmacol 167(6):1287–1300

Hsin IL, Ou CC, Wu MF, Jan MS, Hsiao YM, Lin CH et al (2015) GMI, an immunomodulatory protein from *Ganoderma microsporum*, potentiates cisplatin-induced apoptosis via autophagy in lung cancer cells. Mol Pharm 12(5):1534–1543

Hsu MJ, Lee SS, Lin WW (2002) Polysaccharide purified from *Ganoderma lucidum* inhibits spontaneous and Fas-mediated apoptosis in human neutrophils through activation of the phosphatidylinositol 3 kinase/Akt signaling pathway. J Leukoc Biol 72(1):207–216

Hsu H, Hua K, Wu W, Hsu J, Weng S-T, Lin T-L, et al. (2008) Reishi immun-modulation protein induces interleukin-2 expression via protein kinase-dependent signaling pathways within human T cells. J Cell Physiol 215(1):15-26

Huang J, Nie Q, Liu X, Zhang S, Nie S, Huang D et al (2016) *Ganoderma atrum* polysaccharide modulates TNF0± secretion and mRNA expression in macrophages of S-180 tumor-bearing mice. Food Hydrocolloids 53:24–30

Jeong Y-T, Yang B, Jeong S-C, Kim S-m, Song C-H (2008) *Ganoderma applanatum*: a promising mushroom for antitumor and immunomodulating activity. Phytother Res 22(5):614-9

Kino KYA, Yamaoka K, Watanabe J, Tanaka S, Ko K, Shimizu K, Tsunoo H (1989) Isolation and characterization of a new immunomodulatorry protein, Ling Zhi-8 (LZ-8), from *Ganoderma lucidum*. Biol Chem 264:472–478

Kuo MC, Weng CY, Ha CL, Wu MJ (2006) *Ganoderma lucidum* mycelia enhance innate immunity by activating NF-kappaB. J Ethnopharmacol 103(2):217–222

Li WJ, Chen Y, Nie SP, Xie MY, He M, Zhang SS, et al. (2011) *Ganoderma atrum* polysaccharide induces antitumor activity via the mitochondrial apoptotic pathway related to activation of host immune response. J Cell Biochem 112(3):860–71

Li B, Lee DS, Kang Y, Yao NQ, An RB, Kim YC (2013) Protective effect of ganodermanondiol isolated from the Lingzhi mushroom against tert-butyl hydroperoxide-induced hepatotoxicity through Nrf2-mediated antioxidant enzymes—scienceDirect. Food Chem Toxicol 53(3):317–324

Li X, He LP, Yang Y, Liu F, Cao Y, Zuo J (2015) Effects of extracellular polysaccharides of *Ganoderma lucidum* supplementation on the growth performance, blood profile, and meat quality in finisher pigs. Livest Sci 178:187–194

Liang C, Li H, Zhou H, Zhang S, Liu Z, Zhou Q et al (2012) Recombinant Lz-8 from *Ganoderma lucidum* induces endoplasmic reticulum stress-mediated autophagic cell death in SGC-7901 human gastric cancer cells. Oncol Rep 27:1079–1089

Liang F, Ling Y, Meng D, Yan C, Ming-Hua Z, Jun-Fei G, et al. (2013) Anti-lung cancer activity through enhancement of immunomodulation and induction of cell apoptosis of total triterpenes extracted from *Ganoderma luncidum* (Leyss. ex Fr.) Karst. Molecules 18(8):9966–81

Liao CH, Hsiao YM, Sheu GT, Chang JT, Wang PH, Wu MF et al (2007) Nuclear translocation of telomerase reverse transcriptase and calcium signaling in repression of telomerase activity in human lung cancer cells by fungal immunomodulatory protein from *Ganoderma tsugae*. Biochem Pharmacol 74 (10):1541–1554

Liao CH, Hsiao YM, Lin CH, Yeh CS, Ko JL (2008) Induction of premature senescence in human lung cancer by fungal immunomodulatory protein from *Ganoderma tsugae*. Food Chem Toxicol 46(5):1851–1859

Lin WH, Hung CH, Hsu CI, Lin JY (1997) Dimerization of the N-terminal amphipathic alpha-helix domain of the fungal immunomodulatory protein from *Ganoderma tsugae* (Fip-gts) defined by a yeast two-hybrid system and site-directed mutagenesis. J Biol Chem 272(32):20044–20048

Lin Y, Liang Y, Lee S-S, Chiang B (2005) Polysaccharide purified from *Ganoderma lucidum* induced activation and maturation of human monocyte derived dendritic cells by the NF-kappaB and p38 mitogen activated protein kinase pathways. J Leukocyte Biol 78(2): 533-543

Lin KI, Kao YY, Kuo HK, Yang WB, Chou A, Lin HH et al (2006) Reishi polysaccharides induce immunoglobulin production through the TLR4/TLR2-mediated induction of transcription factor blimp-1. J Biol Chem 281(34):24111–24123

Lin CH, Sheu GT, Lin YW, Yeh CS, Huang YH, Lai YC et al (2010) A new immunomodulatory protein from *Ganoderma microsporum* inhibits epidermal growth factor mediated migration and invasion in A549 lung cancer cells. Process Biochem 45(9):1537–1542

Lin CC, Yu YL, Shih CC, Liu KJ, Ou KL, Hong LZ et al. (2011) A novel adjuvant Ling Zhi-8 enhances the efficacy of DNA cancer vaccine by activating dendritic cells. Cancer Immunol Immunother 60(7):1019–1027

Lin T-Y, Hsu H, Sun W, Wu T-H, Tsao S-M (2017) Induction of Cbl-dependent epidermal growth factor receptor degradation in Ling Zhi 8 suppressed lung cancer. Int J Cancer 140(11):2596-2607

Meng L, Xie J, Lv G, Hu D, Zhao J, Duan J et al (2014) A comparative study on immunomodulatory activity of polysaccharides from two official species of *Ganoderma* (Lingzhi). Nutr Cancer 66:1124–1131

Pang X, Chen Z, Gao X, Liu W, Slavin M, Yao W et al (2007) Potential of a novel polysaccharide preparation (GLPP) from Anhui-grown *Ganoderma lucidum* in tumor treatment and immunostimulation. J Food Sci 72(6):S435–S442

Payne G, Javed W, Lee S et al. (2016) *Ganoderma applanatum*—potential target for stimulating macrophages in immunosuppressive breast cancer microenvironment. Breast Cancer Res Treat 159(1):181–181

Qiang Y, Nie S, Wang J, Huang D, Xie MY (2015) Toll-like receptor 4 mediates the antitumor host response induced by *Ganoderma atrum* polysaccharide. J Agric Food Chem 63(2):517–525

Que Z, Zou F, Zhang A, Zheng Y, Bi L, Zhong J et al (2014) Ganoderic acid Me induces the apoptosis of competent T cells and increases the proportion of Treg cells through enhancing the expression and activation of indoleamine 2,3-dioxygenase in mouse lewis lung cancer cells. Int Immunopharmacol 23(1):192–204

Shing MK, Leung T, Chu Y, Li C, Chik KW, Leung P et al (2008) Randomized, double-blind and placebo-controlled study of the immunomodulatory effects of Lingzhi in children with cancers. J Clin Oncol 26:14021–14021

Song YS, Kim SH, Sa JH, Jin C, Lim CJ, Park EH (2004) Anti-angiogenic and inhibitory activity on inducible nitric oxide production of the mushroom *Ganoderma lucidum*. J Ethnopharmacol 90(1):17–20

Sun L, Lin Z, Li X, Li M, Lu J, Duan X et al (2011a) Promoting effects of *Ganoderma lucidum* polysaccharides on B16F10 cells to activate lymphocytes. Basic Clin Pharmacol Toxicol 108(3):149–154

Sun L, Lin Z, Duan X, Lu J, Ge Z-H, Li X, et al. (2011b) *Ganoderma lucidum* polysaccharides antagonize the suppression on lymphocytes induced by culture supernatants of B16F10 melanoma cells. J Pharm Pharmacol 63(5):725-735

Sun LX, Lin ZB, Duan XS, Lu J, Li WD (2012) Enhanced MHC class I and costimulatory molecules on B16F10 cells by *Ganoderma lucidum* polysaccharides. J Drug Target 20(7):582–592

Sun L, Li W, Lin Z, Duan X, Li X-f, Yang N, et al. (2014) Protection against lung cancer patient plasma-induced lymphocyte suppression by *Ganoderma lucidum* polysaccharides. Cell Physiol Biochem 33:289–99

Wang SY, Hsu ML, Hsu HC, Lee SS, Ho CK (1997) The antitumor effect of *Ganoderma lucidum* is mediated by cytokines released from activated macrophages and T lymphocytes. Int J Cancer 70(6):699–705

Wang Y, Khoo K, Chen S, Lin C, Wong C (2002) Studies on the immuno-modulating and antitumor activities of *Ganoderma lucidum* (Reishi) polysaccharides: functional and proteomic analyses of a fucose-containing glycoprotein fraction responsible for the activities. Bioorg Med Chem 10(4):1057–1062

Wang G, Zhao J, Liu J, Huang Y, Zhong J, Tang W (2007) Enhancement of IL-2 and IFN-gamma expression and NK cells activity involved in the antitumor effect of ganoderic acid Me in vivo. Int Immunopharmacol 7(6):864–870

Wang C-l, Pi C-C, Kuo C-W, Zhuang Y, Khoo K, Liu W-H, et al. (2011) Polysaccharides purified from the submerged culture of *Ganoderma formosanum* stimulate macrophage activation and protect mice against Listeria monocytogenes infection. Biotechnol Lett 33:2271–8

Wang PY, Zhu XL, Lin ZB (2012) Antitumor and immunomodulatory effects of polysaccharides from broken-spore of *Ganoderma lucidum*. Front Pharmacol 13(3):135. https://doi.org/10.3389/fphar.2012.00135

Wang CL, Lu CY, Hsueh YC, Liu WH, Chen CJ (2014) Activation of antitumor immune responses by *Ganoderma formosanum* polysaccharides in tumor-bearing mice. Appl Microbiol Biotechnol 98(22):9389–9398

Wanmuang H, Leopairut J, Kositchaiwat C, Wananukul W, Bunyaratvej S (2007) Fatal fulminant hepatitis associated with *Ganoderma lucidum* (Lingzhi) mushroom powder. J Med Assoc Thai 90 (1):179–181

Xu H, Kong Y-Y, Chen X, Guo M-y, Bai X, Lu Y-J, et al. (2016) Recombinant FIP-gat, a fungal immunomodulatory protein from *Ganoderma atrum*, induces growth inhibition and cell death in breast cancer cells. J Agric Food Chem 64(13):2690–8

Yu Q, Nie S, Wang J-q, Yin P, Huang D, Li W, Xie M (2014) Toll-like receptor 4-mediated ROS signaling pathway involved in *Ganoderma atrum* polysaccharide-induced tumor necrosis factor-± secretion during macrophage activation. Food and chemical toxicology. Int J Br Ind Biol Res Assoc 66:14–22

Yu Q, Nie S, Wang J-q, Huang D, Li W, Xie M (2015) Molecular mechanism underlying chemoprotective effects of *Ganoderma atrum* polysaccharide in cyclophosphamide-induced immunosuppressed mice. J Funct Foods 15:52–60

Zhang JP, Zheng L, Wang JH, Magnusson KE, Liu X (2010) Lipid extract from completely sporoderm-broken germinating *Ganoderma sinensis* spores elicits potent antitumor immune responses in human macrophages. Phytotherapy Research Ptr. 23(6):844–850

Zhang S, Nie S, Huang D, Li W, Xie M (2013) Immunomodulatory effect of *Ganoderma atrum* polysaccharide on CT26 tumor-bearing mice. Food Chem 136(3–4):1213–1219

Zhang S, Nie S, Huang D, Feng Y, Xie M (2014) A novel polysaccharide from *Ganoderma atrum* exerts antitumor activity by activating mitochondria-mediated apoptotic pathway and boosting the immune system. J Agric Food Chem 62(7):1581–1589

Zhang J, Gao X, Pan Y, Xu N, Jia L (2016) Toxicology and immunology of *Ganoderma lucidum* polysaccharides in Kunming mice and Wistar rats. Int J Biol Macromol 85:302–310

Zhao H, Zhang Q, Zhao L, Huang X, Wang J, Kang X (2012) Spore powder of *Ganoderma lucidum* improves cancer-related fatigue in breast cancer patients undergoing endocrine therapy: a pilot clinical trial. Evidence-Based Complementary Alternat Med 2012:809614

Zhu XL, Lin ZB (2005) Effects of *Ganoderma lucidum* polysaccharides on proliferation and cytotoxicity of cytokine-induced killer cells. Acta Pharmacologica Sinica 026(9):1130–7

Zhu X, Lin Z (2006) Modulation of cytokines production, granzyme B and perforin in murine CIK cells by *Ganoderma lucidum* polysaccharides. Carbohyd Polym 63(2):188–197

A Comprehensive Meta-Analysis on the Effectiveness and Safety of Lingzhi

11

Rui Liu, Shan He, Jiaqi Yu, Yueyue Liang, and Po Cao

Abstract

Although many studies have been conducted on the chemistry and pharmacological activities of Lingzhi, there is a lack of comprehensive meta-analysis on its effectiveness and safety. In this chapter, we briefly introduced the concept and history of meta-analysis and described how to conduct a meta-analysis; we also constructed a data model and utilized R modules and PERL language to implement a pipeline capable of carrying out multiple meta-analysis automatically. Using Lingzhi as the study subject, we selected 57 outcome measures as the indicators and carried out meta-analysis on 89 datasets. These datasets address several pharmacological effects including immunomodulation, anti-oxidation, and anti-fatigue. The overall results strongly support the notion that Lingzhi has various health-promoting effects. During the process, we have developed an analytical robust pipeline. The results will help the regulatory agencies for new policies regarding the functional food administration. Furthermore, this information can help the health providers to choose Lingzhi products for particular health-promoting goals.

11.1 Introduction

During the past several years, many research works on the pharmacological effects of Lingzhi have been conducted quantitatively, and reviewing these papers is critical to take advantage of this wealth of information. In this chapter, we used the meta-analysis method to analyze the effectiveness and safety of Lingzhi. This chapter is arranged in the following manner, in the first part, we introduced the concept and history of meta-analysis; in the second part, we described the procedures of how to conduct a meta-analysis; in the third part, we constructed a pipeline for automatic meta-analysis and applied it in the analysis of the effectiveness of Lingzhi in details; in the fourth part, we briefly reviewed the meta-analysis results on the safety of Lingzhi. In conclusion, the pipeline was proved to be efficient for large-scale meta-analysis. The results obtained from this study can help government agencies to establish policies. In addition, the information will be invaluable for the consumers of functional food to make decisions on the utilization Lingzhi for various health-promoting needs.

R. Liu (✉) · S. He · J. Yu · Y. Liang · P. Cao
School of Information Management, Central China Normal University, Wuhan, China
e-mail: liuruiccnu@hotmail.co

C. Liu (ed.), *The Lingzhi Mushroom Genome*, Compendium of Plant Genomes,
https://doi.org/10.1007/978-3-030-75710-6_11

11.2 An Introduction to Meta-Analysis

11.2.1 Basic Terminology

A meta-analysis is a quantitative approach for systematically assessing previous research results to arrive at conclusions about the body of research. The term "meta-analysis" was first coined by Glass in 1976. The Greek prefix "meta" means "transcending" (Glass 1976). It identifies a process of analysis retrospectively performed on available published data on a specific topic. A meta-analysis is a powerful tool for the synthesis of different studies' results. It is an analytical method where both independent and different studies are integrated and their results pooled into a single common result.

11.2.2 A Brief History of Meta-Analysis

The first meta-analysis was attributed to Pearson, who analyzed the data from five studies on the correlation between the vaccination for enteritis fever and its mortality (Pearson 1904). Then several similar approaches were described from the 1930s, Tippett (1931), Fisher (1932), Cochran (1937), and Pearson (1938) each described statistical techniques for combining data from different studies.

In the mid-1970s, the need for techniques to combine data from many studies of the same topic became especially acute in the social sciences, when hundreds of studies existed for some topics. Gene Glass and colleagues first introduced the term "meta-analysis (Glass 1976) and prompted attention to the approach, especially in the field of psychology. In particular, Smith and Glass published a meta-analysis paper on psychotherapy effectiveness based on 375 studies (Smith 1977). They showed that psychotherapy was effective and there was little difference in effectiveness across different therapies. Their convincing approach to the difficult question of psychotherapies' relative effectiveness demonstrated many of the value of meta-analysis. A rapid increase in the use of this approach was observed in the following years. During the late 1970s and early 1980s, social scientists, including Rosenthal (1978), Glass et al. (1981), Hedges and Larry (1982), Hedges (1983), Hunter et al. (1982), Light (1983), and Light and Pillemer (1984), popularized meta-analysis and further developed the statistical methods supporting the meta-analysis.

In the late 1980s, descriptions of the meta-analysis method appeared almost simultaneously in three influential medical journals, the New England Journal of Medicine, Lancet, and the Annals of Internal Medicine. The use of meta-analysis in medicine has been growing since.

Shortly after that, Hedges and Olkin published a book on a meta-analysis that was deeply rooted in traditional statistics (Hedges and Olkin 1985). This rooting was essential in bringing formality and perceived statistical merit to the approach and serving as a starting point for the advances to meta-analytic techniques.

11.3 Procedures of Conducting Meta-Analysis

Undertaking a meta-analysis requires five crucial steps. See procedures of conducting meta-analysis in Fig. 11.1.

11.3.1 Searching the Literature

The sources of information likely to prove most fruitful to researchers are called reference databases.

Even though the reference database is an excellent source of research, there are limitations. First, there may be a time gap between completing the research and displaying it in the reference database. Besides, it is still necessary to prepare and submit a research report, accept its main report, display it in print or online, and then compile it into a reference database. Therefore, the newly completed research will not appear in

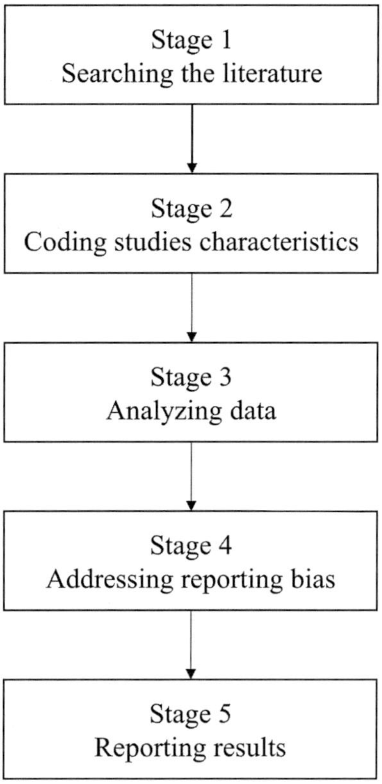

Fig. 11.1 Procedures of conducting meta-analysis

the reference database. Second, each reference database contains some restrictions that limit the scope of entry into the system based on subject or subject boundaries. Therefore, interdisciplinary subject retrieval requires access to multiple reference databases. Third, some reference databases contain only published research, others contain published and unpublished research, and others include only unpublished research. Therefore, if one wants to minimize release bias, it is essential to determine the database's coverage one plans to use and have a database that compiles unreleased and published documents.

The citation index is a unique reference database that can identify and group all published articles citing the same publication, making earlier publications index terms for more articles. Currently, Web of Science provides "Science Citation Index Extension" (including articles published in scientific journals from 1900 to the present), "Social Science Citation Index" (beginning in 1956), and "Art and Humanities Citation Index". Web of Science also provides a reference search for citations, allowing users to move back and forth throughout the document. The citation index's information will limit references to published research including journals and books. Therefore, finding expected references in research reports will bias the null hypotheses in the citations. However, in these categories, the "Social Science Citation Index" coverage is very detailed. It should be kept in mind that citation indexing takes time for indexing documents, so the latest publications will be lacking.

Typically, the reference database search is started by determining the database to be accessed. The database is usually selected based on the database's possibilities to include documents relevant to the search. Besides, the database collection of documents should also be familiar. Searchers can browse the thesaurus attached to different databases to identify terms that might not have been initially considered. No matter what terms to use in the search, use appropriate and detailed terms in the search and query of reference databases and research registries. When people expect the document to contain all keywords, they can use the Boolean operator AND; when people don't want the document to include particular keywords, use the OR operator, and sometimes people can also use the NOT operator. Here are some frequently used methods (Cooper 2017).

(1) Backward Searches
"Backward search", sometimes called "footnote chase", is the process of citing related research in existing works. That is to work from research "backward" to find out previously conducted relevant research. This approach's potential critical bias comes from the possibility that studies yielding favorable results are more likely to be cited than those with unfavorable results. Despite the potential biases of backward searches, they represent a valuable method of searching. The approach is also practical in identifying literature that might have been missed in other search approaches due to failures in using

appropriate keywords or searching literature in other disciplines.

(2) Forward Searches

A forward search tries to find research that references existing research and is usually performed using a particular database. The search is "forwarding" in time and is more likely to find newer articles than searching backward. There are varying degrees of intensity for forward search. The less drastic method identifies the earliest and most important few research topics to be studied and then performs a forward search to find the research citing these seminal papers.

There is no standard answer to which and how many sources of information should be used in a search. The appropriate sources depend on the topic under consideration and the resources available. Nevertheless, searchers should use multiple channels with different entry and access restrictions. It would minimize any systematic differences caused by the differences in the searching results.

11.3.2 Coding Studies Characteristics

11.3.2.1 Inclusion and Exclusion Criteria

There are other possible inclusion or exclusion criteria such as time frame and study design, characteristics of the study's context, participation sample, and outcomes. Sometimes inclusion and exclusion criteria can be applied before the coding of studies begins.

11.3.2.2 Developing a Coding Guide

A coding guide's construction is not easy. Firstly, the characteristics of studies to be collected should be listed; then, each variable's potential research values need to be considered. When the initial set of coding questions and response categories are generated, it is necessary to ask knowledgeable people to suggest additional codes and response categories. And ambiguities in questions and responses will be pointed out. The precision of coding questions and response categories can be improved using a coding guide to encode randomly selected studies.

The information to include in a coding guide can be divided into eight categories. They are the report; the predictor or independent variable; the setting in which the study took place; participant and sample characteristics; the dependent or outcome variables and how they were measured; the type of research design; statistical outcomes and effect sizes; coder and coding process characteristics.

11.3.2.3 Selecting and Training Coders

Some researchers will have every study coded independently by more than one coder, called double coding. The codes for every study are then compared, and discrepancies are resolved in meeting the coders or by a third party. This procedure can significantly reduce potential bias, make evident different interpretations of questions and responses, and catch mechanical errors.

Double coding is not the only way to enhance the reliability of codes. First, Coders who have the right background and interest can be picked. Second, coding sheets can be accompanied by coding guides that define and explain distinctions in each study characteristic. Third, coders should be trained before actual coding. The coders will raise concerns you have not thought of, which will lead to greater clarity in questions, responses, and conventions to use when reports are unclear. Fourth, the coders should gather information for the same few studies independently and share their responses in a group. The clarity of the coding guidelines has increased further with the discussion of errors.

Estimating reliability should be done before coders get started. It is usually essential to obtain numerical estimates of coder reliability. Two indicators are used for this purpose. The first one

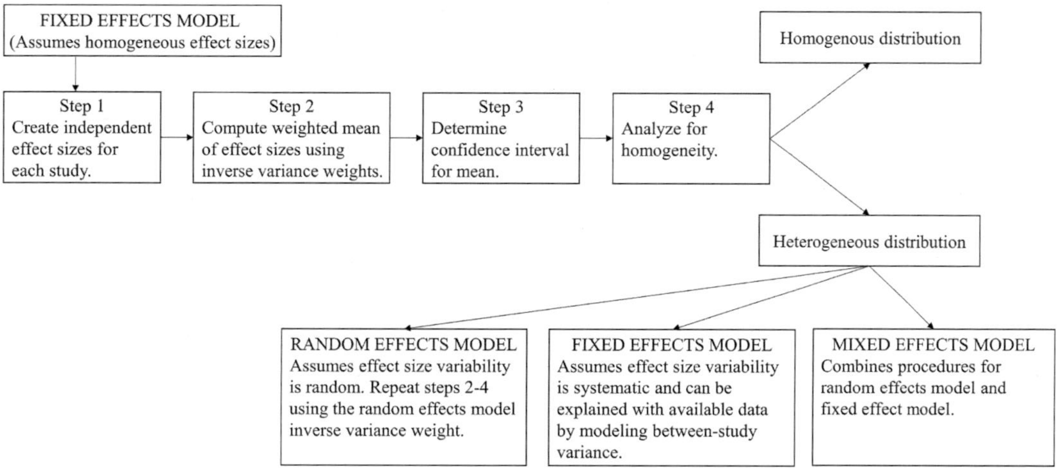

Fig. 11.2 Roadmap of data analysis

is agreement rate between pairs of coders. The agreement rate is defined as "the number of agreed-on codes divided by the total number of coding opportunities". The second is Cohen's kappa. The kappa value is defined as the improvement over chance reached by the coders. Often kappa is presented along with the agreement rate.

11.3.3 Analyzing Data

The meta-analysis's primary analytic goals are to combine and analyze the distribution of effect sizes and examine the relationship between effect sizes and other descriptive variables.

For meta-analysis, a statistic that is used to estimate the intervention effect is crucial for each study. Several terms have been widely used to call this statistics including "intervention effect", "effect estimate" or "effect size" (Deeks and Altman 2011). They are interchangeable and we will use the "effect size" in the following text to represent all the terms. The effect estimates from multiple studies were synthesized based on a particular statistic model such as fixed-effects model and random-effects model. It is critical to select the most appropriate model as it will affect the analysis result dramatically. The data analysis workflow is shown in Fig. 11.2.

11.3.3.1 Effect Size

An effect size is an index of the direction and magnitude of the association between two variables (Wolfe et al. 2016).

This definition includes traditional measures of correlations between two variables, differences between two groups, and contingencies between two dichotomies. In other words, we can compute the correlation (r), standardized mean difference (d or g), or odds ratio (o) from a variety of information commonly reported in primary studies.

(1) R-index

The r-index is simply the Pearson product-moment correlation coefficient. The r-index can be calculated using the following formula:

$$r = \sqrt{\frac{t^2}{t^2 + df_{\text{error}}}}$$

$$v_r = \frac{(1 - r^2)^2}{n - 1}$$

The formula can be used to calculate the 95% confidence interval as r-$1.95v_r \leq r \geq r + 1.95v_r$.

(2) D-index and g-index

D-index. The d-index of effect size is appropriate when the difference between two means is being compared.

$$d = \frac{\overline{X_1} - \overline{X_2}}{SD_{\text{within}}}$$

$\overline{X_1}$ and $\overline{X_2}$ = the two group means;

SD_{within}= the estimated common standard deviation of the two groups.

The standardized mean difference expresses the size of each study's intervention effect relative to the variability observed in that study.

$$\text{SMD} = \frac{\text{difference in mean outcome between groups}}{\text{standard deviation of outcome among participants}}$$

(3) Odds Ratio and Risk Ratio

The third class of effect size metric is applicable when both variables are dichotomous, such as with or without clinical improvement, becoming dead or staying alive. In clinical trials, the most commonly used effect measures for dichotomous data are the risk ratio (RR) and the odds ratio (OR).

Risk is the possibility of a risk event (or an adverse event) occurring. In research, the risk is commonly expressed as a decimal number between 0 and 1, occasionally, it is converted into a percentage. The odds are the ratio of the probability that a particular event will occur to the probability that it will not occur. Odds have been used frequently in other areas.

The risk ratio describes the ratio of two risks. For example, in Table 11.1, assume a study having two groups: treatment and control group. Each group has 100 patients. Ten patients died in the treated group. Twenty patients died in the control group. The risk of death is 10/100 in the treatment group and 20/100 in the control group. So the risk ratio is 0.50.

The odds ratio means the ratio of two odds. As the probability of death in the treatment group is 10/100, and the probability of life is 90/100, the odds of death in the treatment group would be 10/90, or 0.11. The odds of death in the control group would be 20/80, or 0.25. The ratio of the two odds is 0.11/0.25, or 0.44. The odds ratio has statistical properties that often make it the best choice for a meta-analysis.

11.3.3.2 Meta-Analysis

The effect estimates are synthesized to generate a quantitative result after it is calculated for each study. The studies with larger weights will have more influence on the effect estimate in the meta-analysis, and the weights are determined by the selected meta-analysis model. Generally speaking, there are two meta-analysis models: the fixed-effects model and the random-effects model.

(1) Fixed-effects Model

In the fixed-effects model, the method for assigning weights is the "inverse-variance" method. It means that the weight of a study is the inverse of the variance of its effect estimate (Higgins and Green 2008).

Table 11.1 Fictional data for a 2 * 2 table

	Dead	Alive	N
Treated	10	90	100
Control	20	80	100

(2) Random-effects Model
The random-effects model relaxes the assumption of one true intervention effect. Instead, a distribution of true intervention effects is assumed. Many methods are available to calculate this variance. The most common one is the DerSimonian and Laird's method of moments estimator (DerSimonian and Laird 1986).

(3) Heterogeneity
Statistical heterogeneity is a consequence of methodological or clinical diversity (or both) among studies. It is also known as (Higgins et al. 2003) variability in the intervention effects that are evaluated in different studies. Several methods have been developed to quantify inconsistency across studies. They focus on assessing the impact of heterogeneity on the meta-analysis rather than testing the presence of heterogeneity (Higgins et al. 2003). I^2 test (Higgins et al. 2003) can measure the extent of heterogeneity rather than stating if it is present or not. Thresholds for the interpretation of I^2 can be misleading since the importance of inconsistency depends on several factors. A rough guide to the interpretation of thresholds for I^2 is as follows:

① 0–40%: might not be important;
② 30–60%: may represent moderate heterogeneity*;
③ 50–90%: may represent substantial heterogeneity*;
④ 75–100%: considerable heterogeneity*;

(4) Selection of the Meta-analysis Model
The meta-analysis model selection has two approaches. The first one assumes a model and switches to an alternate model if there is statistical evidence that it is more appropriate. The second approach selects the most plausible model. The rationale for the choice of model should be provided, and it is strongly encouraged to fit the alternate model for the sensitivity analyses.

11.3.4 Addressing Reporting Bias

11.3.4.1 Reporting Bias

Reporting biases arise when the nature and direction of results influence the dissemination of research findings. Seven significant reporting biases are listed in Table 11.2.

Table 11.2 Different types of reporting biases

Type of reporting bias	Definition
Publication bias	The publication or non-publication of research findings, depending on the nature and direction of results
Time lag bias	The rapid or delayed publication of research findings, depending on the nature and direction of the results
Multiple (duplicate) publication bias	The multiple or singular publication of research findings, depending on the nature and direction of the results
Location bias	The publication of research findings in journals with different ease of access or indexing levels in standard databases, depending on the nature and direction of results
Citation bias	The citation or non-citation of research findings, depending on the nature and direction of the results
Language bias	The publication of research findings in a particular language, depending on the nature and direction of the results
Outcome reporting bias	The selective reporting of some outcomes but not others, depending on the nature and direction of the results

Modified from Cochrane Handbook for Systematic Reviews of Interventions (Higgins and Green 2008)

11.3.4.2 Detecting Reporting Bias

The reporting bias can be detected using Funnel plot and sensitivity analysis.

(1) Funnel Plot
A funnel plot is a simple scatter plot of the intervention effect estimates from individual studies against certain measure of each study's size or precision. Since the estimated intervention effect's precision increases as the study's size increases, the plot looks like a "funnel". Effect estimates from small studies tend to scatter more widely at the bottom of the graph. In the absence of bias, the plot should approximately resemble an asymmetrical (inverted) funnel.

(2) Sensitivity Analysis
Another way to detect reporting bias is sensitivity analysis. There are three kinds of sensitivity analysis.
The first kind of sensitivity analysis focuses on the impact of decisions that lead to the analysis' different data. A typical example of sensitivity analysis is to ask how results might have changed if different study inclusion rules had been used.
The second kind of sensitivity analysis focuses on the impact of the statistical methods. Statistical sensitivity analyses are to determine whether and how the analyses' conclusions might differ using different statistical procedures or different assumptions about the data.
The third kind of sensitivity analysis focuses on how missing data were addressed. It can be used to investigate whether the conclusions would differ substantially across a range of plausible imputed values.

11.3.5 Reporting Results

In general, the results should contain three parts: (1) analysis methods; (2) list of studies; and (3) statistical analysis results.

At the beginning of the analysis methods, information on the procedures, sources, keywords, and years covered by the literature search is needed for the readers to assess the thoroughness of the search and, therefore, how much credibility to place in the research conclusions. Researchers can use a flow diagram suggested by PRISMA to present the results of a search. It displays the number of records identified, included, and excluded as well as the reasons for exclusions.

Then, the analysis methods should describe the characteristics of the people who retrieved information from the studies, the procedures used to train them, how the retrieved information's reliability was assessed, and what this assessment revealed. It is also essential to discuss in the coding procedures section how missing data were handled.

Lastly, the analysis methods should describe the software and procedures used to carry out any quantitative analyses.

For the second part, researchers should present a table listing all the studies included in the meta-analysis. Certain aggregate descriptive statistics about the literature should be reported as well, such as the number of studies, effect sizes, and samples that went into the meta-analysis, the range of years in which reports appeared, the total number of participants across all studies, as well as the range, median, mean and variance of sample sizes within studies.

Forest plot often appears in the third part. A forest plot shows effect estimates and confidence intervals for individual studies and meta-analysis (Lewis and Clarke 2001). Each study is represented by a block at the point estimate of intervention effect with a horizontal line extending either side of the block. The block area indicates the weight assigned to the study. The block's size indicates the study with a larger weight (usually with narrower confidence intervals), which dominates the calculation of the pooled result. The horizontal line depicts the confidence interval, which shows the intervention effects range compatible with the study's results.

11.4 Meta-Analysis on the Effectiveness of Lingzhi

The Lingzhi studied used the following names: Reishi Mushroom Powder, Reishi Spore Powder, Lingzhi spore powder, *Ganoderma lucidum* spores (GLS), *Ganoderma* spore powder. The readers should be aware that the chemical equivalence of those materials remains unclear.

11.4.1 Procedure of Meta-Analysis on the Effectiveness of Lingzhi

(1) Criteria for Including Studies
The selection of the dataset was based on the following aspects. Firstly, China Food and Drug Administration (CFDA) has issued 27 health-promoting functions for health food. The claim functions of Lingzhi must be one of it. Secondly, the study subject is mouse.

(2) Search Methods
The following databases were used as the sources: The China National Knowledge Infrastructure (CNKI, 1979 to October 2020) and the PubMed (1950 to October 2020). Two strategies were applied when searching literature. For the first strategy, 27 health-promoting functions were used as a specification of functions. The specific function was then combined with the material name to form the query terms. An example of the search query is "Lingzhi spore powder" AND "immunomodulation". For the second strategy, we simply combined the material names with the keyword "function". In particular, we searched PubMed using the query of (((((*Ganoderma lucidum* spores) OR Lingzhi spore powder) OR Reishi Spore Powder) OR Reishi Mushroom Powder) OR Ganoderma spore powder) AND function.

(3) Data Model
We first constructed a data model shown in Fig. 11.3a. The data model included 23 columns. We also used material_id, name, source, note, and create_time to describe the material used in the meta-analysis. See the material models in Fig. 11.3b. The sources of the included studies were recorded using the source model in Fig. 11.3c. The key attributes included source_id, source_type, and citation. A total 5863 studies were selected from above. Their contents were extracted from these studies and used to fill the data model.

(4) Database
We constructed a relational database using MySQL (version 5.14). The database schema contained three tables matching the data model. The names of these tables were exp_data, materials, and sources, matching the data model. The definitions of the three tables were shown in Fig. 11.3a, b, c, respectively. The database was running on a CentOS 7 operating system.

(5) Meta-analysis Pipeline
The meta-analyses were performed following the protocol recommended by the Cochrane Collaboration (Higgins and Green 2008). The exact procedure was implemented as a computation pipeline using the Perl programming language (5.23). The pipeline called the meta library from the R package (version 3.2.0) (Schwarzer 2007).

11.4.2 Results of Meta-Analysis on the Effectiveness of Lingzhi

(1) Searching Results
A total of 288 articles were found using the two search strategies. Eleven of the 288 articles cannot be downloaded. 151 of the 288 articles were excluded for the reason that their titles, abstracts, or both were irrelevant. In addition, we also excluded 32 articles analyzing complex health-promoting functions and 27 records lacking experimental data. Lastly, 57 articles were retained for downstream analysis. See the PRISMA workflow diagram in Fig. 11.4.

(a)

```
mysql> desc exp_data;
+-------------------------+-------------------------------------------+------+-----+---------+----------------+
| Field                   | Type                                      | Null | Key | Default | Extra          |
+-------------------------+-------------------------------------------+------+-----+---------+----------------+
| id_record               | int(10) unsigned                          | NO   | PRI | NULL    | auto_increment |
| id_material             | int(11)                                   | YES  |     | NULL    |                |
| id_source               | int(11)                                   | YES  |     | NULL    |                |
| test_category           | enum('function','toxicity')               | YES  |     | NULL    |                |
| test_type               | varchar(256)                              | YES  |     | NULL    |                |
| exp_name                | varchar(256)                              | YES  |     | NULL    |                |
| exp_group               | enum('normal','model','exp')              | YES  |     | NULL    |                |
| replicate_id            | int(11)                                   | YES  |     | NULL    |                |
| treatment_subject_name  | varchar(45)                               | YES  |     | NULL    |                |
| treatment_subject_sex   | enum('male','female','not_specified')     | YES  |     | NULL    |                |
| treatment_name          | varchar(45)                               | YES  |     | NULL    |                |
| treatment_dose          | float                                     | YES  |     | NULL    |                |
| treatment_unit          | varchar(45)                               | YES  |     | NULL    |                |
| treatment_time          | float                                     | YES  |     | NULL    |                |
| treatment_time_unit     | varchar(45)                               | YES  |     | NULL    |                |
| indicator_name          | varchar(45)                               | YES  |     | NULL    |                |
| indicator_value_unit    | varchar(45)                               | YES  |     | NULL    |                |
| indicator_value_mean    | float                                     | YES  |     | NULL    |                |
| indicator_value_std     | float                                     | YES  |     | NULL    |                |
| sample_size             | int(11)                                   | YES  |     | NULL    |                |
| desc                    | varchar(1024)                             | YES  |     | NULL    |                |
| create_time             | datetime                                  | YES  |     | NULL    |                |
| update_time             | datetime                                  | YES  |     | NULL    |                |
+-------------------------+-------------------------------------------+------+-----+---------+----------------+
23 rows in set
```

(b)

```
mysql> desc material;
+-------------+--------------+------+-----+---------+----------------+
| Field       | Type         | Null | Key | Default | Extra          |
+-------------+--------------+------+-----+---------+----------------+
| id_material | int(11)      | NO   | PRI | NULL    | auto_increment |
| name        | varchar(45)  | YES  |     | NULL    |                |
| source      | varchar(45)  | YES  |     | NULL    |                |
| note        | varchar(256) | YES  |     | NULL    |                |
| create_time | datetime     | YES  |     | NULL    |                |
+-------------+--------------+------+-----+---------+----------------+
```

(c)

```
mysql> desc source;
+----------------+---------------+------+-----+---------+-------+
| Field          | Type          | Null | Key | Default | Extra |
+----------------+---------------+------+-----+---------+-------+
| id             | int(64)       | NO   | PRI | 0       |       |
| source_type    | varchar(45)   | YES  |     | NULL    |       |
| id_noteexpress | varchar(45)   | YES  |     | NULL    |       |
| citation       | varchar(1024) | YES  |     | NULL    |       |
| create_time    | datetime      | YES  |     | NULL    |       |
| update_time    | datetime      | YES  |     | NULL    |       |
+----------------+---------------+------+-----+---------+-------+
```

Fig. 11.3 Schema for the database supporting the automatic meta-analysis. **a** exp_data. **b** material. **c** source

(2) Effects of Treatments

These 57 articles contained a total of 89 datasets addressing three health-promoting effects of Lingzhi: immunomodulation, anti-oxidation, and anti-fatigue. In some analyses, multiple datasets from the same study were used.

① Immunomodulation

Twenty-eight studies were selected for meta-analysis of NK-cell (Natural Killer Cell) activity (Fig. 11.5). We first carried out heterogeneity analysis. The I^2

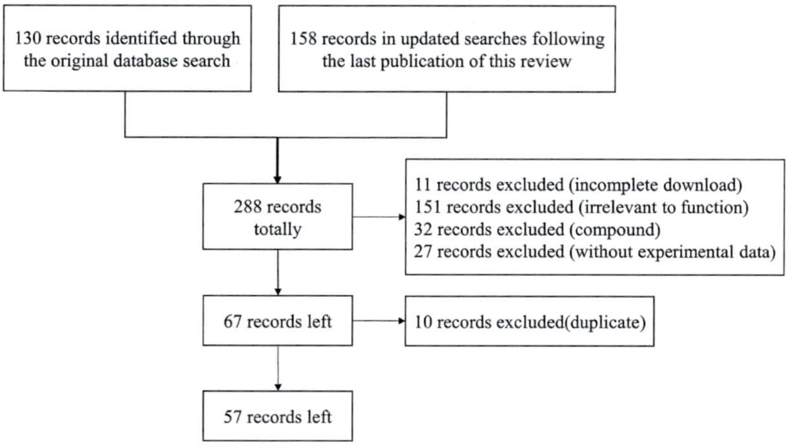

Fig. 11.4 PRISMA study flow diagram

Study		Experimental Mean	Experimental SD	Total	Control Mean	Control SD	Total	Weight	Std. Mean Difference IV,Random,95%CI
20140079	subgroup	38.74	8.7000	11	32.49	5.1100	10	3.5%	0.83 [-0.07;1.73]
20150269	subgroup	22.54	12.1400	10	14.47	9.0600	10	3.4%	0.72 [-0.19;1.63]
20040848	subgroup	31.50	6.8000	10	27.00	8.9000	10	3.5%	0.54 [-0.35;1.44]
20150137	subgroup	41.96	4.0100	11	38.78	3.3600	11	3.7%	0.83 [-0.05;1.71]
20150719	subgroup	30.54	6.7300	10	31.59	5.6600	10	3.7%	-0.16 [-1.04;0.72]
20150877	subgroup	31.20	6.2000	10	30.80	3.8000	10	3.7%	0.07 [-0.80;0.95]
20040559	subgroup	55.31	11.2600	10	32.29	4.8700	10	1.8%	2.54 [1.30;3.78]
20160280	subgroup	37.70	7.4000	10	27.90	7.1000	10	2.9%	1.29 [0.31;2.28]
20160417	subgroup	49.20	5.4000	10	41.50	5.6000	10	2.9%	1.34 [0.35;2.33]
20140754	subgroup	27.13	7.1200	10	17.80	5.4400	10	2.8%	1.41 [0.41;2.41]
20140453	subgroup	47.63	4.3400	12	33.13	3.0300	12	1.4%	3.74 [2.33;5.15]
20140697	subgroup	33.10	10.1000	10	36.30	16.3000	10	3.7%	-0.23 [-1.11;0.65]
20150030	subgroup	42.83	28.9900	10	62.36	46.2900	10	3.6%	-0.48 [-1.38;0.41]
20150525	subgroup	25.90	5.4000	10	28.80	5.5000	10	3.6%	-0.51 [-1.40;0.38]
20160346	subgroup	16.54	3.1100	10	10.63	1.3900	10	2.0%	2.35 [1.15;3.55]
20160424	subgroup	58.70	8.6000	12	37.10	7.1000	12	2.2%	2.64 [1.50;3.79]
20160374	subgroup	37.70	7.4000	10	27.90	7.1000	10	2.9%	1.29 [0.31;2.28]
20040791	subgroup	51.98	6.2400	10	29.61	2.4500	10	0.9%	4.52 [2.73;6.31]
20041380	subgroup	49.57	17.0600	12	42.52	9.0400	12	4.3%	0.50 [-0.32;1.31]
15400	subgroup	79.20	9.6000	10	69.10	11.4000	10	3.3%	0.92 [-0.01;1.85]
15485	subgroup	39.55	10.1400	10	22.61	7.1600	10	2.4%	1.85 [0.76;2.93]
15485	subgroup	39.55	10.1400	10	28.14	4.9600	10	2.9%	1.37 [0.37;2.37]
15485	subgroup	38.87	5.9500	10	28.14	4.9600	10	2.4%	1.88 [0.79;2.97]
15485	subgroup	38.87	5.9500	10	22.61	7.1600	10	2.0%	2.37 [1.17;3.56]
15477	subgroup	0.98	0.3300	12	0.94	0.2200	12	4.4%	0.14 [-0.66;0.94]
15470	subgroup	23.48	3.1000	10	19.99	2.6300	10	3.1%	1.16 [0.20;2.13]
15492	subgroup	35.10	5.0400	10	23.80	6.7100	10	2.4%	1.82 [0.74;2.90]
15492	subgroup	35.10	5.0400	10	20.25	5.7000	10	1.8%	2.64 [1.38;3.91]
15492	subgroup	0.00	0.0000	10	23.80	6.7100	10	0.0%	
15492	subgroup	0.00	0.0000	10	20.25	5.7000	10	0.0%	
15487	subgroup	37.52	14.6300	30	30.36	10.3500	30	10.7%	0.56 [0.04;1.07]
15542	subgroup	24.40	2.1600	10	7.90	6.8100	10	1.5%	3.13 [1.75;4.52]
15519	subgroup	32.00	10.9000	10	20.60	9.0000	10	3.1%	1.09 [0.14;2.05]
15141	subgroup	13.09	3.4500	10	10.79	2.9900	10	3.5%	0.69 [-0.23;1.59]
Total(95%CI)				**370**			**369**	**100.0%**	**0.98 [0.81;1.15]**

Heterogeneity: $Tau^2 = 0.7558$; $Chi^2 = 128.94$, df = 31 ($P<0.01$); $I^2 = 76\%$

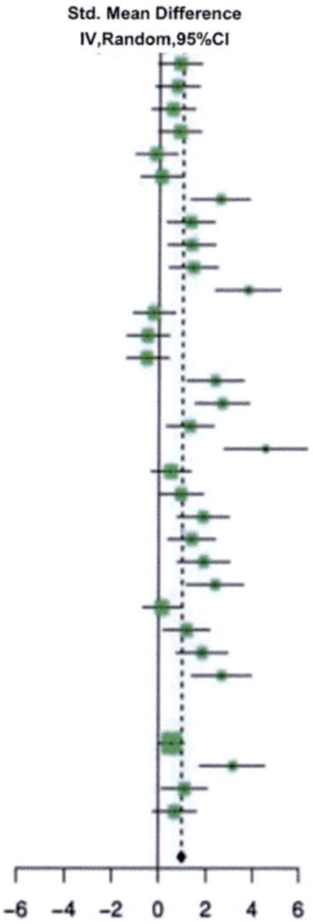

Fig. 11.5 Meta-analysis for NK-cell activity

was larger than 50%, indicating the presence of significant heterogeneity among the studies. Consequently, we selected the random-effects model for downstream analysis. The standard mean difference (SMD) was 0.9804 (p-value < 0.01). The 95% confidence interval (CI) ranged from 0.8116 to 1.1491. In conclusion, the treatment with Lingzhi Spore significantly increased the NK-cell (Natural Killer Cell) activity.

Two studies were selected for meta-analysis of OD (Optical Density) value increase of ConA (Concanavalin A)

Study		Experimental Mean	SD	Total	Control Mean	SD	Total	Weight	Std. Mean Difference IV,Fixed,95%CI
20040559	subgroup	0.18	0.0061	10	0.14	0.0044	10	43.4%	8.00 [5.09;10.90]
20040866	subgroup	0.17	0.0023	10	0.14	0.0054	10	56.6%	6.90 [4.36;9.44]
Total(95%CI)				20			20	100.00%	7.38 [5.46;9.29]

Heterogeneity: $Tau^2 = 0$; $Chi^2 = 0.31$, df = 1 (P=0.58); $I^2 = 0\%$

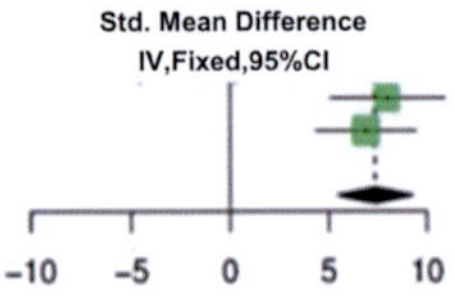

Fig. 11.6 Meta-analysis for the OD value increase of ConA

(Fig. 11.6). We firstly carried out heterogeneity analysis. The I^2 was less than 50%. Consequently, we selected the fixed-effects model for downstream analysis. The SMD was 7.3759 (p-value < 0.0001). The 95% CI ranged from 5.4641 to 9.2877. In conclusion, the treatment with Lingzhi Spore significantly increased the OD value of ConA.

Twenty-three studies were selected for meta-analysis of body weight (Fig. 11.7). We firstly carried out heterogeneity analysis. The I^2 was larger than 50%. Consequently, we selected the random-effects model for downstream analysis. The SMD was −0.1744 (p-value = 0.0009). The 95% CI ranged from −0.2774 to −0.0714. In conclusion, the treatment with Lingzhi Spore significantly reduced the body weight.

Two studies were selected for meta-analysis of half hemolysis number (Fig. 11.8). We firstly carried out heterogeneity analysis. The I^2 was larger than 50%. Consequently, we selected the random-effects model for downstream analysis. The SMD was 1.2934 (p-value = 0.0003). The 95% CI ranged from 0.5862 to 2.0005. In conclusion, the treatment with Lingzhi Spore significantly increased the half hemolysis number.

Twenty-four studies were selected for meta-analysis of phagocytic index (Fig. 11.9). We firstly carried out heterogeneity analysis. The I^2 was larger than 50%. Consequently, we selected the random-effects model for downstream analysis. The SMD was 0.9599 (p-value < 0.0001). The 95% CI ranged from 0.7789 to 1.1410. In conclusion, the treatment with Lingzhi Spore significantly increased the phagocytic index.

Thirty studies were selected for meta-analysis of phagocytic index a (Fig. 11.10). We firstly carried out heterogeneity analysis. The I^2 was larger than 50%. Consequently, we selected the random-effects model for downstream analysis. The SMD was 0.4704 (p-value < 0.0001). The 95% CI ranged from 0.3216 to 0.6193. In conclusion, the treatment with Lingzhi Spore significantly increased the phagocytic index a.

Twenty-three studies were selected for meta-analysis of phagocytic rate (Fig. 11.11). We firstly carried out heterogeneity analysis. The I^2 was larger than 50%. Consequently, we selected the random-effects model for downstream analysis. The SMD was 0.8648 (p-value < 0.0001). The 95% CI ranged from 0.6736 to 1.0559. In conclusion, the treatment with Lingzhi Spore significantly increased the phagocytic rate.

Seventeen studies were selected for meta-analysis of increase of body weight (Fig. 11.12). We firstly carried out heterogeneity analysis. The I^2 was less than 50%. Consequently, we selected the fixed-effects model for downstream analysis. The SMD was 0.1970 (p-value = 0.0020). The 95% CI ranged from 0.0719 to 0.3222. In conclusion, the treatment with Lingzhi Spore significantly increased the body weight.

Two studies were selected for meta-analysis of the number of antibody-producing cells (Fig. 11.13). We firstly carried out heterogeneity analysis. The I^2

Study		Experimental Mean	SD	Total	Control Mean	SD	Total	Weight	Std. Mean Difference IV,Random,95%CI
20140079	subgroup	26310.00	830.0000	11	26420.00	800.0000	10	1.4%	-0.13 [-0.99;0.73]
20140079	subgroup	26200.00	620.0000	11	26320.00	700.0000	10	1.4%	-0.17 [-1.03;0.68]
20140079	subgroup	26040.00	780.0000	11	25760.00	780.0000	10	1.4%	0.34 [-0.52;1.21]
20140079	subgroup	26220.00	830.0000	11	26000.00	890.0000	10	1.4%	0.25 [-0.61;1.11]
20150269	subgroup	37700.00	2000.0000	10	37700.00	1500.0000	10	1.4%	0.00 [-0.88;0.88]
20150269	subgroup	37500.00	1600.0000	10	37900.00	1500.0000	10	1.4%	-0.25 [-1.13;0.63]
20040848	subgroup	35000.00	1600.0000	10	35300.00	1800.0000	10	1.4%	-0.17 [-1.05;0.71]
20150137	subgroup	25600.00	1580.0000	10	25140.00	1660.0000	11	1.4%	0.27 [-0.59;1.13]
20150137	subgroup	25510.00	890.0000	11	25770.00	760.0000	11	1.5%	-0.30 [-1.14;0.54]
20150137	subgroup	26080.00	1020.0000	11	25950.00	890.0000	11	1.5%	0.14 [-0.71;0.97]
20150137	subgroup	25530.00	610.0000	10	25130.00	580.0000	10	1.3%	0.64 [-0.26;1.55]
20150088	subgroup	34200.00	1900.0000	10	34900.00	2100.0000	10	1.4%	-0.33 [-1.22;0.55]
20150719	subgroup	32200.00	1320.0000	10	30800.00	2440.0000	10	1.3%	0.68 [-0.22;1.59]
20150877	subgroup	31400.00	2700.0000	10	30400.00	1900.0000	10	1.3%	0.41 [-0.48;1.30]
20150877	subgroup	29200.00	2300.0000	10	30100.00	2000.0000	10	1.3%	-0.40 [-1.29;0.49]
20150877	subgroup	33500.00	1500.0000	10	31200.00	2700.0000	10	1.2%	1.01 [0.07;1.95]
20150877	subgroup	32500.00	2100.0000	10	31000.00	1900.0000	10	1.3%	0.72 [-0.19;1.63]
20160422	subgroup	32700.00	2800.0000	12	34000.00	2100.0000	12	1.6%	-0.51 [-1.32;0.31]
20040559	subgroup	34000.00	1100.0000	10	24000.00	1000.0000	10	0.1%	9.11 [5.84;12.38]
20040866	subgroup	35000.00	1300.0000	10	35000.00	1400.0000	10	1.4%	0.00 [-0.88;0.88]
20160280	subgroup	24790.00	1530.0000	10	24990.00	1280.0000	10	1.4%	-0.14 [-1.01;0.74]
20160417	subgroup	30200.00	2000.0000	10	29100.00	1800.0000	10	1.3%	0.55 [-0.34;1.45]
20160417	subgroup	29900.00	1200.0000	10	30400.00	1700.0000	10	1.4%	-0.33 [-1.21;0.56]
20160417	subgroup	30600.00	2300.0000	10	32400.00	1800.0000	10	1.2%	-0.83 [-1.76; 0.09]
20160417	subgroup	29600.00	3200.0000	10	29000.00	1600.0000	10	1.4%	0.23 [-0.65;1.11]
20140453	subgroup	25850.00	1310.0000	12	25650.00	1010.0000	12	1.7%	0.17 [-0.64;0.97]
20140453	subgroup	26150.00	920.0000	12	26250.00	940.0000	12	1.7%	-0.10 [-0.90;0.70]
20140453	subgroup	25930.00	1180.0000	12	26000.00	1280.0000	12	1.7%	-0.05 [-0.86;0.75]
20140453	subgroup	26330.00	1020.0000	12	25870.00	950.0000	12	1.6%	0.45 [-0.36;1.26]
20140697	subgroup	27900.00	1100.0000	10	29400.00	1400.0000	10	1.1%	-1.14 [-2.10;-0.18]
20140697	subgroup	27900.00	1100.0000	10	28400.00	2000.0000	10	1.4%	-0.30 [-1.18;0.59]
20140697	subgroup	27900.00	1100.0000	10	29300.00	2000.0000	10	1.2%	-0.83 [-1.75;0.09]
20140697	subgroup	27900.00	1100.0000	10	28800.00	2800.0000	10	1.3%	-0.41 [-1.29;0.48]
20140697	subgroup	28200.00	2500.0000	10	28800.00	2800.0000	10	1.4%	-0.22 [-1.10;0.66]
20140697	subgroup	28200.00	2500.0000	10	29400.00	1400.0000	10	1.3%	-0.57 [-1.47;0.33]
20140697	subgroup	28200.00	2500.0000	10	28400.00	2000.0000	10	1.4%	-0.08 [-0.96;0.79]
20140697	subgroup	28200.00	2500.0000	10	29300.00	2000.0000	10	1.3%	-0.47 [-1.36;0.43]
20140697	subgroup	28600.00	1500.0000	10	28400.00	2000.0000	10	1.4%	0.11 [-0.77;0.99]
20140697	subgroup	28600.00	1500.0000	10	29400.00	1400.0000	10	1.3%	-0.53 [-1.42;0.37]
20140697	subgroup	28600.00	1500.0000	10	28800.00	2800.0000	10	1.4%	-0.09 [-0.96;0.79]
20140697	subgroup	28600.00	1500.0000	10	29300.00	2000.0000	10	1.4%	-0.38 [-1.27;0.51]
20140697	subgroup	26300.00	2200.0000	10	29400.00	1400.0000	10	1.0%	-1.61 [-2.65;-0.57]
20140697	subgroup	26300.00	2200.0000	10	28400.00	2000.0000	10	1.2%	-0.96 [-1.89;-0.02]
20140697	subgroup	26300.00	2200.0000	10	28800.00	2800.0000	10	1.2%	-0.95 [-1.89;-0.01]
20140697	subgroup	26300.00	2200.0000	10	29300.00	2000.0000	10	1.1%	-1.37 [-2.36;-0.37]
20150030	subgroup	24900.00	1000.0000	10	25200.00	1200.0000	10	1.4%	-0.26 [-1.14;0.62]
20150030	subgroup	25000.00	400.0000	10	25400.00	1000.0000	10	1.3%	-0.50 [-1.40;0.39]
20150030	subgroup	25700.00	800.0000	10	25700.00	1100.0000	10	1.4%	0.00 [-0.88;0.88]
20150030	subgroup	24600.00	1100.0000	10	25100.00	900.0000	10	1.3%	-0.48 [-1.37;0.42]
20150525	subgroup	28500.00	2400.0000	10	30400.00	2300.0000	10	1.3%	-0.77 [-1.69;0.14]
20150525	subgroup	28500.00	2400.0000	10	28600.00	2500.0000	10	1.4%	-0.04 [-0.92;0.84]
20150525	subgroup	28500.00	2400.0000	10	29400.00	2100.0000	10	1.4%	-0.38 [-1.27;0.50]
20150525	subgroup	28500.00	2400.0000	10	30800.00	2400.0000	10	1.2%	-0.92 [-1.85;0.01]
20150525	subgroup	29200.00	1700.0000	10	30400.00	2300.0000	10	1.3%	-0.57 [-1.47;0.33]
20150525	subgroup	29200.00	1700.0000	10	29400.00	2100.0000	10	1.4%	-0.10[-0.98;0.78]
20150525	subgroup	29200.00	1700.0000	10	30800.00	2400.0000	10	1.3%	-0.74 [-1.65;0.18]
20150525	subgroup	29200.00	1700.0000	10	28600.00	2500.0000	10	1.4%	0.27 [-0.61;1.15]
20150525	subgroup	28700.00	1600.0000	10	30400.00	2300.0000	10	1.2%	-0.82 [-1.74;0.10]
20150525	subgroup	28700.00	1600.0000	10	28600.00	2500.0000	10	1.4%	0.05 [-0.83;0.92]
20150525	subgroup	28700.00	1600.0000	10	30800.00	2400.0000	10	1.2%	-0.99 [-1.93;-0.05]
20150525	subgroup	28700.00	1600.0000	10	29400.00	2100.0000	10	1.4%	-0.36 [-1.24;0.53]
20150525	subgroup	27500.00	2000.0000	10	28600.00	2500.0000	10	1.3%	-0.47 [-1.36;0.43]
20150525	subgroup	27500.00	2000.0000	10	30400.00	2300.0000	10	1.1%	-1.29 [-2.27;-0.31]
20150525	subgroup	27500.00	2000.0000	10	30800.00	2400.0000	10	1.0%	-1.43 [-2.44;-0.42]
20150525	subgroup	27500.00	2000.0000	10	29400.00	2100.0000	10	1.2%	-0.89 [-1.82;0.04]
20160424	subgroup	34500.00	2000.0000	10	33800.00	2300.0000	10	1.6%	0.31 [-0.49;1.12]
20160374	subgroup	24790.00	1530.0000	10	24990.00	1280.0000	10	1.4%	-0.14 [-1.01;0.74]
20040791	subgroup	35000.00	1400.0000	10	35000.00	1200.0000	10	1.4%	0.00 [-0.88;0.88]
20041380	subgroup	32190.00	1750.0000	12	31670.00	1830.0000	12	1.6%	0.28 [-0.52;1.09]
11127	subgroup	26400.00	1300.0000	10	26000.00	2400.0000	10	1.4%	0.20 [-0.68;1.08]
11174	subgroup	29090.00	1190.0000	10	27590.00	660.0000	10	1.0%	1.46 [0.45;2.48]
11174	subgroup	28190.00	1370.0000	10	27590.00	660.0000	10	1.3%	0.53 [-0.36;1.43]
11159	subgroup	21400.00	1800.0000	10	21300.00	2300.0000	10	1.4%	0.05 [-0.83;0.92]
11159	subgroup	21400.00	1800.0000	10	26000.00	2400.0000	10	0.8%	-2.08 [-3.21;-0.94]
11159	subgroup	26400.00	1300.0000	10	26000.00	2400.0000	10	1.4%	0.20 [-0.68;1.08]
11159	subgroup	26400.00	1300.0000	10	21300.00	2300.0000	10	0.7%	2.61 [1.36;3.87]
Total(95%CI)				**780**			**777**	**100.0%**	**-0.17 [-0.28;-0.07]**

Heterogeneity: Tau^2= 0.2479; Chi^2= 163.53, df = 75 (P<0.01); I^2= 54%

Fig. 11.7 Meta-analysis for body weight

Study		Experimental Mean	SD	Total	Control Mean	SD	Total	Weight	Std. Mean Difference IV,Random,95%CI
11154	subgroup	128.53	11.1600	10	97.00	19.4500	10	41.6%	1.90 [0.81;3.00]
11154	subgroup	113.87	18.1900	10	97.00	19.4500	10	58.4%	0.86 [-0.07;1.78]
Total(95%CI)				**20**			**20**	**100.00%**	**1.29 [0.59;2.00]**

Heterogeneity: Tau²= 0.2796; Chi²= 2.04, df = 1 (P=0.15); I²= 51%

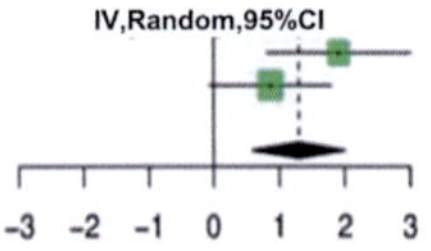

Fig. 11.8 Meta-analysis for half hemolysis number

Study		Experimental Mean	SD	Total	Control Mean	SD	Total	Weight	Std. Mean Difference IV,Random,95%CI
20150269	subgroup	0.65	0.0600	10	0.54	0.1100	10	3.5%	1.19 [0.22;2.16]
20040848	subgroup	0.22	0.1200	10	0.19	0.0600	10	4.2%	0.30 [-0.58;1.19]
20150137	subgroup	0.60	0.0800	11	0.55	0.0800	11	4.4%	0.60 [-0.26;1.46]
20150877	subgroup	0.77	0.2900	10	0.61	0.2600	10	4.1%	0.56 [-0.34;1.45]
20040559	subgroup	0.29	0.0400	10	0.16	0.0200	10	1.3%	3.94 [2.32;5.56]
20160280	subgroup	0.30	0.0200	10	0.27	0.0400	10	3.8%	0.91 [-0.02;1.84]
20160417	subgroup	0.83	0.3300	10	0.78	0.3400	10	4.3%	0.14 [-0.73;1.02]
20140453	subgroup	0.48	0.1100	12	0.33	0.1200	12	4.1%	1.26 [0.37;2.15]
20150030	subgroup	0.33	9.0000	10	0.30	0.0900	10	4.3%	0.00 [-0.87;0.88]
20150525	subgroup	0.77	0.3400	10	0.84	0.2500	10	4.2%	-0.22 [-1.10;0.66]
20160424	subgroup	0.93	0.1700	12	0.71	0.1170	12	3.9%	1.42 [0.51;2.34]
20160374	subgroup	0.30	0.0200	10	0.27	0.0400	10	3.8%	0.91 [-0.02;1.84]
20040791	subgroup	0.28	0.0400	10	0.17	0.0200	10	1.6%	3.33 [1.88;4.78]
20041380	subgroup	0.47	0.1800	12	0.26	0.0900	12	3.9%	1.42 [0.51;2.34]
11153	subgroup	0.52	0.0433	10	0.20	0.0402	10	0.4%	7.50 [4.76;10.24]
15490	subgroup	0.26	0.1210	10	0.15	0.0850	10	3.7%	0.99 [0.05;1.93]
15539	subgroup	0.12	0.0100	10	0.12	0.0100	10	4.3%	0.00 [-0.88;0.88]
15480	subgroup	0.08	0.0127	10	0.05	0.0109	10	2.6%	2.04 [0.91;3.16]
15470	subgroup	0.85	0.2800	10	0.64	0.1300	10	3.8%	0.92 [-0.01;1.85]
15543	subgroup	0.99	0.1600	10	0.82	0.1600	10	3.7%	1.02 [0.07;1.96]
15487	subgroup	0.38	0.1630	30	0.23	0.0960	30	11.0%	1.10 [0.55;1.64]
15484	subgroup	0.76	0.3500	10	0.27	0.1100	9	2.7%	1.76 [0.66;2.86]
15484	subgroup	0.76	0.3500	10	0.36	0.1200	7	2.7%	1.35 [0.25;2.44]
15484	subgroup	0.62	0.3700	9	0.36	0.1200	7	3.0%	0.85 [-0.20;1.89]
15484	subgroup	0.62	0.3700	9	0.27	0.1100	9	3.1%	1.22 [0.19;2.25]
15519	subgroup	0.78	0.2900	10	0.64	0.2100	10	4.1%	0.53 [-0.37;1.43]
15141	subgroup	0.12	0.0130	10	0.10	0.0060	10	3.6%	1.14 [0.17;2.10]
Total(95%CI)				**295**			**289**	**100.0%**	**0.96 [0.78;1.14]**

Heterogeneity: Tau²= 0.4697; Chi²= 78.56, df = 26 (P<0.01); I²= 67%

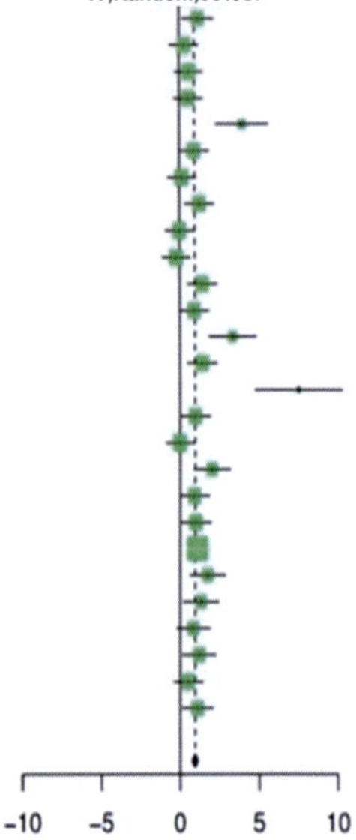

Fig. 11.9 Meta-analysis for phagocytic index

was less than 50%. Consequently, we selected the fixed-effects model for downstream analysis. The SMD was 1.1940 (p-value = 0.0005). The 95% CI ranged from 0.5254 to 1.8627. In conclusion, the treatment with Lingzhi Spore significantly increased the number of antibody-producing cells.

Three datasets were selected for meta-analysis of lymphocyte transformation rate (Fig. 11.14). We firstly carried out heterogeneity analysis. The I^2 was less than 50%. Consequently, we selected the fixed-effects model for downstream analysis. The SMD was 3.6258 (p-value < 0.0001). The 95% CI ranged from 2.7334 to 4.5183. In conclusion, the treatment with Lingzhi Spore significantly increased lymphocyte transformation rate.

Fifteen studies were selected for meta-analysis of the number of hemolytic plaques (Fig. 11.15). We firstly carried out heterogeneity analysis. The I^2 was larger than 50%. Consequently, we selected the random-effects model for downstream analysis. The SMD was 1.4462 (p-value < 0.0001). The 95% CI ranged from 1.1907 to 1.7017. In conclusion, the treatment with Lingzhi Spore significantly increased the number of hemolytic plaques.

Fourteen studies were selected for meta-analysis of auricle swelling (Fig. 11.16). We firstly carried out heterogeneity analysis. The I^2 was larger than 50%.

Study		Experimental Mean	SD	Total	Control Mean	SD	Total	Weight	Std. Mean Difference IV,Random,95%CI
20140079	subgroup	1.32	0.2300	11	1.06	0.1700	10	0.3%	1.22[0.27;2.17]
20140079	subgroup	6.88	0.8200	10	6.54	0.9700	10	2.8%	0.36[-0.52;1.25]
20150269	subgroup	10.31	0.6300	10	9.12	0.9600	10	2.2%	1.40[0.40;2.41]
20040848	subgroup	5.05	0.8200	10	4.68	0.8700	10	2.8%	0.42[-0.47;1.31]
20150137	subgroup	4.18	0.3900	10	4.06	0.5100	10	2.9%	0.25[-0.63;1.13]
20150719	subgroup	5.72	1.1200	10	5.55	1.2300	10	2.9%	0.14[-0.74;1.02]
20150719	subgroup	1.25	0.0500	10	1.19	0.0600	10	2.5%	1.04[0.09;1.99]
20150877	subgroup	5.65	0.7600	10	6.00	0.4800	10	2.8%	-0.53[-1.42;0.37]
20040559	subgroup	5.28	1.0700	10	3.17	0.4300	10	1.5%	2.48[1.25;3.70]
20040866	subgroup	4.48	0.3800	10	2.80	0.2600	10	0.6%	4.94[3.02;6.86]
20160280	subgroup	5.24	0.7300	10	4.95	0.6300	10	2.8%	0.41[-0.48;1.30]
20160417	subgroup	6.27	0.7300	10	6.20	0.5800	10	2.6%	0.76[-0.16;1.67]
20140754	subgroup	5.37	0.7800	10	4.58	0.6600	10	2.5%	1.05[0.10;2.00]
20140754	subgroup	1.23	0.1000	10	1.18	0.1100	10	2.8%	0.46[-0.44;1.35]
20140453	subgroup	6.54	1.1600	12	6.02	0.9000	12	3.3%	0.48[-0.33;1.30]
20140697	subgroup	1.02	0.2900	10	1.06	0.3600	10	2.9%	-0.12[-0.99;0.76]
20140697	subgroup	8.37	1.1800	10	7.38	0.7000	10	2.5%	0.98[0.04;1.92]
20150030	subgroup	6.8	0.9300	10	6.34	0.9000	10	2.8%	0.48[-0.41;1.37]
20150525	subgroup	5.86	0.6200	10	6.06	0.8700	10	2.9%	-0.25[-1.13;0.63]
20160346	subgroup	5.73	1.1200	10	5.59	1.3300	10	2.9%	0.11[-0.77;0.99]
20160346	subgroup	1.25	0.0400	10	1.20	0.0700	10	2.6%	0.84[-0.08;1.76]
20160424	subgroup	5.32	0.8400	12	5.84	0.6400	12	3.2%	-0.67[-1.50;0.15]
20160374	subgroup	5.24	0.7300	10	4.95	0.6300	10	2.8%	0.41[-0.48;1.30]
20040791	subgroup	4.76	1.1200	10	2.82	0.3300	10	1.6%	2.25[1.08;3.42]
20041380	subgroup	5.37	0.8100	12	5.47	0.8400	12	3.5%	-0.12[-0.92;0.68]
11127	subgroup	6.95	0.7700	10	6.26	0.6900	10	2.6%	0.90[-0.03;1.83]
11174	subgroup	3.43	0.2770	10	3.81	0.8730	10	2.7%	-0.56[-1.46;0.34]
11174	subgroup	2.93	0.4180	10	3.81	0.8730	10	2.3%	-1.23[-2.20;-0.25]
11159	subgroup	6.95	0.7700	10	6.26	0.6900	10	2.6%	0.90[-0.03;1.83]
15490	subgroup	5.12	0.7300	10	4.88	0.7200	10	2.8%	0.32[-0.57;1.20]
15539	subgroup	4.74	0.8000	10	4.74	0.7000	10	2.9%	0.00[-0.88;0.88]
15477	subgroup	4.51	1.0400	12	5.51	0.9700	12	3.0%	-0.96[-1.81;-0.11]
15492	subgroup	6.34	2.3800	10	5.17	0.4800	10	2.7%	0.65[-0.25;1.56]
15492	subgroup	5.23	0.5300	10	5.17	0.4800	10	2.9%	0.11[-0.76;0.99]
15487	subgroup	1.56	1.3400	30	7.62	1.0600	30	4.9%	2.40[1.73;3.08]
15519	subgroup	5.82	0.5500	10	5.79	0.4000	10	2.9%	0.06[-0.82;0.94]
15141	subgroup	4.54	0.4700	10	4.22	0.5200	10	2.7%	0.62[-0.28;1.52]
Total(95%CI)				**399**			**398**	**100.0%**	**0.47[0.32;0.62]**

Heterogeneity: $Tau^2 = 0.5840$; $Chi^2 = 134.42$, df = 36 (P<0.01); $I^2 = 73\%$

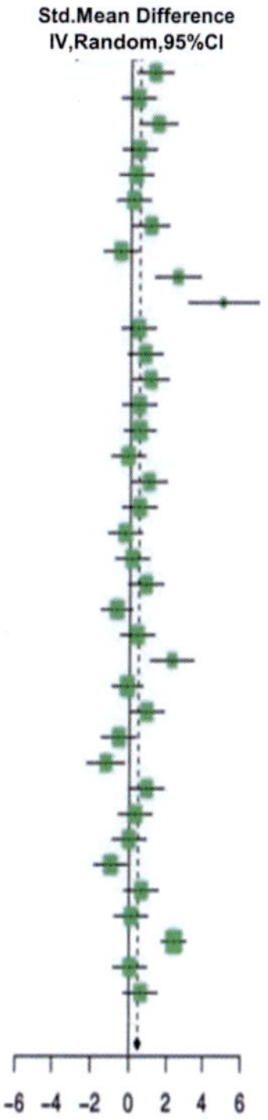

Fig. 11.10 Meta-analysis for phagocytic index a

Consequently, we selected the random-effects model for downstream analysis. The SMD was 1.1865 (p-value < 0.01). The 95% CI ranged from 0.9487 to 1.4242. In conclusion, the treatment with Lingzhi Spore significantly increased the auricle swelling.

Four studies were selected for meta-analysis of spleen index (Fig. 11.17). We firstly carried out heterogeneity analysis. The I^2 was less than 50%. Consequently, we selected the fixed-effects model for downstream analysis. The SMD was 0.4302 (p-value = 0.0364). The 95% CI ranged from 0.0273 to 0.8331. In conclusion, the treatment with Lingzhi Spore significantly increased the spleen index.

Fourteen studies were selected for meta-analysis of toe swelling (Fig. 11.18). We firstly carried out heterogeneity analysis. The I^2 was larger than 50%. Consequently, we selected the random-effects model for downstream analysis. The SMD was 0.5304 (p-value < 0.01). The 95% CI ranged from 0.2907 to 0.7701. In conclusion, the treatment with Lingzhi Spore significantly increased the toe swelling.

② Anti-oxidation

Two studies were selected for meta-analysis of GSH-PX (Glutathione Peroxidase) (Fig. 11.19). We firstly carried out heterogeneity analysis. The I^2 was less than 50%. Consequently, we selected the fixed-effects model for downstream analysis. The SMD was −2.9387 (p-value < 0.01). The 95% CI ranged from −4.0234 to −1.8540. In conclusion, the treatment with Lingzhi Spore significantly reduced the GSH-PX.

Five studies were selected for meta-analysis of MDA (Malondialdehyde) (Fig. 11.20). We firstly carried out heterogeneity analysis. The I^2 was larger

Study		Experimental Mean	SD	Total	Control Mean	SD	Total	Weight	Std. Mean Difference IV,Random,95%CI
20140079	subgroup	36.68	3.5500	11	31.55	3.6500	10	3.9%	1.37[0.40;2.34]
20150269	subgroup	30.20	3.4000	10	25.20	4.2000	10	3.8%	1.25[0.28;2.23]
20040848	subgroup	15.20	6.9000	10	15.90	4.9000	10	4.7%	-0.11[-0.99;0.77]
20150137	subgroup	34.53	4.9200	11	29.11	4.0500	11	4.3%	1.16[0.24;2.07]
20150719	subgroup	27.25	1.4400	10	25.15	2.0600	10	4.0%	1.13[0.17;2.09]
20150877	subgroup	27.60	10.7000	10	23.10	7.0000	10	4.6%	0.48[-0.42;1.37]
20040559	subgroup	16.00	1.5000	10	10.00	1.3000	10	1.3%	4.09[2.43;5.76]
20160280	subgroup	25.70	2.6000	10	24.50	4.3000	10	4.7%	0.32[-0.56;1.21]
20160417	subgroup	42.00	11.2000	10	34.60	11.7000	10	4.5%	0.62[-0.28;1.52]
20140754	subgroup	28.80	3.9100	10	25.50	3.7500	10	4.3%	0.83[-0.10;1.75]
20140453	subgroup	23.29	5.4100	12	15.99	4.1000	12	4.3%	1.47[0.55;2.39]
20140697	subgroup	36.40	6.9000	10	38.20	10.8000	10	4.7%	-0.19[-1.07;0.69]
20150030	subgroup	27.40	6.9000	10	25.90	7.7000	10	4.7%	0.20[-0.68;1.08]
20150525	subgroup	32.40	10.8000	10	36.30	7.7000	10	4.6%	-0.40[-1.29;0.49]
20160346	subgroup	27.75	2.1200	10	26.30	2.0800	10	4.5%	0.66[-0.24;1.57]
20160424	subgroup	49.90	7.9000	12	41.30	7.7000	12	4.9%	1.06[0.20;1.93]
15490	subgroup	16.30	6.1000	10	10.40	1.4000	10	3.8%	1.28[0.30;2.26]
15539	subgroup	10.40	1.0500	10	10.25	1.2300	10	4.7%	0.13[-0.75;1.00]
15543	subgroup	37.20	3.1500	10	31.20	4.4600	10	3.5%	1.49[0.47;2.50]
15484	subgroup	43.48	13.5500	10	20.27	7.9600	9	2.8%	1.97[0.83;3.11]
15484	subgroup	43.48	13.5500	10	23.51	4.3300	7	2.6%	1.75[0.57;2.92]
15484	subgroup	39.28	13.7900	9	23.51	4.3300	7	2.9%	1.38[0.25;2.51]
15484	subgroup	39.28	13.7900	9	20.27	7.9600	9	3.0%	1.61[0.51;2.71]
15542	subgroup	24.00	1.1100	10	13.05	0.2700	10	0.2%	12.98[-0.07;1.78]
15519	subgroup	39.50	8.4000	10	32.40	7.4000	10	4.3%	0.86[-0.07;1.78]
15141	subgroup	9.45	1.1650	10	8.55	0.7620	10	4.3%	0.88[-0.05;1.80]
Total(95%CI)				**264**			**257**	**100.0%**	**0.86[0.67;1.06]**

Heterogeneity: $Tau^2 = 0.5629$; $Chi^2 = 81.69$, df =25 (P<0.01); $I^2 = 69\%$

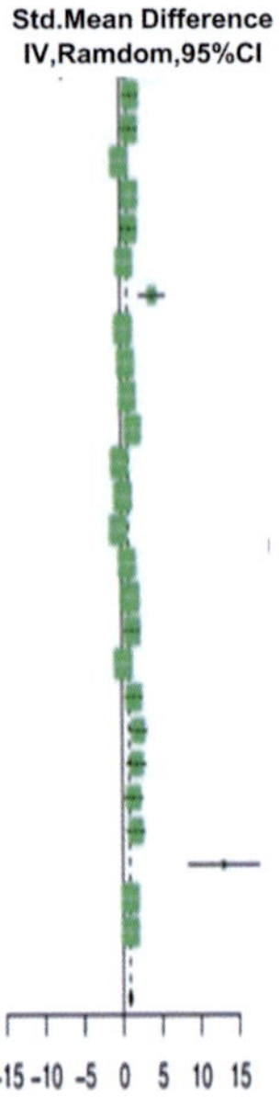

Fig. 11.11 Meta-analysis for phagocytic rate

than 50%. Consequently, we selected the random-effects model for downstream analysis. The SMD was −1.2082 (*p*-value < 0.01). The 95% CI ranged from −1.6465 to −0.7699. In conclusion, the treatment with Lingzhi Spore significantly reduced the MDA.

Two datasets was selected for meta-analysis of thymus/body ratio (Fig. 11.21). We firstly carried out heterogeneity analysis. The I^2 was larger than 50%. Consequently, we selected the random-effects model for downstream analysis. The SMD was 3.2705 (*p*-value < 0.01). The 95% CI ranged from 2.5709 to 3.9702. In conclusion, the treatment with Lingzhi Spore significantly increased the thymus/body ratio.

Five studies were selected for meta-analysis for SOD (Superoxide Dismutase) (Fig. 11.22). We firstly carried out heterogeneity analysis. The I^2 was larger than 50%. Consequently, we selected the random-effects model for downstream analysis. The SMD was 0.4401 (*p*-value = 0.0373). The 95% CI ranged from 0.0258 to 0.8543. In conclusion, the treatment with Lingzhi Spore significantly increased the SOD.

③ Anti-fatigue

Two studies were selected for meta-analysis of CK (Creatine Kinase) Activity (Fig. 11.23). We firstly carried out heterogeneity analysis. The I^2 was less than 50%. Consequently, we selected the fixed-effects model for downstream analysis. The SMD was 5.6671 (*p*-value < 0.0001). The 95% CI ranged from 3.9182 to 7.4161. In conclusion, the treatment with Lingzhi Spore significantly increased the CK.

Two studies were selected for meta-analysis of GSH-PX (Glutathione Peroxidase) Activity (Fig. 11.24). We firstly carried out heterogeneity analysis. The I^2 was larger than 50%. Consequently, we selected the random-effects model for downstream analysis. The SMD was −0.9854 (*p*-value = 0.0033). The 95% CI ranged from −1.6420 to −0.3289. In conclusion, the treatment with Lingzhi Spore significantly reduced the GSH-PX.

Study		Experimental Mean	Experimental SD	Experimental Total	Control Mean	Control SD	Control Total	Weight	Std. Mean Difference IV,Fixed,95%CI
20140079	subgroup	5430.00	390.0000	11	5550.00	480.0000	10	2.1%	-0.26[-1.13;0.60]
20140079	subgroup	5440.00	390.0000	11	5520.00	320.0000	10	2.1%	-0.21[-1.07;0.65]
20140079	subgroup	5570.00	340.0000	11	5490.00	420.0000	10	2.1%	0.20[-0.66;1.06]
20140079	subgroup	5530.00	320.0000	11	5360.00	260.0000	10	2.0%	0.56[-0.32;1.06]
20150269	subgroup	18000.00	1100.0000	11	17600.00	1500.0000	10	2.0%	0.29[-0.59;1.17]
20150269	subgroup	18000.00	1600.0000	10	17100.00	1400.0000	10	1.9%	0.57[-0.33;1.47]
20150269	subgroup	17300.00	1200.0000	10	17900.00	1700.0000	10	2.0%	-0.39[-1.28;0.50]
20150269	subgroup	18400.00	1500.0000	10	18400.00	2100.0000	10	2.0%	0.00[-0.88;0.88]
20150269	subgroup	17600.00	1500.0000	10	17300.00	1700.0000	10	2.0%	0.18[-0.72;1.06]
20150269	subgroup	17900.00	1500.0000	10	17800.00	1600.0000	10	2.0%	0.06[-0.82;0.94]
20040848	subgroup	14900.00	1800.0000	10	15300.00	2200.0000	10	2.0%	-0.19[-1.07;0.69]
20150137	subgroup	5430.00	1220.0000	10	5050.00	610.0000	11	2.1%	0.38[-0.48;1.25]
20150137	subgroup	5350.00	990.0000	11	5350.00	990.0000	11	2.2%	0.00[-0.48;0.84]
20150137	subgroup	5560.00	450.0000	11	5880.00	660.0000	11	2.1%	-0.55[-1.40;0.31]
20150137	subgroup	5220.00	610.0000	10	5240.00	630.0000	10	2.0%	-0.03[-0.91;0.85]
20150088	subgroup	14200.00	2000.0000	10	15000.00	2900.0000	10	2.0%	-0.31[-1.19;0.58]
20150088	subgroup	14200.00	1700.0000	10	14100.00	1600.0000	10	2.0%	0.06[-0.82;0.93]
20150088	subgroup	14700.00	1800.0000	10	14700.00	1800.0000	10	2.0%	0.00[-0.88;0.88]
20150088	subgroup	15100.00	1400.0000	10	14800.00	1700.0000	10	2.0%	0.18[-0.69;1.06]
20150719	subgroup	11800.00	1990.0000	10	10600.00	2880.0000	10	2.0%	0.46[-0.43;1.36]
20040559	subgroup	14000.00	500.0000	10	14000.00	1700.0000	10	2.0%	0.00[-0.88;0.88]
20040559	subgroup	14000.00	1000.0000	10	14000.00	1200.0000	10	2.0%	0.00[-0.88;0.89]
20040559	subgroup	14000.00	1000.0000	10	14000.00	1800.0000	10	2.0%	0.00[-0.88;0.90]
20040559	subgroup	14000.00	1200.0000	10	14000.00	1200.0000	10	2.0%	0.00[-0.88;0.91]
20040866	subgroup	15000.00	1000.0000	10	15000.00	1000.0000	10	2.0%	0.00[-0.88;0.92]
20040866	subgroup	15000.00	600.0000	10	15000.00	800.0000	10	2.0%	0.00[-0.88;0.93]
20040866	subgroup	16000.00	500.0000	10	15000.00	500.0000	10	1.3%	1.92[0.82;3.01]
20040866	subgroup	15000.00	700.0000	10	15000.00	800.0000	10	2.0%	0.00[-0.88;0.88]
20160280	subgroup	4840.00	1420.0000	10	5030.00	1460.0000	10	2.0%	-0.13[-1.00;0.75]
20160280	subgroup	5450.00	1580.0000	10	5450.00	2200.0000	10	2.0%	0.00[-0.88;0.88]
20160280	subgroup	6150.00	1680.0000	10	5440.00	1570.0000	10	2.0%	0.42[-0.47;1.31]
20160280	subgroup	5770.00	1550.0000	10	4780.00	1600.0000	10	1.9%	0.60[-0.30;1.50]
20160280	subgroup	5700.00	1680.0000	10	5150.00	1400.0000	10	2.0%	0.34[-0.54;1.23]
20140453	subgroup	5250.00	740.0000	12	5430.00	720.0000	12	2.4%	-0.24[-1.04;0.57]
20140453	subgroup	5550.00	670.0000	12	5590.00	770.0000	12	2.4%	-0.05[-0.85;0.75]
20140453	subgroup	5450.00	450.0000	12	5340.00	460.0000	12	2.4%	0.23[-0.57;1.04]
20140453	subgroup	5570.00	410.0000	12	5320.00	480.0000	12	2.3%	0.54[-0.28;1.36]
20160346	subgroup	11800.00	1870.0000	10	11300.00	2060.0000	10	2.0%	0.24[-0.64;1.12]
20160424	subgroup	13800.00	2000.0000	12	13200.00	3000.0000	12	2.4%	0.23[-0.58;1.03]
20160374	subgroup	4840.00	1420.0000	10	2150.00	400.0000	10	1.0%	2.47[1.25;3.69]
20160374	subgroup	4840.00	1420.0000	10	5030.00	1460.0000	10	2.0%	-0.13[-1.00;0.75]
20160374	subgroup	5700.00	1680.0000	10	2150.00	400.0000	10	0.9%	2.78[1.48;4.09]
20160374	subgroup	5700.00	1680.0000	10	5030.00	1460.0000	10	2.0%	0.41[-0.48;1.30]
20040791	subgroup	15000.00	900.0000	10	15000.00	700.0000	10	2.0%	0.00[-0.88;0.88]
20040791	subgroup	15000.00	700.0000	10	15000.00	500.0000	10	2.0%	0.00[-0.88;0.89]
20040791	subgroup	15000.00	700.0000	10	15000.00	700.0000	10	2.0%	0.00[-0.88;0.90]
20040791	subgroup	15000.00	500.0000	10	15000.00	700.0000	10	2.0%	0.00[-0.88;0.91]
20041380	subgroup	11830.00	1030.0000	12	11250.00	1140.0000	12	2.4%	0.52[-0.30;1.33]
11127	subgroup	6000.00	400.0000	10	4700.00	500.0000	10	0.9%	2.75[1.46;4.04]
11159	subgroup	5000.00	400.0000	10	4700.00	500.0000	10	1.9%	0.63[-0.27;1.54]
Total(95%CI)				**518**			**515**	**100.0%**	**0.20[0.07;0.32]**

Heterogeneity: Tau^2= 0.0985; Chi^2=72.69, df =49(P=0.02); I^2= 33%

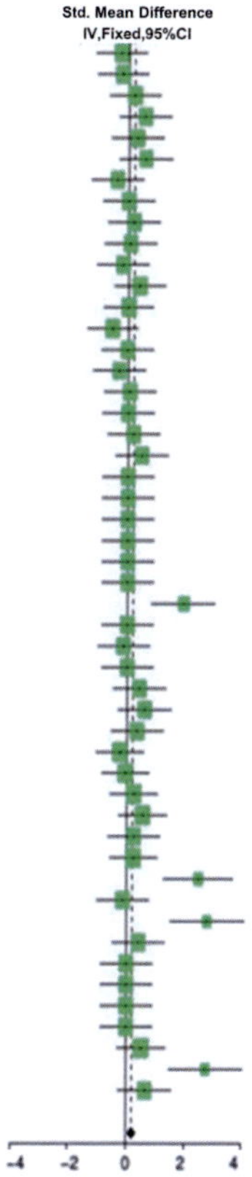

Fig. 11.12 Meta-analysis for the increase of body weight

Study		Experimental Mean	Experimental SD	Experimental Total	Control Mean	Control SD	Control Total	Weight	Std. Mean Difference IV,Fixed,95%CI
20140079	subgroup	510.45	127.56	11	360.50	101.34	10	49.30%	1.24[0.29; 2.19]
20150137	subgroup	510.00	165.93	10	318.64	154.73	11	50.70%	1.15[0.21; 2.09]
Total(95%CI)				**21**			**21**	**100.00%**	**1.19[0.53; 1.86]**

Heterogeneity: Tau^2= 0; Chi^2= 0.02, df = 1 (P=0.89); I^2= 0%

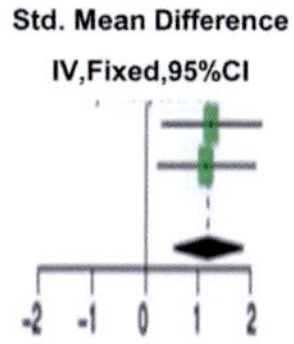

Fig. 11.13 Meta-analysis for number of antibody-producing cells

Study		Experimental Mean	Experimental SD	Experimental Total	Control Mean	Control SD	Control Total	Weight	Std. Mean Difference IV,Fixed,95%CI
11156	subgroup	70.33	3.9600	10	54.26	2.2400	10	22.7%	4.78[2.91;6.66]
11156	subgroup	76.62	8.5600	10	54.26	2.2400	10	36.7%	3.42[1.95;4.90]
11156	subgroup	81.35	11.3900	10	54.26	2.2400	10	40.6%	3.16[1.76;4.56]
Total(95%CI)				**30**			**30**	**100.0%**	**3.63[2.73;4.52]**

Heterogeneity: Tau^2= 0; Chi^2= 1.97, df =2 (P=0.37); I^2= 0%

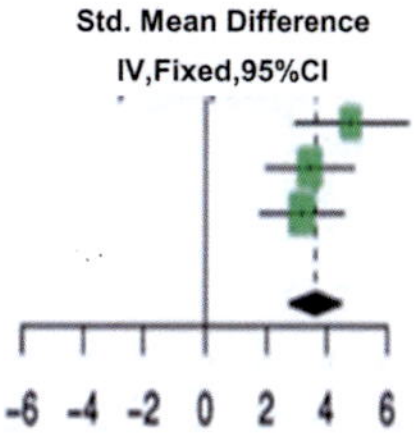

Fig. 11.14 Meta-analysis for the increase of lymphocyte transformation rate

Study	Experimental Mean	SD	Total	Control Mean	SD	Total	Weight	Std. Mean Difference IV,Random,95%CI
20150269 subgroup	2131 .00	162.0000	10	1769.00	151 .0000	10	4.8%	2.21 [1.05; 3.38]
20040848 subgroup	23.70	8.2000	10	16.30	7.4000	10	7.5%	0.91 [-0.02; 1.84]
20150719 subgroup	140.60	19.7200	10	108.10	10.7400	10	5.3%	1.96 [0.85; 3.07]
20150877 subgroup	49.00	11.0000	10	48.00	6.0000	10	8.5%	0.11 [-0.77; 0.99]
20040559 subgroup	24.50	3.4100	10	9.50	2.1700	10	1.7%	5.03 [3.08; 6.97]
20040866 subgroup	23.90	2.9200	10	9.60	2.4100	10	1.7%	5.12 [3.14; 7.09]
20160280 subgroup	132.00	23.0000	10	95.00	28.0000	10	6.5%	1.38 [0.38; 2.38]
20140754 subgroup	275.00	54.0600	10	235.50	69.2200	10	8.0%	0.61 [-0.29; 1.51]
20150030 subgroup	539.00	165.7000	10	341.00	166.3000	10	7.1%	1.14 [0.18; 2.10]
20160374 subgroup	132.00	23.0000	10	95.00	28.0000	10	6.5%	1.38 [0.38; 2.38]
15485 subgroup	193.00	75.0000	10	116.00	51.0000	10	7.0%	1.15 [0.19;2.11]
15539 subgroup	0.00	0.0005	10	0.00	0.0005	10	7.7%	0.81 [-0.11; 1.74]
15487 subgroup	242.50	40.2000	10	167.30	38.1000	10	17.2%	1.90 [1.28;2.51]
15519 subgroup	135.00	28.0000	10	61.00	28.0000	10	4.3%	2.53 [1.29; 3.77]
15141 subgroup	102.30	19.4600	10	75.81	11.0000	10	6.1%	1.61 [0.57; 2.64]
Total(95%CI)			**170**			**170**	**100%**	**1.45 [1.19; 1.70]**

Heterogeneity: Tau²= 0.6675; Chi²= 49.96, df = 14 (P<0.01); I²= 72%

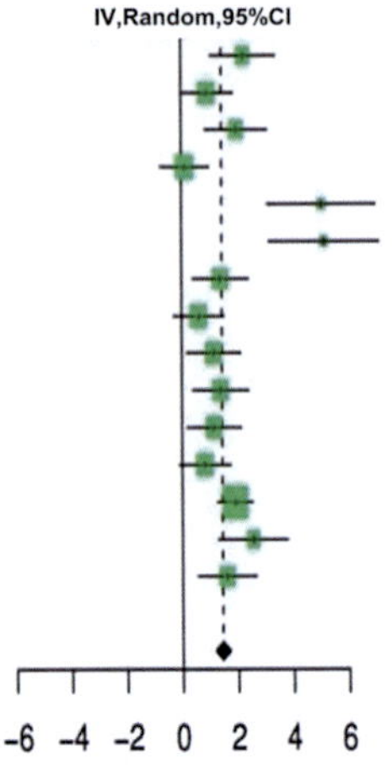

Fig. 11.15 Meta-analysis for number of hemolytic plaques

Study	Experimental Mean	SD	Total	Control Mean	SD	Total	Weight	Std. Mean Difference IV,Random,95%CI
20150269 subgroup	0.21	0.0380	10	0.14	0.0430	10	4.9%	1.79 [0.72; 2.87]
20040559 subgroup	19.10	2.7000	10	17.60	2.6000	10	7.0%	0.54 [-0.35; 1.44]
20040866 subgroup	23.10	4.1000	10	15.80	3.4000	10	4.8%	1.86 [0.77; 2.94]
20160280 subgroup	15.70	3.1000	10	11.60	3.0000	10	5.8%	1.29 [0.30; 2.27]
20160374 subgroup	15.70	3.1000	10	11.60	3.0000	10	5.8%	1.29 [0.30; 2.27]
20040791 subgroup	23.00	4.6000	10	18.40	2.7000	10	6.1%	1.17 [0.20; 2.13]
11127 subgroup	19.82	5.5600	10	9.29	2.9200	10	4.1%	2.27 [1.09; 3.45]
11153 subgroup	22.36	1.7900	10	18.00	1.5100	10	3.7%	2.52 [1.29; 3.76]
11154 subgroup	10.98	1.8500	10	8.09	1.5300	10	5.2%	1.63 [0.59; 2.67]
11154 subgroup	8.35	0.9200	10	8.09	1.5300	10	7.3%	0.20 [-0.68; 1.08]
11159 subgroup	19.82	5.5600	10	9.29	2.9200	10	4.1%	2.27 [1.09; 3.45]
15464 subgroup	2.11	0.7900	10	4.87	1.0100	10	3.2%	-2.92 [-4.25; -1.58]
15464 subgroup	2.11	0.7900	10	1.02	0.5300	10	5.4%	1.55 [0.52; 2.58]
15464 subgroup	3.84	1.7500	10	4.87	1.0100	10	6.8%	-0.69 [-1.60; 0.22]
15464 subgroup	3.84	1.7500	10	1.02	0.5300	10	4.4%	2.09 [0.95; 3.22]
15487 subgroup	16.74	2.0800	10	10.88	3.7800	10	14.9%	1.90 [1.28; 2.51]
15542 subgroup	8.81	4.7400	10	4.93	3.0600	10	6.5%	0.93 [0.00; 1.87]
15141 subgroup	24.10	2.8000	10	211.90	1.1000	10	0.0%	-84.55 [-113.81; -55.30]
Total(95%CI)			**200**			**200**	**100%**	**1.19 [0.95; 1.42]**

Heterogeneity: Tau²= 1.5537; Chi²= 115.23, df = 17 (P<0.01); I²= 85%

Fig. 11.16 Meta-analysis for auricle swelling

Study	Experimental Mean	SD	Total	Control Mean	SD	Total	Weight	Std. Mean Difference IV,Fixed,95%CI
20040886 subgroup	0.58	0.8500	10	0.56	0.1200	10	21.1%	0.03 [-0.85: 0.91]
20040559 subgroup	0.56	0.1200	10	0.50	0.0800	10	20.1%	0.56 [-0.33; 1.46]
15539 subgroup	0.32	0.0400	10	0.32	0.0200	10	21.1%	0.00 [-0.88; 0.88]
15484 subgroup	0.75	0.1390	20	0.64	0.1240	19	37.6%	0.82 [0.17; 1.48]
Total(95%CI)			**50**			**49**	**100%**	**0.43 [0.03; 0.83]**

Heterogeneity: Tau²= 0.0108; Chi²= 3.19, df = 3 (P=0.36); I²= 6%

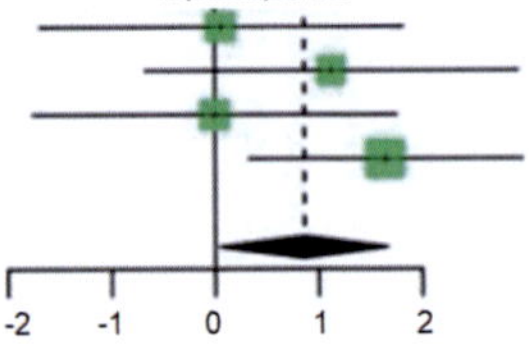

Fig. 11.17 Meta-analysis for spleen index

Study	Experimental Mean	Experimental SD	Experimental Total	Control Mean	Control SD	Control Total	Weight	Std. Mean Difference IV,Random,95%CI
20140079 subgroup	0.04	0.0080	10	0.03	0.0080	10	6.6%	1.12 [0.18; 2.05]
20040848 subgroup	0.06	0.0170	10	0.04	0.0170	10	6.5%	1.01 [0.07; 1.95]
20150137 subgroup	0.05	0.0070	10	0.04	0.0070	10	5.7%	1.54 [0.54; 2.54]
20150088 subgroup	0.10	0.0270	10	0.06	0.0270	10	5.8%	1.34 [0.35; 2.33]
20150719 subgroup	0.11	0.0320	10	0.07	0.0320	10	6.4%	1.05 [0.10; 2.00]
20150877 subgroup	0.05	0.0100	10	0.04	0.0100	10	6.1%	1.21 [0.24; 2.19]
20160422 subgroup	0.06	0.0160	10	0.04	0.0160	10	7.8%	0.99 [0.14; 1.85]
20160417 subgroup	0.04	0.0070	10	0.03	0.0070	10	6.5%	0.96 [0.02; 1.89]
20150030 subgroup	0.10	11.9000	10	0.08	11.9000	10	7.5%	0.00 [-0.87; 0.88]
330 subgroup	0.05	0.0170	10	0.06	0.0170	10	7.4%	-0.24 [-1.12; 0.64]
11174 subgroup	0.04	0.0260	10	0.06	0.0260	10	6.7%	-0.84 [-1.76; 0.08]
11174 subgroup	0.04	0.0100	10	0.06	0.0100	10	6.4%	-1.02 [-1.97; -0.08]
15480 subgroup	0.05	0.0190	10	0.02	0.0190	10	5.4%	1.55 [0.52; 2.57]
16319 subgroup	0.05	0.0150	10	0.02	0.0150	10	5.0%	1.78 [0.71; 2.85]
15488 subgroup	0.03	0.0104	15	0.04	0.0104	15	10.1%	-0.89 [-1.65; -0.14]
Total(95%CI)			**158**			**158**	**100%**	**0.53 [0.29; 0.77]**

Heterogeneity: Tau^2= 0.7344; Chi^2= 59.74, df = 14 (P<0.01); I^2= 72%

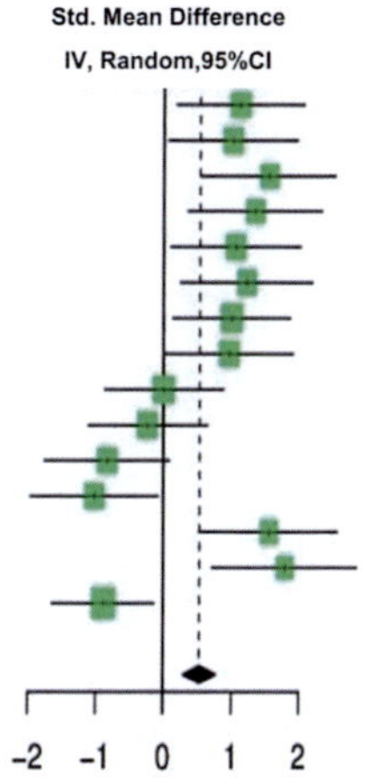

Fig. 11.18 Meta-analysis for toe swelling

Study		Experimental Mean	Experimental SD	Experimental Total	Control Mean	Control SD	Control Total	Weight	Std. Mean Difference IV,Fixed,95%CI
15471	subgroup	61875000.00	14364000.0000	8	109534000.00	19289000.0000	8	56.6%	-2.65 [-4.09; -1.21]
11152	subgroup	65470000.00	7010000.0000	8	93780000.00	9010000.0000	8	43.4%	-3.32 [-4.96; -1.67]
Total(95%CI)				**16**			**16**	**100%**	**-2.94 [-4.02; -1.85]**

Heterogeneity: Tau^2= 0; Chi^2= 0.36, df = 1 (P=0.55); I^2= 0%

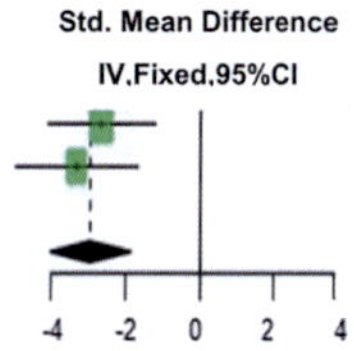

Fig. 11.19 Meta-analysis for GSH-PX

Study		Experimental Mean	Experimental SD	Experimental Total	Control Mean	Control SD	Control Total	Weight	Std. Mean Difference IV,Random,95%CI
15473	subgroup	0.00	0.0004	10	0.00	0.0004	10	24.4%	-0.40[-1.28; 0.49]
15489	subgroup	5.06	0.9100	10	6.10	0.2600	10	18.6%	-1.49[-2.50; 0.47]
15471	subgroup	15.31	5.3250	8	10.28	5.3720	8	17.7%	0.89[-0.15; 1.93]
11152	subgroup	17.26	1.2600	8	13.40	1.3000	8	8.5%	2.85[1.35; 4.35]
11155	subgroup	392800.00	109600.0000	20	919600.0000	123100.0000	20	13.4%	-4.43[-5.63; -3.23]
11155	subgroup	562500.00	53900.0000	20	919600.0000	123100.0000	20	17.4%	-3.68[-4.74; -2.63]
Total(95%CI)				**76**			**76**	**100.00%**	**-1.21[-1.65; -0.77]**

Heterogeneity: Tau^2= 5.5729; Chi^2= 96.28, df = 5 (P<0.01); I^2= 95%

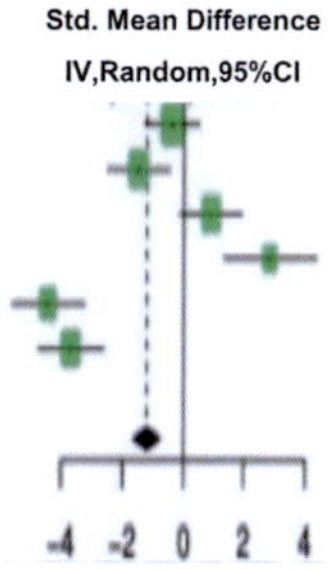

Fig. 11.20 Meta-analysis for MDA

Study		Experimental Mean	Experimental SD	Experimental Total	Control Mean	Control SD	Control Total	Weight	Std. Mean Difference IV,Random,95%CI
11155	subgroup	3274.00	527.0000	20	1051.00	522.0000	20	37.6%	4.15[3.01; 5.30]
11155	subgroup	2320.00	374.00	20	1051.00	522.0000	20	62.4%	2.74[1.85; 3.62]
Total(95%CI)				**40**			**40**	**100.00%**	**3.27[2.57; 3.97]**

Heterogeneity: Tau^2= 0.7293; Chi^2= 3.68, df = 1 (P=0.05); I^2= 73%

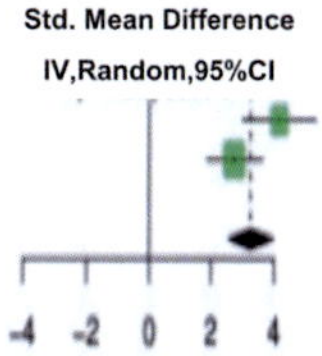

Fig. 11.21 Meta-analysis for thymus/body ratio

Study		Experimental Mean	Experimental SD	Experimental Total	Control Mean	Control SD	Control Total	Weight	Std. Mean Difference IV,Random,95%CI	Std. Mean Difference IV,Random,95%CI
15473	subgroup	145280.00	9370.0000	10	151210.00	12010.0000	10	21.4%	-0.53[-1.42; 0.37]	
15489	subgroup	22730000.00	1190000.0000	10	26480000.00	4910000.0000	10	19.3%	-1.01[-1.95; -0.06]	
15471	subgroup	83506000.00	10484000.0000	8	116572000.00	17977000.0000	8	10.2%	-2.12[-3.42; -0.83]	
11152	subgroup	132800000.00	12040000.0000	8	158470000.00	13920000.0000	8	11.3%	-1.86[-3.10; -0.63]	
11155	subgroup	0.15	0.0509	20	0.02	0.0047	20	16.1%	3.58[2.55; 4.61]	
11155	subgroup	0.06	0.0202	20	0.02	0.0047	20	21.6%	2.77[1.88; 3.66]	
Total(95%CI)				**76**			**76**	**100.00%**	**0.44[0.03; 0.85]**	

Heterogeneity: Tau²= 5.3776; Chi²= 103.82, df = 5 (P<0.01); I²= 95%

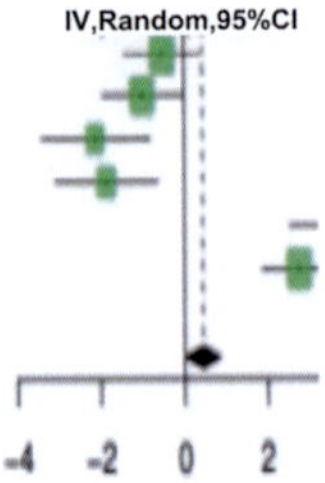

Fig. 11.22 Meta-analysis for SOD

Study		Experimental Mean	Experimental SD	Experimental Total	Control Mean	Control SD	Control Total	Weight	Std. Mean Difference IV,Fixed,95%CI	Std. Mean Difference IV,Fixed,95%CI
15278	subgroup	517.00	39.0000	8	342.00	22.0000	8	57.6%	5.23[2.92; 7.53]	
15474	subgroup	242260.00	19790.0000	8	129760.00	13580.0000	8	42.4%	6.27[3.58; 8.95]	
Total(95%CI)				**16**			**16**	**100.00%**	**5.67[3.92; 7.42]**	

Heterogeneity: Tau²= 0; Chi²= 0.33, df = 1 (P=0.56); I²= 0%

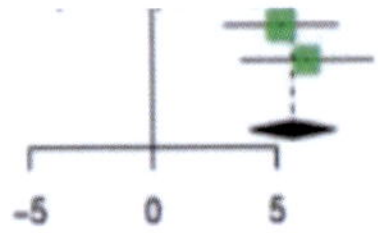

Fig. 11.23 Meta-analysis for CK

Study		Experimental Mean	Experimental SD	Experimental Total	Control Mean	Control SD	Control Total	Weight	Std. Mean Difference IV,Random,95%CI	Std. Mean Difference IV,Random,95%CI
15478	subgroup	207200.00	30100.0000	8	216600.00	45200.0000	8	44.5%	-0.23[-1.22; 0.75]	
15474	subgroup	65470000.00	7010000.0000	8	93780000.00	9010000.0000	8	15.9%	-3.32[-4.96; -1.67]	
15474	subgroup	34870000.00	7750000.0000	8	41230000.00	5460000.0000	8	39.6%	-0.90[-1.94; 0.15]	
Total(95%CI)				**24**			**24**	**100.00%**	**-0.99[-1.64; -0.33]**	

Heterogeneity: Tau²= 1.4439; Chi²= 9.98, df = 2 (P<0.01); I²= 80%

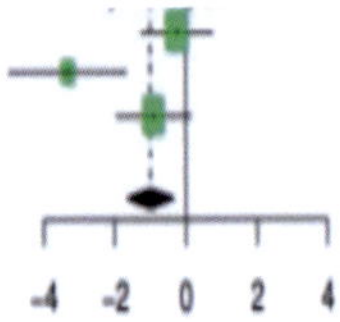

Fig. 11.24 Meta-analysis for GSH-PX

Two studies were selected for meta-analysis of MDA (Malondialdehyde) (Fig. 11.25). We firstly carried out heterogeneity analysis. The I^2 was larger than 50%. Consequently, we selected the random-effects model for downstream analysis. The SMD was 2.3269 (p-value < 0.0001). The 95% CI ranged from 1.4942 to 3.1595. In conclusion, the treatment with Lingzhi Spore significantly increased the MDA.

Two studies were selected for meta-analysis of SOD (Superoxide Dismutase) (Fig. 11.26). We firstly carried out heterogeneity analysis. The I^2 was less than 50%. Consequently, we selected the fixed-effects model for downstream analysis. The SMD was −2.4471 (p-value < 0.01). The 95% CI ranged from −3.2574 to −1.6369. In conclusion, the treatment with Lingzhi Spore significantly reduced the SOD.

(3) Summary

We analyzed 89 datasets for 57 outcome measures concerning the health-promoting functions of Lingzhi. Forty-nine outcome measures were related to the immunomodulatory effects, and 14 meta-analysis results showed significant differences between the control group and the experimental group. For the anti-oxidation meta-analysis, 18 outcome measures were analyzed, and 4 meta-analysis results showed significant differences between the control group and the experimental group. There were 19 outcome

Study		Experimental Mean	SD	Total	Control Mean	SD	Total	Weight	Std. Mean Difference IV,Random,95%CI
15478	subgroup	0.01	0.0007	8	0.01	0.0006	8	55.3%	1.36[0.24; 2.48]
15474	subgroup	17.26	1.2600	8	13.40	1.3000	8	30.7%	2.85[1.35; 4.35]
15474	subgroup	11.19	1.2900	8	6.01	0.4900	8	13.9%	5.02[2.79; 7.25]
Total(95%CI)				**24**			**24**	**100.00%**	**2.33[1.49; 3.16]**

Heterogeneity: Tau^2= 2.1625; Chi^2= 8.95, df = 2 (P=0.01); I^2= 78%

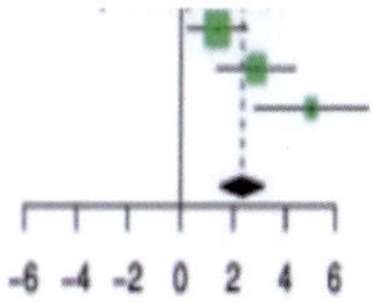

Fig. 11.25 Meta-analysis for MDA

Study		Experimental Mean	SD	Total	Control Mean	SD	Total	Weight	Std. Mean Difference IV,Fixed,95%CI
15478	subgroup	9789.97	1041.0500	8	19134.85	4777.1100	8	32.8%	-2.56[-3.97; -1.14]
15474	subgroup	132800000.00	12040000.0000	8	158470000.00	158470000.00	8	43.3%	-1.86[-3.10; -0.63]
15474	subgroup	88500000.00	9210000.0000	8	128770000.00	128770000.00	8	23.8%	-3.36[-5.02; -1.70]
Total(95%CI)				**24**			**24**	**100.00%**	**-2.45[-3.26; -1.64]**

Heterogeneity: Tau^2= 0.0099; Chi^2= 2.04, df = 2 (P=0.36); I^2= 2%

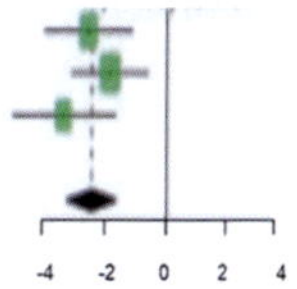

Fig. 11.26 Meta-analysis for SOD

measures related to anti-fatigue, 4 meta-analysis results showed significant differences between the control group and the experimental group. The descriptive statistics of the meta-analysis were listed in Table 11.3.

11.5 Meta-Analysis on the Safety of Lingzhi

There are relatively few studies focused on the safety of products containing Lingzhi. Wicks and Tong investigated the safety and tolerability of 16 healthy participants with a daily intake of 2 g of Lingzhi extract for 10 consecutive days (Wicks et al. 2007). No clinical adverse effects were observed. And no significant changes were found in the blood for proteins including CD4, CD8, CD9, and CD56. Wicks and Tong concluded that Lingzhi extract was well-tolerated without obvious adverse effects. Based on the American Herbal Pharmacopoeia, Lingzhi is a class 1 herb. It can be consumed safely in healthy individuals when used appropriately. However, there have been some reports that sensitive individuals may experience skin rashes or mild gastrointestinal discomfort such as diarrhea after consuming Lingzhi.

11.6 Concluding Remarks

Lingzhi has been used extensively to treat various human diseases, such as hepatitis, hypertension, hypercholesterolemia, and cancer (Yun 1999; Liu et al. 2007; Sliva et al. 2012). Two systematic reviews and meta-analyses on Lingzhi have been reported. Both of them focused on improving cancer-related fatigue (Jin et al. 2012; Zhao et al. 2012). Here, we selected 89 datasets from 57 studies for meta-analysis on the health-promoting effects of Lingzhi related to immunomodulation, anti-oxidation, and anti-fatigue. The results strongly support the notion that Lingzhi has beneficial immunomodulatory effects. A pipeline was developed to facilitate

Table 11.3 Descriptive statistics of meta-analysis results

Function	Outcome measure	Sample size	I^2 (%)	*p*-value
Immunomodulation	NK-cell activity	28	76	0.0001
Immunomodulation	OD value increase of ConA	2	0	0.58
Immunomodulation	Bodyweight	23	54	0.0001
Immunomodulation	Half hemolysis number	2	51	0.15
Immunomodulation	Phagocytic index	24	67	0.0001
Immunomodulation	Phagocytic index a	30	73	0.0001
Immunomodulation	Phagocytic rate	23	69	0.0001
Immunomodulation	Bodyweight	17	33	0.02
Immunomodulation	Antibody-Producing cells	2	0	0.89
Immunomodulation	Lymphocyte Transformation rate	1	0	0.37
Immunomodulation	Number of Hemolytic plaques	15	72	0.0001
Immunomodulation	Auricle swelling	14	85	0.0001
Immunomodulation	Spleen index	4	6	0.0364
Immunomodulation	Toe swelling	14	72	0.0001
Anti-oxidation	GSH-PX	2	0	0.55
Anti-oxidation	MDA	5	95	0.0001
Anti-oxidation	Thymus/body ratio	1	73	0.05
Anti-oxidation	SOD	5	95	0.0001
Anti-fatigue	CK	2	0	0.56
Anti-fatigue	GSH-PX	2	80	0.0001
Anti-fatigue	MDA	2	78	0.01
Anti-fatigue	SOD	2	2	0.36

high-throughput meta-analysis. The results will help the regulatory agencies for new policies regarding the functional food administration. Furthermore, this information can help the healthcare providers to choose Lingzhi products for particular health-promoting goal.

References

Cochran W (1937) Problems arising in the analysis of a series of similar experiments. J Royal Stat Soc B 4:102–118

Cooper HM (2017) Research synthesis and meta-analysis. Sage Publications

Deeks JJ, H. J., Altman DG, (editors). (March 2011). “Chapter 9: analysing data and undertaking meta-analyses. In: Higgins JPT, Green S, (eds.) Cochrane handbook for systematic reviews of interventions. Version 5.1.0.The Cochrane Collaboration, 2011.” 2020, from www.cochrane-handbook.org

DerSimonian R, Laird N (1986) Meta-analysis in clinical trials. Control Clin Trials 7:177–188

Fisher R (1932) Statistical methods for research workers. Oliver and Boyd, LondonFisher R.A. (1992) Statistical Methods for Research Workers. In: Kotz S., Johnson N.L. (eds) Breakthroughs in Statistics. Springer Series in Statistics (Perspectives in Statistics). Springer

Glass GV (1976) Primary, secondary, and meta-analysis of research. Educ Res 5(10):3–8

Glass GV, Mcgaw B, Smith ML (1981) Meta-analysis in social research. Sage Publications

Hedges, Larry V (1982) Estimation of effect size from a series of independent experiments. Psychol Bull. 92 (2): 490–499

Hedges LV (1983) Combining independent estimators in research synthesis. Br J Math Stat Psychol 36(1): 123-131

Hedges LV, Olkin I (1985) Statistical methods for meta-analysis. New Directions for Program Evaluation. 1984(24): 25–42

Higgins J, Thompson S, Deeks J, Altman D (2003) Measuring inconsistency in meta-analyses. BMJ 327: 557–600

Higgins JPT, Green S, Cochrane Collaboration (2008) Cochrane handbook for systematic reviews of interventions. Wiley-Blackwel

Hunter JE, Schmidt FL, Jackson GB (1982) Meta-analysis: cumulating research findings across studies. Sage Publications

Jin X, Beguerie J, Sze D, Chan G (2012) Ganoderma lucidum (Reishi mushroom) for cancer treatment. Cochrane Database Syst Rev 6(6): CD007731

Lewis S, Clarke M (2001) Forest plots: trying to see the wood and the trees. BMJ 322(7300): 1479–1480

Light R (1983) Evaluation studies review annual, vol. 8. Sage Publications

Light RJ, Pillemer DB (1984) Summing up: the science of reviewing research. Harvard University Press

Liu J, Shimizu K, Konishi F, Sato M, Noda K et al (2007) Anti-androgenic activities of the triterpenoids fraction of Ganoderma lucidum. Food Chem 100:1691–1696

Pearson E (1938) The probability integral transformation for testing goodness of fit and combining independent tests of significance. Biometrika 30:134–148

Pearson K (1904) Report on certain enteric fever inoculation statistics. BMJ 2(2288):1243–1246

Rosenthal R (1978) Combining results of independent studies. Psychol Bull 85:185–193

Schwarzer G (2007) meta: An R package for meta-analysis. R News. From https:// cran.r-project.org/doc/Rnews/Rnews_2007-3.pdf

Schwarzer G, Carpenter JR, Rücker G (2007) Meta-Analysis with R (Use-R!). Springer International Publishing, Switzerland. From http://www.springer.com/gp/book/9783319214153

Sliva D, Loganathan J, Jiang J, Jedinak A, Lamb JG et al (2012) Mushroom ganoderma lucidum prevents colitis-associated carcinogenesis in mice. PLoS ONE 7:e47873

Smith ML, Glass GV (1977) Meta-analysis of psychotherapy outcome studies. Am Psychol 32: 752–760

Tippett LHC (1931) The methods of statistics, William & Norgate Ltd

Wicks SM, Tong R, Wang CZ, O'Connor M, Yuan CS (2007) Safety and tolerability of Ganoderma lucidum in healthy subjects: a double-blind randomized placebo-controlled trial. Am J Chin Med 35(3): 407–414

Wolfe R, Abramson M, Mckenzie JE, Beller EM, Forbes AB (2016) Introduction to systematic reviews and meta-analysis. Respirology 21(4):626–637

Yun TK (1999) Update from Asia. Asian studies on cancer chemoprevention. Ann N Y Acad Sci 889: 157–192

Zhao H, Zhang Q, Zhao L, et al (2012) Spore powder of Ganoderma lucidum improves cancer-related fatigue in breast cancer patients undergoing endocrine therapy: a pilot clinical trial. Evid Based Complement Alternat Med. 2012:809614

Survey of Lingzhi Health Foods and Drugs

12

Liqiang Wang

Abstract

Lingzhi, a top-grade medicinal material used in traditional medicine, has been used for thousands of years to treat various diseases, particularly in China. In recent years, different tissue parts of Lingzhi have been developed into approximately 180 drugs and 1270 health foods in China. This chapter analyzed these drugs and health foods in terms of their source tissues, dosage forms, major chemical components, maker compounds, health effects, national management policies, etc. We discuss the deficiencies of Lingzhi in the development of health foods and drugs and attempt to identify the trend of health foods and drugs. The results can guide the development of health foods and drugs containing medicinal materials like Lingzhi.

Electronic supplementary material The online version of this chapter (https://doi.org/10.1007/978-3-030-75710-6_12) contains supplementary material, which is available to authorized users.

L. Wang (✉)
College of Pharmacy, Heze University, 274015
Heze, Shandong Province, People's Republic of China
e-mail: lys832000@163.com

12.1 Introduction

The Lingzhi, a medicinal material used in traditional medicine, was revered in ancient China and hailed as the "mushroom of immortality" and the "medicine of kings" (Babu and Subhasree 2008). The earliest written records of Lingzhi can be traced back to Shan Hai Jing ("Mountains and Seas"). And six medicinal qualities of Lingzhi are recorded in Shen Nong Ben Cao Jing (100–200 A.D), which is the earliest pharmaceutical monograph in China. Since ancient times, Lingzhi has been seen as a precious raw material for nourishing and strengthening the body (Yuan et al. 2018). This superb tonic offers many extraordinary health-promoting benefits, including potent immune-enhancing properties, stress relief, emotional equilibrium, elevates mood and happiness, increases intellectual capacity and memory, promotes agility, lengthens the life span, and supports the major organ systems: the heart, kidneys, liver, and lungs (Babu and Subhasree 2008).

The source organisms of Lingzhi are fungi from the basidiomycete, belonging to the family of Polyporaceae. In the Pharmacopoeia of the People's Republic of China (2020 edition), red Lingzhi is considered the most superior variety of Lingzhi. Red Lingzhi is the primary raw material used to make drugs and health foods. Its scientific name is *Ganoderma lucidum* (Leyss. ex Fr.) Karst. Purple Lingzhi is the second most

C. Liu (ed.), *The Lingzhi Mushroom Genome*, Compendium of Plant Genomes,
https://doi.org/10.1007/978-3-030-75710-6_12

widely used Lingzhi species, and its scientific name is *G. sinenes* Zhao, Xu et Zhang. Besides, several other species, like *G. tsugae* Murr., are also used as medicinal material.

Traditionally, the fruit body of the *Ganoderma* fungi is the most used tissue. However, with the development of modern technology, additional tissues are used for drugs and health food. These include mycelium (Fig. 12.1A–C), fruit body (Fig. 12.1D–F), spore powder (Fig. 12.1G–I), and spore oil (Fig. 12.1J–L). Modern research has identified the main active components of Lingzhi as polysaccharides and triterpenoids (Zhou et al. 2018; Zhao and Chang 2007). Various pharmacological effects have been identified and studied (Sanodiya et al. 2009; Boh et al. 2007). In practice, the polysaccharides and triterpenoids are usually used as the marker components for the quality of related drugs and health foods.

Studies on Lingzhi range from new cultivar breeding, large-scale standardized plantation, processing, manufacturing, to quality control (Li et al. 2019a). Lingzhi has been continuously developed into various products, including herbal drugs and health foods, cosmetics, ornaments, and agricultural products. Since 1996, more than 1270 Lingzhi health food and 180 Lingzhi drugs have been approved and reached the market. These products are packaged medicinal materials in various dosage forms, including hard capsule, soft capsule, granule, tea preparation, oral agents, liquor, tablet, mixture, syrups, injection, drop pill, etc. (Chen et al. 2016). However, to our knowledge, there has been no general introduction on Lingzhi health foods and drugs. As the last chapter of this book, we reviewed this subject to give the reader a clear message about the future applications of Lingzhi.

This chapter analyzes Lingzhi health foods and Lingzhi drugs in terms of their source tissues, dosage forms, major chemical components, maker compounds, health effects, national management policies, etc. We describe the trends of Lingzhi products' development and identify potential problems. Lastly, using Lingzhi as an example, we briefly discuss the current development landscape for herbal drugs and health foods.

12.2 Data and Methods

12.2.1 Data Source

The data of the approved Lingzhi health foods were retrieved from the Health Food Registration Management Information System Database, Center for Food Evaluation, State Administration for Market Regulation (http://tsspxx.gsxt.gov.cn/gcbjp/tsspindex.xhtml). The registration time was set to be before December 31st, 2018. To avoid the duplicated data, only health foods declared for the first time were counted, and the health food that were re-registered were not considered. Finally, information for 1270 approved health foods was collected and used for analyses. Among them, information for 538 health foods approved after 2010 is shown in Table 12.1. The data of the approved Lingzhi drugs were retrieved from the National Medical Products Administration database (http://app1.nmpa.gov.cn/data_nmpa/face3/dir.html). The registration time was set to be before December 31st, 2020. And 180 approved drug information was collected (Table 12.2).

12.2.2 Data Analysis

The names of raw materials of the collected health foods and drugs were standardized as follows. Firstly, the materials named "Lingzhi", "Lingzhi extraction", "Lingzhi powder", and "Lingzhi fruit body" were all renamed "Lingzhi fruit body" uniformly. Secondly, the materials named "Lingzhi spore powder" and "wall-breaking spore powder" were renamed "Lingzhi spore powder" uniformly. According to the health foods and drugs' formula and process characteristics, the accessory materials were removed. Subsequently, the health foods and drugs were counted according to the approval years, the tissue parts, the claimed product's

Fig. 12.1 Lingzhi materials used in health food and drugs. The mycelium (**A–C**), fruit body (**D–F**), spore powder (**G–I**), and spore oil (**J–L**) of Lingzhi are shown. **A** Preservation of Lingzhi strains in the form of myce lium (http://img.yulucn.com/upload/f/2c/f2ce7817141bb 4dc60cfbc66c4b5e9e2.jpg). **B** Expanded mycelium in a Petri dish (http://img.xjishu.com/img/zl/2017/10/261059 54926335.gif). **C** Mycelium cultured in liquid medium for health food production. **D** Lingzhi cultured in bags with sawdust, wheat bran, etc. as a medium (http://news.cnhubei.com/xw/jj/201211/W020121104185436615726.jpg). **E** Lingzhi cultured in short wood segments suitable for the growth of Lingzhi (http://www.371zy.com/uploads/allimg/170825/1-1FR5104J6404.jpg). **F** Lingzhi in the wild (http://cbu01.alicdn.com/img/ibank/2017/096/647/4416746690_260384149.jpg). **G** Lingzhi spore powder. **H** The Lingzhi spore powder that can be consumed directly (http://5b0988e595225.cdn.sohucs.com/images/201 71220/8370acdd99354f0a85182fef2d01ddf6.jpeg). **I** The Lingzhi spore powder brewed with warm water (http://img.bj003.com/AreaSoft/26120/1/21.jpg).

Table 12.1 Statistics of approved single and multiple composition health food and drug products

Materials	Type of products	Proportion of single composition (%)	Proportion of multiple compositions (%)
Fruit body	Health foods	6.8	93.2
	Drugs	67.1	32.9
Fruit body extracts	Health foods	3.6	96.4
	Drugs	12.5	87.5
Spore oil	Health foods	80.8	19.2
	Drugs	–	–
Spore powder	Health foods	49.7	50.3
	Drugs	100	0
Spore powder extracts	Health foods	25.0	75.0
	Drugs	–	–
Mycelium	Health foods	28.6	71.4
	Drugs	100	0
Mycelium extracts	Health foods	20.0	80.0
	Drugs	100	0

Note –, no data

Table 12.2 The marker components in the health foods with different parts of Lingzhi as raw materials (Sa 2020)

Raw materials	Crude polysaccharides (%)	Triterpenoids (%)
Fruit body	80	10
Spore powder	84	50
Spore oil	0	100
Mycelium	100	0

health benefits, the marker ingredients, the dosage form of products, and the composition of the product multiple compositions. The histogram and scatter plots were drawn using Excel. The pie chart and network diagram were drawn using SPSS Modeler (v.18.0, IBM, New York).

12.3 Results

12.3.1 Number Increase of Health Foods and Drugs by Time

Since the approval of the first Lingzhi health foods in 1996, the heat of Lingzhi as health food raw materials has not decreased (Fig. 12.2A). The health foods with edible parts of Lingzhi as raw materials account for about 8% of all approved health foods. In 2002, all drugs approved by local governments before 2002 were put under the CFDA administration. So there is a temporary peak in the number of approved drugs in 2002. Up to now, the approved number of Lingzhi drugs accounts for about 1% of all approved drugs (Fig. 12.2B).

12.3.2 Statistics of the Approved Lingzhi Health Foods and Drugs

Common raw materials are usually from the fruit body extracts, fruit body fine powder, mycelia, spores, and spore oil. As shown in Fig. 12.3A, the health foods with Lingzhi fruit body (including extracts) as raw materials are the most

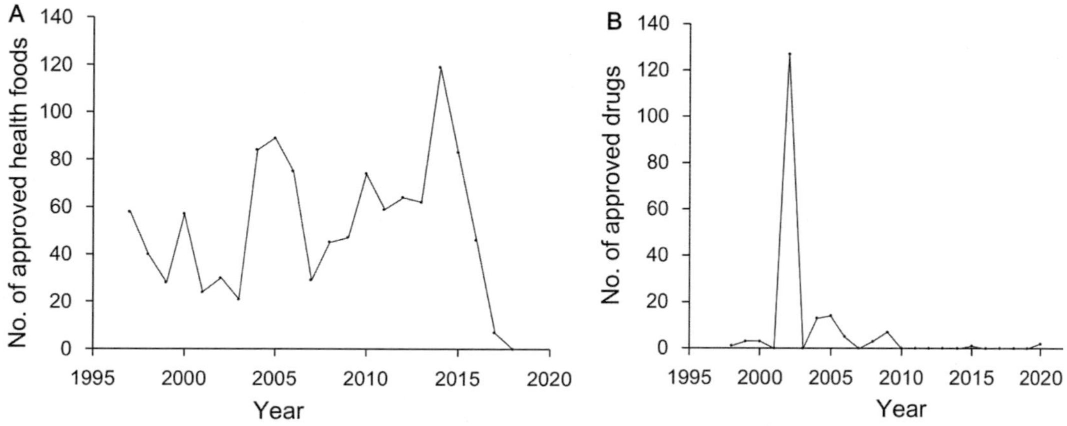

Fig. 12.2 The number of approved Lingzhi health foods (**A**) and drugs (**B**) over the years

common, which is the same as the traditional medicinal parts of Lingzhi (Fig. 12.3B). The frequency of approved health foods with Lingzhi mycelium as raw materials is the lowest. In the Pharmacopoeia of the People's Republic of China (2020 edition), the available parts of Lingzhi that are used as raw materials of drugs are mainly from the fruit body. In contrast, few drugs use spore powder and mycelium as raw materials. And no drugs use the spore oil as its raw material (Fig. 12.3B).

12.3.2.1 Tissue Parts Used in Lingzhi Single and Multiple Composition Health Foods and Drugs

The statistics of the single and multiple composition health foods and drugs are shown in Table 12.1. Lingzhi fruit body, including its extracts used as medicinal materials, is used in more than 90% of multiple composition health food products. The extracted Lingzhi spore oil with definite ingredients is usually used in single

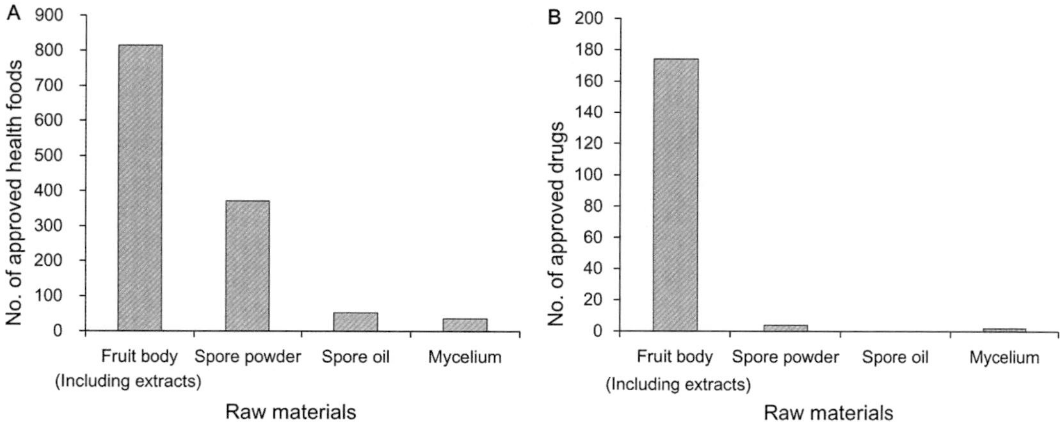

Fig. 12.3 Approved numbers of health foods (**A**) and drugs (**B**) with different tissue parts of Lingzhi as raw materials

composition products. More recently, Lingzhi spore powder and mycelium are generally used in combination with other medicinal materials. In contrast, two-thirds of the Lingzhi drugs containing Lingzhi fruit body are single composition products. The numbers of the drugs with fruit body extracts, spore oil, spore powder, spore powder extracts, mycelium, and mycelium extracts as raw materials are too small to draw any statistically significant conclusion.

12.3.2.2 Health Effects of Lingzhi Health Foods and Drugs

To guide the usage of health foods in the market, CFDA established a catalog of health effects that the health foods can claim to have. The catalog contains 27 health effects, including enhancing immune systems, sleep improvement, alleviating physical fatigue, enhancing anoxia endurance, irradiation hazard protection function, increasing bone density, assisting liver protection against injury, alleviating eye fatigue, eliminating acne, eliminating skin pigmentation, improving skin ability to retain moisture, improving skin oil content function, assisting blood lipids reduction, assisting blood sugar reduction, anti-oxidative function, assisting memory improvement, alleviating lead excretion, improving throat function, assisting blood pressure reduction, facilitating milk secretion, assisting weight control, improving child growth, improving nutritional anemia, regulating gastrointestinal flora, facilitating digestion, facilitating bowel movement, and protection of gastric mucosa (http://sfdachina.com/info/86-1.htm). The health foods can only claim to have health effects listed in the catalog.

Lingzhi is the most commonly used medicinal materials in all registered health foods. Its approved health foods are claimed to involved 20 of the 27 health effects, which are enhancing immune systems (Kuo et al. 2006; Cao et al. 2018), alleviating physical fatigue (Luo et al. 2014; Chen et al. 2003), assisting blood lipids reduction (Yi and Xu 2001; Tang et al. 2003), anti-oxidation function (Sun and Liu 2017; Vu et al. 2019; Yang et al. 2010a), assisting blood sugar reduction (Liu et al. 2007; Ma et al. 2015; Zhang et al. 1999), assisting memory improvement (Zhang et al. 2009), assisting blood pressure reduction (Zhang et al. 1999), sleep improvement (Sun et al. 2019; Yu et al. 2019), enhancing anoxia endurance (Mao and You 2015; Lan and Xu 2009), irradiation hazard protection function (Kunmu et al. 2020; Ding et al. 2014; Miao et al. 2019), increasing bone density (Li et al. 2008), assisting weight control (Ning et al. 2004), improving child growth (Xia et al. 2014), assisting liver protection against injury (Sun and Zhang 2020; Chen et al. 2019; Chen and Chen 2017), alleviating eye fatigue (Guo et al. 2008; Deng et al. 2009), eliminating skin pigmentation (Zhang et al. 2002), improving skin ability to retain moisture (Jiang et al. 2016; Huang et al. 2015), regulating gastrointestinal flora (Yang et al. 2020; Zhang et al. 2008), facilitating bowel movement (Feng et al. 2008), and protection of gastric mucosa (Yang et al. 2009, 2010b).

Among the approved health food products, 66.8% of them declared the health effect of enhancing immune systems, and 29% claimed two health effects at the same time. Among the products with Lingzhi spore powder as raw materials, 91% of such products declared the Health effect of enhancing immune systems, and 24% of such products claimed two Health effects. Among the products with Lingzhi spore oil as raw materials, 92% declared the Health effect of enhancing immune systems, and 10% claimed two Health effects. Among the products with Lingzhi mycelium as raw material, 68% reported the Health effect of enhancing immune systems, and 43% of the products declared two health effects. All Lingzhi health foods claim to enhance immune systems as their primary health effect (Fig. 12.4).

In the Pharmacopoeia of the People's Republic of China (2020 edition), the health effects of Lingzhi are invigorating "Chi" and soothing the nerves, relieving cough, and relieving asthma. The indication of Lingzhi is suitable for restlessness, insomnia, palpitations, cough, and asthma due to lung deficiency, fatigue, shortness of breath, and loss of appetite. For the drugs with any tissue parts of Lingzhi as raw materials, they usually contain multiple compositions. So, the pharmacological effects of drugs usually differ

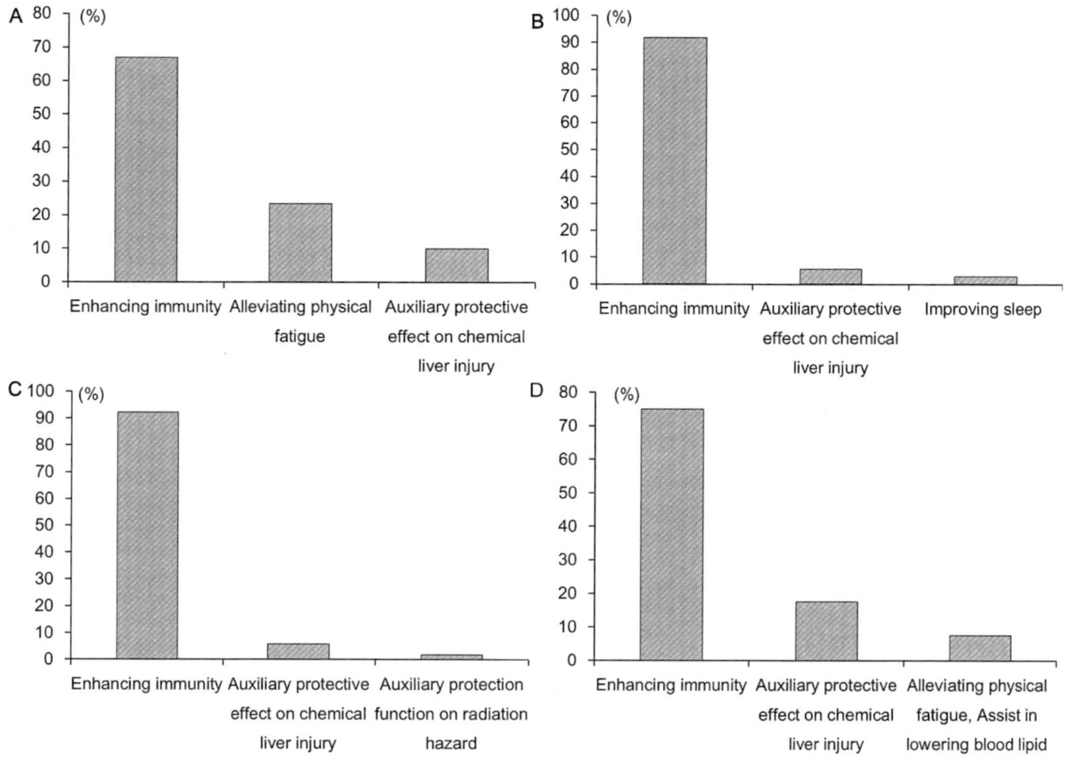

Fig. 12.4 The proportion of the health foods with different health effects. The health foods shown in panels (**A**), (**B**), (**C**), and (**D**) were made of fruit body, spore powder, spore oil, and mycelium of Lingzhi, respectively.

with different compositions (Table 12.2). In conclusion, the pharmacological effects of Lingzhi drugs mainly involve immune regulation, cardiovascular disease treatment, gastrointestinal disease treatment, tumor treatment, anti-inflammatory, hepatitis treatment, calming the mind and nerves, treating insomnia and forgetfulness, nourishing liver and kidney, nourishing "Chi" and spleen, nourishing "Chi" and blood, promoting blood circulation and relieving pain, relieving cough and resolving phlegm, etc.

12.3.2.3 Marker Components of Lingzhi Health Foods and Drugs

In single composition health foods with different tissue parts of Lingzhi as raw materials, the marker components are different (Table 12.2). Lingzhi health foods' marker components are mostly crude polysaccharides and triterpenoids, which are the bioactive ingredients in the health foods. The content of marker components is an important index to control the product quality. Its content in health foods is generally between 1/3 and 1/2 of the drug usage amount specified in the Pharmacopoeia of the People's Republic of China (2020 edition). For the drug, using anhydrous glucose as the indicator of polysaccharides, the amount of glucose should be more than 0.90% of total dry products by weight. Similarly, using oleanolic acid as an indicator of triterpenoids, the amount of oleanolic acid should be more than 0.50% of the dry products by weight.

12.3.2.4 Types of the Dosage Forms

Hard capsules are used in most health foods (51%), followed by soft capsules (8%) and tablets (8%). Lingzhi spore powder is used in 72% of powder products; Lingzhi fruit bodies are used in 95% of liquor products and 98% of tea products. These dosage forms are also described in the Pharmacopoeia of the People's Republic of

China (2020 edition). The distribution of Lingzhi drugs' dosage forms is consistent with those of Lingzhi's health food products (Fig. 12.5). For the drugs, the most used dosage forms are also hard capsules (45%), followed by tablets (25%) and granules (10%). Compared with health foods, the drugs have 12 dosage forms. The diversity of dosage forms is conducive to the usage of drugs for different types of patients, improving the efficiency of drug usage.

12.3.3 Usage Frequency of Raw Materials in the Approved Lingzhi Health Foods and Drugs

Traditionally, the source tissue of Lingzhi used for health foods and drugs is the fruit body. Other raw materials used to make health foods are also commonly used medicinal materials. The top ten and top eight ingredients combined with Lingzhi in the health foods and drugs are shown in Tables 12.3 and 12.4, respectively. The medicinal materials frequently emerged in health foods' formula can by themselves enhance immune systems. The properties of Lingzhi are sweet and neutral. Most of its combined ingredients in health foods have a sweet taste. The book "Materia Medica Preparation General Meaning of Medicinal Properties" describes that "the medicinal materials with sweet taste can tonify and slow down". Therefore, this kind of health food is mainly tonic and moderating. The medicinal materials combined with Lingzhi used in the drugs are different from those used in the health foods. This may result from the different goals of health foods and drugs. The drugs are expected to treat disease, and health foods are not.

12.3.4 Analysis of the Combination of Lingzhi and Other Medicinal Materials

The 760 approved multiple composition health foods collected from the 1270 approved health foods are analyzed to explain the relevance of the raw materials used in combination with Lingzhi. The support degree of the SPSS Modeler software was set to 30. The support degree indicates the number of one material that appeared in all health foods. A connection network is shown in Fig. 12.6A for each raw material in the 760 approved multiple composition health foods. The thickness of lines is proportional to the correlation frequency. The top five raw materials most commonly used with Lingzhi are wolfberry, American ginseng, Astragalus, Ginseng, and Lingzhi spore powder (Fig. 12.6A). Here Lingzhi spore powder is considered a medicinal material that is different from Lingzhi.

Similarly, the raw materials commonly used in combination with Lingzhi in drugs were also analyzed. A total of 70 approved multiple

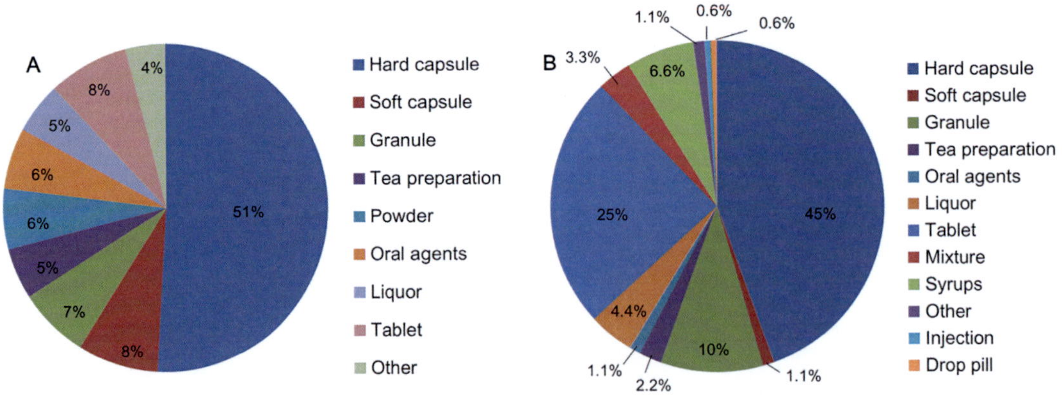

Fig. 12.5 Dosage form distribution of Lingzhi health foods (**A**) and drugs (**B**)

Table 12.3 Top ten raw materials most commonly used in combination with Lingzhi in health foods

No.	Raw materials		Source plant of raw materials	No. of health foods
	Common name	Latin name		
1	Barbary wolfberry fruit	Lycii Fructus	*Lycium barbarum* L.	182
2	American ginseng	Panacis Quinquefolii Radix	*Panax quinquefolium* L.	139
3	Milkvetch root	Astragali Radix	*Astragalus membranaceus* (Fisch.) Bge. var. *mongholicus* (Bge.) Hsiao; *Astragalus membranaceus* (Fisch.) Bge	101
4	Ginseng	Ginseng Radix et Rhizoma	*Panax ginseng* C.A.Mey	95
5	Lingzhi spore powder	Ganoderma Spore Powder	*Ganoderma lucidum* (Leyss. ex Fr.) Karst.; *Ganoderma sinense* Zhao, Xu et Zhang	78
6	Poria	Poria	*Poria cocos* (Schw.) Wolf	68
7	Zizyphus Vulgaris	Ziziphispinosaesemen	*Ziziphus jujube* Mill. var. *Spinosa* (Bunge) Huex H. F. Chou	62
8	Polygonatum	Polygonati Rhizoma	*Polygonatum kingianum* Coll. et Hemsl.; *Polygonatum sibiricum* Red.; *Polygonatum cyrtonema* Hua	54
9	Rhodiola root	Rhodiola Crenulatae Radix et Rhizoma	*Rhodiola crenulata* (Hook. f. et Thoms.) H. Ohba	45
10	Chinese magnolia vine fruit	Schisandrae Chinensis Fructus	*Schisandra chinensis* (Turcz.) Baill	37

composition drugs are collected from the 180 approved drugs. Because the number of drugs is less than that of health foods, the support degree of SPSS Modeler was set to 10. The thickness of lines in Fig. 12.6B is also proportional to the correlation frequency. The top two raw materials commonly used combined with Lingzhi in drugs are Chinese magnolia vine fruit and Curcuma root (Fig. 12.6B).

12.4 Discussions

12.4.1 Current Status and Deficiencies of Lingzhi Health Foods and Drugs

In the approved health foods, the polysaccharide of Lingzhi fruit body is usually prepared using multiple-time hot water extraction. This method supposes to have a high extraction rate of Lingzhi polysaccharides. The content of water-soluble crude polysaccharides is relatively stable, although some chemical components and contents differ because of the different origin of Lingzhi (Yan et al. 2019). The quality of most approved health foods prepared with Lingzhi fruit body was usually measured using "polysaccharide" or "triterpenoids and sterols" as the marker components. The detection methods of the approved health foods were spectrophotometry. The drawback of this method is its low specificity, and the measured content of polysaccharide is also easily affected by other auxiliary materials. Therefore, detection indexes with more specificity need to be developed further.

The approved health foods prepared with Lingzhi spore powder also contained

Table 12.4 Top eight raw materials most commonly in combination with Lingzhi in drugs

No.	Raw materials		Source species of raw materials	No. of drugs
	Common name	Latin name		
1	Chinese magnoliavine fruit	Schisandrae Chinensis Fructus	*Schisandra chinensis* (Turcz.) Baill.	30
2	Curcuma root	Curcumae Radix	*Curcuma wenyujin* Y. H. Chen et C. Ling *Curcuma Longa* L. *Curcuma kwangsiensis* S. G. Lee et C. F. Liang *Curcuma phaeocaulis* Valeton	29
3	Bupleurum root	Bupleuri Radix	*Bupleurum chinense* DC. *Bupleurum scorzonerifolium* Willd.	20
4	Tangerine peel	Citri Reticulate Pericarpium	*Citrus reticulata* Blanco	16
5	Prepared fleece flower root	Polygoni Multiflori Radix Praeparata	*Polygonum multiflorum* Thunb.	15
6	Ginseng	Ginseng Radix et Rhizoma	*Panax ginseng* C. A. Mey	10
7	Cyperus root	Cyperi Rhizoma	*Cyperus rotundus* L.	10
8	Margarita	Margarita	*Pteria martensii* (Dunker) *Hyriopsis cumingii* (Lea 1852) *Cristaria plicata* (Leach 1814)	10

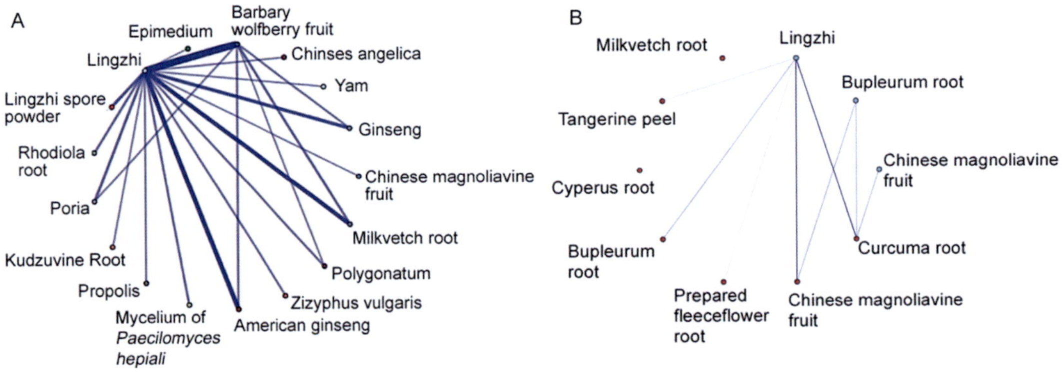

Fig. 12.6 Raw material interaction networks of Lingzhi health foods (**A**) (Sa 2020) and drugs (**B**). The thickness of lines in the network is proportional to the correlation frequency of each pair of raw materials in the multiple composition health foods and drugs. The correlation frequency is the appearing number of each pair of raw materials in the multiple composition health foods and drugs. The line is not shown if its corresponding correlation frequency is less than 30 in panel A and less than 10 in panel B

polysaccharides and triterpenoids. The Lingzhi spore is usually undergone the wall-broken process. More active components in the spore are released. The contents of triterpenoids, water-soluble polysaccharides (Yang and Zhu 2010), and crude fat (Zhao et al. 2011) in wall-broken Lingzhi spore powder were significantly higher than those in non-broken Lingzhi spore powder. More than 95% of the products directly use the wall-broken Lingzhi spore powder. However, if the wall is broken, many unsaturated fatty acids will oxidize and become rancid. Therefore, to ensure the quality of health foods, there are generally higher requirements for the inner packaging materials that directly contact the raw materials. The wall-broken Lingzhi spores are not processed further in most of the approved health foods.

The primary function of the approved Lingzhi spore oil and mycelium health foods is to enhance immune systems, consistent with the results in existing literature (Yi et al. 2012; Chen and Chen 2016). For the health foods made from Lingzhi spore oil, the marker component is the total Lingzhi triterpene contents. Since 2006, Lingzhi spore oil has been approved as the raw material of health foods. It contains many active components, such as triterpenoid, Lingzhi acid, and unsaturated fatty acid. However, Lingzhi spore oil has almost no antioxidant components, which can be oxidized when exposed to the air. It is easy to crystallize at low temperatures. The acid value and peroxide value will change significantly after long-term storage (Zeng et al. 2014).

In our view, several deficiencies existed in the Lingzhi health foods and drugs, which affected the products' further development. First of all, the dosage forms of health foods are too limited. The capsule and tablet are the most commonly used dosage form; they are also the typical forms of drugs and may confuse the consumers as drugs, while they should be consumed as food. Also, the health benefits of health foods are too limited. The health foods claimed that "enhancing immune system" are more than 50% of the total number of the approved health foods. The CFDA has only approved 27 functions for health foods.

12.4.2 The Future Development of Medicinal Materials

12.4.2.1 Catalogue of Exempt Raw Material for Health Foods

In August 2019, the State Administration for Market Regulation issued the "Health Food Raw Material Catalog and Health Effect Catalog Management Policies", which provided legal support for accelerating the management of the "Raw Material Catalog of Health Foods" (abbreviated as the "Raw Material Catalog") and the "Health Effect Catalog of Health Foods" (abbreviated as the "Health Effect Catalog"). The raw materials included in the "Raw Material Catalog"have well-defined health benefits, dosage, and suitable consumer population. Health food enterprises can produce and market products that meet the registration requirements instead of going through a lengthy approval process. In 2018, the Center for Food Evaluation of the State Administration for Market Regulation released in the "Raw Material Catalog", including wall-broken Lingzhi spore powder (Liu et al. 2020). As a result, the products containing wall-broken Lingzhi spore powder can be developed and marketed quickly.

12.4.2.2 Development of New Health Effects Catalog

The Health effects of approved health foods are based on modern studies on nutrition, pharmacology, clinical medicine, etc. Functional evaluation tests are used to verify whether the health foods have the claimed health effects. The health functions included in the "Health Effect Catalog" clearly define that the health care functions should align with the TCM health care theory. This provision opens up a new path for the development of health care functions in the future. Taking Lingzhi as an example, the efficacy of Lingzhi recorded in the Pharmacopoeia of the People's Republic of China (2020 edition) is "invigorating Chi and calming nerves, relieving cough and asthma". It is possible to add "calming nerves, relieving cough and asthma" as additional health effects of Lingzhi in the "Health Effect Catalog".

12.5 Concluding Remarks

Lingzhi is one of the most well-known medicinal materials because of its apparent medical efficacy and the absence of unfavorable side effects. This chapter analyzed the characteristics of Lingzhi health foods and drugs and the relevant national management policies. The Lingzhi health foods and drugs on the market are made from fruit bodies, spores, spore oil, and mycelia through sophisticated or straightforward processes. Lingzhi has been made into final health foods or drugs either alone or in combination with other medicinal materials and in the forms of hard capsule, soft capsule, granule, tea preparation, oral agents, liquor, tablet, mixture, syrups, injection, drop pill, etc. (Chen et al. 2016). Overall, there is a great potential to develop new Lingzhi health foods and drugs worldwide.

References

Babu PD, Subhasree RS (2008) The sacred mushroom "Reishi"-a review. Am Eurasian J Bot 1(3):107–110

Boh B, Berovic M, Zhang J, Zhi-Bin L (2007) *Ganoderma lucidum* and its pharmaceutically active compounds. Biotechnol Annu Rev 13:265–301

Cao Y, Xu X, Liu S, Huang L, Gu J (2018) *Ganoderma*: a cancer immunotherapy review. Front Pharmacol 9:1217

Chen T, Chen R (2016) Experimental study on effects of immune function in mice by *Ganoderma ludidum* spore oil capsule. Anhui J Preventive Med 22(6):367–370

Chen Y, Chen Q (2017) Anti-inflammatory and hepatoprotective effects of *Ganoderma lucidum* polysaccharides on carbon tetrachloride-induced acute liver injury in mice. Food Sci 38(17):210–215

Chen J, Zheng Y, Lin X, Lin R, Cai H (2003) Effects of *Ganoderma-lucidum*-fermented-tea on hemolysin and cell immunity of the male mice. Fujian J Agric Sci 18 (2):127–128

Chen Z, Huang W, Jin X, Liu Z, Huang Y, Li P, Zheng L (2016) Research progress on *Ganoderma lucidum* intensive processing in China. J Food Saf Qual 7 (2):639–644

Chen D, Tang Y, Piao W, Wang O, Huang J, Wang L (2019) Protective effect of *Ganoderma* fruit body concentrated capsule on acute alcoholic liver injury in mice. Food Nutr China 25(2):54–58

Deng X, Lin X, Gao Y, Sun Q, Zhong Z (2009) Effect of oil in *Ganoderma lucidum* spores on retina photoreceptor damage of rats induced by N-methyl-N-nitrosourea. Chin Tradit Herbal Drugs 40(3):415–420

Ding Y, Zhou X, Cui L, Chen H, Zhang Y, Guo G, Sun R, Chen B (2014) Radioprotective effect of *Ganoderma lucidum* polysaccharides on irradiated mice. J Med Postgrad 27(11):1152–1155

Wu F, Zhao X, Xue H, Qiao J, Xu J (2008) The cathartic action of aloes *Ganoderma* capsule. Med J Commun 22(4):338–339

Guo S, Li Y, Zhang Y, Tang S, Zhang C, Cui X (2008) *Ganoderma lucidum* on eye drops radioactive protective role of rabbit study. Mod Chin Med 10(11):34–36

Huang S, Jiao C, Liang H, Yang Y, Chen J, Zhang M, Hong Y, Xie Y (2015) Study on the anti-aging skin efficacy of the active ingredient of *Ganoderma lucidum* water extract. J Anhui Agric Sci 6:27–29

Jiang N, Xu X, Wei W, Yu M, Luo X (2016) *Ganoderma lucidum* extract can resist skin tissue senility. Sichuan J Zool 35(4):585–587

Kunmu DLY, Fu J, Yu D, Yu C (2020) Thymic metabolomics for effect of *Ganoderma* polysaccharides on radiation-injured mice. Chin J Exp Tradit Med Formulae 26(3):102–109

Kuo MC, Weng CY, Ha CL, Wu MJ (2006) *Ganoderma lucidum* mycelia enhance innate immunity by activating NF-κB. J Ethnopharmacol 103(2):217–222

Lan Y, Xu Q (2009) Effects of the *Ganoderma* spores oil in Changbai mountain on hypoxia-tolerance of mice. J Med Sci Yanbian Univ 32(4):251–252

Liu D, Sun H, Li S (2007) Study on antihypertensive effect of triterpenes from *Ganoderma lucidum in vivo*. Lishizhen Med and Mater Med Res 18(2):307–309

Liu R, Chen Z, Hua G, Wang L, Sa Y, Liu C (2020) Construction and analysis of evaluation system for functional food crude materials based on AHP. Chin Tradit Herbal Drugs 51(18):14–21

Li Z, Guo J, Zeng Y, Shen Z, Zhan C, Li Y, He S, Li X, Zhong Z (2008) Effects of *Ganoderma lucidum* spores on endocrine function of ovariectomized rats. Chin J Clin Anat 26(4):419–422

Li Z, Zhou J, Lin Z (2019) Development and innovation of *Ganoderma* industry and products in China. Adv Exp Med Biol 1181:187–204

Li Q, Wang Q, Li L, Sun P, Zhang L, Zhao W, Wang Y, Li Z, Wang J (2019b) Application of TCM constitution identification in "preventive treatment of disease'. Acta Chin Med 34(255):1586–1589

Li H, He K, Zheng H, Li W, Zeng Y, Yang Y (2020) Health-care mechanism and medicine and food resources of invigorating spleen in traditional Chinese medicine. Chin Tradit Herbal Drugs 51(3):780–787

Luo L, Cai L, Hu X (2014) Evaluation of the anti-hypoxia and anti-fatigue effects of *Ganoderma lucidum* polysaccharides. Appl Mech Mater 522–524:303–306

Mao H, You Y (2015) Protective effect of *Ganoderma lucidum* polysaccharides peptide on cultured neonatal rat cardiomyocytes injured by hypoxia/reoxygenation. Chin J Pharmacol Toxicol 29(3):398–403

Ma HT, Hsieh JF, Chen ST (2015) Anti-diabetic effects of *Ganoderma lucidum*. Phytochemistry 114:109–113

Yu C, Dong W, Lian L, Fu J, Guo L, Yu D (2019) Experimental study on anti-ionizing radiation of *Ganoderma Lucidum* polysaccharide. J Liaoning Univ Tradit Chin Med 21(9):40–43

Ning A, Huang M, Dai B, Cao J, Li J (2004) The preliminary study of C96-z mycelium fermentative liquid on the effect of losing weight. Chin J Microecol 16(2):77–79

Sa Y (2020) Development trend of traditional Chinese medicine dietary supplement from approval of *Ganoderma* dietary supplements. Mod Chin Med 22 (7):1124–1129

Sanodiya BS, Thakur GS, Baghel RK, Prasad GB, Bisen PS (2009) *Ganoderma lucidum*: a potent pharmacological macrofungus. Curr Phar Biotechnol 10(8): 717–742.

Sun Y, Liu Q (2017) Extraction technology research of *Ganoderma lucidum* antioxidation. Rinchen Yanjiu 3:19–20

Sun Y, Zhang H (2020) Effects of *Ganoderma* polysaccharide on liver fat deposits and NLRP3 inflammatory corpuscle expression of mice with acute liver injury. World Chin Med 16(6):842–845

Sun L, Zhou B, Sui Z, Meng L, Zhang J (2019) Study on the efficacy of *Ganoderma lucidum* compound granules in improving sleep. Chin J Public Health Eng 18 (5):685–687

Tang F, Guo Y, Zhang J (2003) Study on the reducing the blood lipids of the *Ganoderma luidum*. Food Sci 24 (4):145–149

Vu T, Cuong W, Chen J, Shi M, Zhang H (2019) The anti-oxidation and anti-aging effects of *Ganoderma lucidum* in *Caenorhabditis elegans*. Exp Gerontol 117:99–105

Xia G, Yao H, Dong L, Liu W (2014) Effect of *Ganoderma* polysaccharide on survival, development and senescence of zebrafish. Chin J Pharmacol Toxicol 28(4):491–496

Yang D, Zhou M, Cheng W, Li L, Gao H, Shi D, Du X, Chen M (2010a) Separation, purification and antioxidation character of the *Ganoderma lucidu*m polysaccharides. Hubei Agric Sci 49(11):2883–2886

Yang H, Chen J, Li Y (2009) Protective effect of GLP on intestinal mucosa injury in hemorrhagic shock reperfusion rabbit. Chin J Misdiagnosis 9(3):514–516

Yang H, Chen J, Wang L (2010b) Protective effects of GLP on intestinal mucosa injury in hemorrhagic shock reperfusion and its mechanism. Shaanxi Med J 39 (2):134–136

Yang X, Zhu J (2010) Comparison study on the contents of triterpene between the *G. lucidum* spore powder with broken and unbroken cellular wall comparison study on the content of triterpenes. Chin J Pharm Anal 30(11): 2227–2228

Yang M, Hu Y, Wang Y, Xiong Y, Yang Y, Zheng Q, Huang X (2017) Design ideas of Chinese materia medica health products based on theory and advantages of traditional Chinese medicine. Chin Tradit Herbal Drugs 48(3):419–423

Yang K, Zhang Y, Zhang S, Cai M, Pi X, Hu J, Guan R, Sun P (2020) Preparation of *Ganoderma lucidum* spore oligosaccharide and its regulation on gut microbiota. Food Ferment Ind 46(9):37–42

Yan Z, Hao L, Zhang L, Kang C, Ma T, Cui Y, Zheng Z (2019) Comparison of main chemical constituents in *Ganoderma lucidum* fruiting bodies collected from three producing regions. Food Sci 40(6):240–246

Yi Y, Xu C (2001) Experimental research on lowering the serum lipid effect of *Ganoderma lucidium*. China Inf Rev 24(1):52–53

Yi Y, Hu S, Xiong X, Liu D, Zhong Y, Yi C (2012) Immunoregulatory function of *Ganoderma* spore oil in immunocompromised mice. J Zhejiang Univ 39 (2):161–166

Yuan Y, Wang Y, Su G, Wan Y, Ca L, Shen Y, Yuan B, Han D, Huang L (2018) Archaeological evidence suggests earlier use of *Ganoderma* in Neolithic China. Chin Sci Bull 63(13):1180–1188

Yu Y, Wu X, Xiong K, Bao G, Xu H, Xu J, Wu Q, Li Q, Yao C (2019) Hypnotic effects of *Ganoderma lucidum* triterpene acids in mice. Sci Technol Food Ind 40 (1):297–301

Zeng R, Li Q, Liu C, Song X (2014) Stability of *Ganoderma lucidum* spore oil. Chin J Exp Tradit Med Formulae 20(18):23–25

Zhang G, Jin H, Long J, Qian R, Cao X, Zhang M, Luo B, Wang Z, Sen C (1999) Effects of *Ganoderma lucidum* (Ling Zhi) combined with hypotension on blood sugar, plasma NO, microcirculation and hemorheology in treatment of refractory hypertension. J Chin Microcirc 3(2):75–78

Zhang A, Ge W, Wang Y, Chen J (2002) Preparation of compound mythic fungus cream and observation of clinical effect for chloasma. Chin J Dermatovenereol 16(4):235–236

Zhang Y, Dong X, Tang Z, Han Y (2008) Effect of *Ganoderma lucidum* spore powder on sugar tolerance and intestinal flora of diabetes rats. Chin J Microbiol Immunol 28(6):519–520

Zhang W, Zhang Q, Deng W, Qiu Y, Xing G (2009) Effect of fungus mixture of *Ganoderma lucidum* on the memory and hippocampus neurocells protection in the brain of cerebral ischemic/reperfusion Mongolia gerbils. China J Pharm Econ 4:45–51

Zhao Y, Chang J (2007) Comparison of phenol sulfuric acid method and indirect iodometry method in the determination of *Ganoderma lucidum* polysaccharides. Edible Fungi 29(3):58–59

Zhao X, Ni W, Xing Z, Men D, Bai B (2011) Effect of sporoderm breakage on the quality of *Ganoderma lucidum* spore powders. Acta Edulis Fungi 18(3):71–73

Zhou Y, Wang Z, Feng P (2018) Research progress on polysaccharide from *Ganoderma lucidum* spore powder. Chin Tradit Herbal Drugs 49(21):5205–5210

The manufacturer's authorised representative in the EU is Springer Nature Customer Service Centre GmbH, Europaplatz 3, 69115 Heidelberg, Germany. If you have any concerns regarding our products, please contact ProductSafety@springernature.com

Printed and bound by CPI Group (UK) Ltd, Croydon, CR0 4YY
07/10/2026
02229168-0005